*Systems: Concepts,
Methodologies, and
Applications*

Systems: Concepts, Methodologies, and Applications

2ND EDITION

Brian Wilson
Department of Systems and Information Management,
School of Management,
University of Lancaster, Lancaster, UK

JOHN WILEY & SONS
Chichester • New York • Brisbane • Toronto • Singapore

Other Wiley Editorial Offices

John Wiley & Sons, Inc., 605 Third Avenue,
New York, NY 10158-0012, USA

Jacaranda Wiley Ltd, G.P.O. Box 859, Brisbane,
Queensland 4001, Australia

John Wiley & Sons (Canada) Ltd, 22 Worcester Road,
Rexdale, Ontario M9W 1L1, Canada

John Wiley & Sons (SEA) Pte Ltd, 37 Jalan Pemimpin 05-04,
Block B, Union Industrial Building, Singapore 2057

Library of Congress Cataloging-in-Publication Data:
Wilson, B. (Brian)
  Systems : concepts, methodologies, and applications / Brian
Wilson.—2nd ed.
    p. cm.
  Includes bibliographical references.
  ISBN 0 471 92716 3
  1. System theory. 2. Problem solving. 3. Linguistic models.
4. System analysis. I. Title.
Q295.W54 1990
003—dc20   90-12021
                                                          CIP

British Library Cataloguing in Publication data:
Wilson. B. (Brian) *1933–*
  Systems : concepts, methodologies and applications.
  2nd ed.
  1. Systems theory
  I. Title
  003

  ISBN 0 471 92716 3

Typeset by Thomson Press (India) Ltd, New Delhi

To My Family

In a world that is rapidly changing, organizations need to become more adaptable and to better learn to manage change. From a systems point of view, change is enormously complex and can come from inside or outside the boundaries of the system. A major key to managing change is proper diagnosis of problems and situations, keeping in mind that the performance of the whole is not the sum of the individual parts, but is a consequence of the relationship of the performance between the parts. Thus problems cannot be solved separately, since they are interdependent.

Huse (1980)

Contents

Preface

The word 'system' is open to many interpretations. It is used in everyday language to describe 'the Establishment', our particular part of the Universe, the arrangement of pipework that extracts exhaust gases from a car engine, and even a procedure for placing bets on the race-course. In the technical literature, it appears in book titles as systems engineering, systems analysis, and systems dynamics. Any one of these may refer to the subjects of communications, control theory, computers, or a particular modelling language.

It is clear, given the above confusing range of meanings, that, if the concept 'system' is to be used as the basis for a modelling language, as is the case here, its interpretation must be made explicit and a definition must be derived which makes sense when related to all other interpretations.

In 1981, Peter Checkland produced a book entitled *Systems Thinking, Systems Practice* (Checkland, 1981). This represented an account of his learning experience since joining the Department of Systems at the University of Lancaster in 1969. His aim was to make clear the interpretation of the word 'system' and the activity of 'systems analysis' at it had emerged from the 'action research' programme within the Department, and also to place the work done at Lancaster within the context of the various underlying philosophies of science and the systems movement as a whole.

It is not my aim to reproduce the same material in different words but, in presenting the results of my own experience in using systems ideas, I hope to assemble a coherent account of their development and application together with the relationship of particular systems methodologies and concepts to the nature of the kind of problems encountered. Thus the context in which this work is cast is not one of philosophy but is that of problem-solving in general. Since the basic concepts used are the same as those discussed by Professor Checkland, there will inevitably be some duplication of the ground already covered, but that is necessary if this account is to be coherent.

The audience for this material is the student, or would-be practitioner, of systems concepts and methodologies and it is my hope that this reflection on the particular experiences that I have enjoyed since joining the Department in 1966 will be of interest and of use to them. It is my style of writing to avoid, whenever possible, the use of the impersonal pronoun 'one'. I must add lest I be accused of male chauvinism that, except where a specific reference is being made, 'he' should be read as 'he or she'.

I am indebted to the many organizations and students with whom I have collaborated over the past 24 years, but they are too numerous to mention individually. However, I would like to express my appreciation to British Airways, to the Central Electricity Generating Board (South West Region), to BP, to Glaxochem, to Ponderosa Industrias SA of Chihuahua, and to Materias Primas Monterrey SA, who gave me permission to refer specifically to projects carried out within their organizations. I gratefully acknowledge the contribution made to the development of systems ideas by all the members of the staff of the Department of Systems at Lancaster and its associated consultancy company, ISCOL Ltd. The action research activity has been a collaborative effort and, without continuing critical debate, many of the lessons would have been lost. I am particularly grateful to Gwilym Jenkins, who founded the Department and who provided me with the opportunity of taking part in such a rich learning experience, and to Peter Checkland, with whom I have collaborated in teaching systems ideas to both masters degree students at Lancaster and to representatives of many companies in the UK and abroad.

I would like to thank Mr M. J. Whitmarsh-Everiss for permission to reproduce Figures A10 and A11 of Appendix I, and the McGraw-Hill Book Company for permission to reproduce Figures 24 and 25; also the *Journal of Systems Engineering* for permission to reproduce the paper 'A systems study of a petrochemical plant' which appears as Appendix II. I also wish to acknowledge the contributions of Dr T. R. Barnett and Chris Pogson who produced the Albion Exercise in Appendix III. My grateful thanks go to Mrs Sue Jarman, who converted the original manuscript into legible typescript with both accuracy and speed. Finally, I wish to acknowledge the help and encouragement provided by Mrs Sylvia Johnson in completing this revised edition.

Overview

The material in this book is presented in seven chapters together with three appendices. Following an introduction, which describes the kind of research that has led to the particular systems ideas used and some brief comments on the nature of problems and the organizations in which they reside, Chapter 1 seeks to survey the kind of modelling languages appropriate to various parts of a problem spectrum. This spectrum extends from the well defined (hard) problems, in which the modelling language may be mathematically oriented, to soft, ill structured problems in which a modelling language is required which is capable of a richer description of the real world than mathematics can provide. Such a language is that based upon the concept of a human activity system. In a supporting appendix (Appendix I), examples are given which merely aim to illustrate the many techniques available for modelling, for problems which can be precisely defined, though the reader is referred to selected references if a deeper understanding is required. These may be of interest to the reader who is concerned with developing a capability for modelling over the entire problem spectrum. The emphasis in the book, however, is towards the development of a modelling language, and its application, which is relevant to the 'soft' end of the problem spectrum. Chapter 1 aims to provide the context in which this further development can be placed. Chapter 2 therefore concentrates wholly on a description of the human activity system, which is the basic element within this modelling language. Modelling in these terms consists essentially of two phases. Firstly, it is necessary to derive a definition of the system to be modelled, and secondly, to develop a model of the system so defined. Ways of doing both phases are described in detail in this chapter. Although such models may look very different to those developed as descriptions of situations at the 'hard' end of the problem spectrum (in that they are in words rather than symbols), in principle they are no different. A differential equation is an intellectual construct used to describe the phenomena underlying the behaviour of a physical system involving hardware. In the same way a human activity system is an intellectual construct used to investigate the behaviour of a situation involving people. Chapters 1 and 2 therefore concentrate on the activity of modelling prior to a discussion of how such models can be used within the general activity of problem-solving. In discussing these ideas I have maintained a complete separation between their description and their use. In an analogous situation,

I would argue that before a serious analysis of French literature can be undertaken the analyst should, at least, understand the French language.

Chapter 3 begins to examine methodology using systems ideas but, as in Chapter 1, a brief survey of methodologies in general is included to provide the context for this discussion. The development of systems methodology at Lancaster started from a particular set of ideas and Appendix II is included to illustrate an application of our initial problem-solving approach to a situation that could be described as relatively 'hard', in relation to its position within the problem spectrum.

Chapter 4 continues a discussion of systems methodologies and their relation to specific kinds of problem situations within the spectrum as a whole. This discussion takes place through the description of a number of actual cases in which the specific methodologies and associated concepts were used. They concern a problem of the design of a services complex, an operational system, service systems, and a problem related to testing an hypothesis (more usually associated with scientific investigation).

Chapter 5 continues the discussion but is related entirely to the area of management control. Within this chapter a number of concepts are described which illustrate, and emphasize, the particular distinction being made between management and process control.

A study of management control is included through an illustration of the activity of reorganization. This is a frequent response to the need to undertake some kind of control action, and a systems-based methodology is presented as a rational alternative to the 'ad hoc' tinkering with structure that is the more usual. Within this approach a particular mapping technique is illustrated which uses a variety of transparent overlays.

Within the area of information management considerable impetus has been derived from the developments in computer-based data processing, and Chapter 6 examines the way in which soft-systems methodology can be used to address the initial stages of requirements analysis. Although this chapter is entitled 'Analysis of Business Information' the ideas are generally applicable to any organized purposeful activity.

A frequent consequence of the introduction of new technology is the creation of new roles. These can be either at an individual level (such as information managers, data administrators etc.), or at an organizational level (such as the creation of Information Systems Departments to exploit the new technology for the benefit of the business in general). Once these roles have been created, questions need to be answered such as: what is the area of authority of the role?, what structure is appropriate?, what expertise and information support is necessary? and so on. Chapter 7 discusses role exploration and illustrates approaches to answering the above questions through the description of a number of actual projects.

Finally, Appendix III contains a number of exercises. The reader may wish to tackle these as a means of starting to develop a capability in using this particular set of systems ideas. What I have described in the book is really a way of thinking about problem situations. As such it cannot be taught, it can only be learnt. I would liken it to swimming. You may read a book which describes to you how to swim, but having read it, you will not be an accomplished swimmer. It will be necessary to get into the water. Similarly, an accomplished systems thinker is one who practises it. The exercises are graded and extend from giving practice in picture building (a necessary activity if you are to understand relationships) to problem-solving assignments. All the exercises are open-ended; i.e. there are a number of possible answers and no one answer can be regarded as the correct one. For this reason, and also because the exercises are used within our own teaching programme, specimen answers are not given.

I hope that you find the contents of the book interesting and thought-provoking. I have found the ideas invaluable in my own consultancy activity and I hope that you will also find them of use.

Introduction

Problems and Problem-solving

The most basic statement about the content of this book is that it is about problem-solving. To make that statement, however, is to make a gross over-simplification. The idea of a problem is, itself, highly complex and the notion that a solution can be found which removes the problem represents a naive view of the activity of problem-solving. There are, of course, situations in which a problem can be defined in simple terms; for example, if your car has a flat tyre, then your problem can be defined very easily and a solution will be recognized as such when the desired pressure is maintained in the tyre. This kind of easily defined problem represents one extreme of a problem spectrum which extends to the kind of problem facing the British Government at the present time, i.e. what to do in Northern Ireland? It is difficult to envisage a solution to that situation which will be recognized as a solution by all of the concerned parties.

The emphasis in this book will be on that class of problems generally termed 'management problems' which lie towards the latter end of the above spectrum. Later discussion will aim to clarify what is meant by the term 'management problems'.

The activity of problem-solving consists of, first of all, finding out about the situation in which the problem is believed to lie and then, through some analysis leading to decisions about what to do, taking action to alleviate the perceived problems. Our concern here is with the terms in which such analysis can be undertaken.

The questions, 'What are management problems?' and 'In what terms can management problem-solving be undertaken?' have been the stimulus for a major research programme carried out by the staff and postgraduate students of the Department of Systems and Information Management at Lancaster University, and this has led to a particular process of analysis based upon the use of a particular concept: the human activity system. This is a crucial concept in the analysis of the highly complex area of management problem-solving, and considerable emphasis will be given to its derivation and use later in the book. First, it is worth describing a little of the background to the development process undertaken in the research programme.

1

The Department of Systems and Information Management at Lancaster (formerly Systems Engineering and, latterly, Systems) accepted its first intake of students in 1966. It was established as a postgraduate department and its central activity was a twelve month course leading to a master's degree. This is still the case, though a master's course in Information Management has been introduced and postgraduate activity extends to a doctoral programme. Some undergraduate teaching of systems forms an input to other courses within the university.

The specific aim of the Department was to extend the methodologies of engineering design and 'hard' systems engineering (Goode and Machol, 1957; Williams, 1961; Hall, 1962; Chestnut, 1965) into the area of management problem-solving in general. The philosophy of the Department has always been to undertake this development through involvement in real-world problems. Here the distinction is made between problems in the real world and problems in the laboratory. In the laboratory the analyst has the freedom to define the problem and to control the environment. Thus he can allow certain variables to affect the process under investigation and to constrain others so that they do not. In the real world such freedom does not exist. The analyst has to accept whatever influences the situation under investigation. The mechanism chosen to carry out this development therefore was 'action research' (Foster, 1972; Warmington, 1980). It has been our practice to do both the teaching and the research in this mode and this has demanded that our students should, in the main, be mature postgraduates (to date the average age has been 30), as it is difficult to be effective in this process of learning and development if the students do not already have an appreciation of how organizations function and of the kinds of problems faced by real managers when undertaking the task of real management. In addition to the use of joint staff–student teams on client-sponsored studies, the Department formed a management consultancy company so that this work could be supported through studies unconstrained by the time and resource limitations of the master's degree programme; this was registered as a limited liability company, ISCOL Ltd, in 1970. Experience from both sources will be used to illustrate the development of concepts and methodology in Chapters 3, 4, 5, 6 and 7.

Action Research

The concept of action research is that of simultaneously bringing about change in the project situation (the action) while learning from the process of deriving the change (the research). Over 200 such studies have now been completed in both the Department and ISCOL Ltd and these have led to the particular systems language and methodologies which form the basis of our teaching and consultancy work.

The process of action research can best be seen as a learning cycle illustrated by Figure 1. The process is initiated by the existence of situations in outside organizations in which there are preceived to be problems. In order to undertake

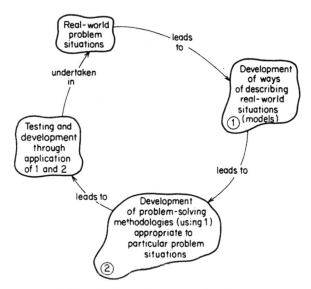

Figure 1. An action research cycle

analysis of these problem situations, the analyst must have available, or develop, ways of describing them (i.e. models) together with modelling languages. The initial expertise within the Department (related to the previous industrial experience of the members of staff) was in the area of process modelling and hence the modelling language was, in the main, based upon the use of mathematics.

The methodologies available for problem-solving within the Department, at that time, were those related to process design, process control, and optimization and those of the Hall and Rand 'systems engineering' variety (Quade and Boucher, 1968). Thus, armed with this initial expertise, the learning cycle of Figure 1 was operated and the modelling process and methodologies extended, adapted, and created to cope with the variety of problem situations encountered. Since the cycle is completed by application, the development remains problem oriented and the output is usable problem-solving methodologies capable of practical application.

The Process of Inquiry

The action research activity described above can only be successful if both the 'action' and the 'research' take place. The action without the research could be seen as no more than consultancy where 'what is gained' can be described as experience. For the research to be present it is necessary to know *what has been learned* from the experience. Such learning requires intellectual reflection on the experience and that in turn requires the establishment of concepts so that 'what has been learned' can be known and made explicit. Without this the knowledge gained cannot be made transferable. For this intellectual reflection to be successful a particular distinction needs to be made and maintained throughout the action research process. Failure to maintain this distinction will lead to confusion and an incoherent description of the research.

Our concern throughout the book is with real-world problem solving (i.e. external to the laboratory) and hence we need to know, at any time, whether what we are describing is activity which is in the real world (i.e. that activity undertaken by real managers when carrying out the real tasks associated with management) or whether what we are describing is a way of thinking about the real world, i.e. intellectual acivity. This is such an important distinction that it is worth emphasizing it further.

We (as the human animal) inhabit the real world but, more significantly, we inhabit the particular bit of the real world represented by the situation in which we wish to intervene. Unlike other animals we have the ability to withdraw intellectually from the real world and think about ourselves thinking about the real world. This notion is illustrated by the following diagram in which we have removed ourselves from the real world and are viewing it through a number of intellectual constructs. Each construct could be likened to a filter so that the image of the real world is particular to the characteristics of the filter and is determined by them.

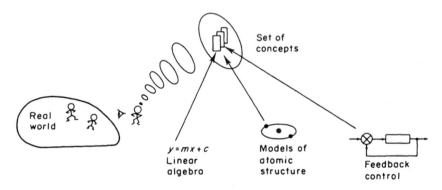

Real-world/intellectual world distinction

To illustrate this notion further consider a number of examples. In the figure opposite particular real-world concerns are shown together with some of the intellectual constructs which are made use of in thinking about and understanding those bits of the real world. The final example leaves as a question, those intellectual constructs which may be made use of when our bit of the real world is taken to be management activity in general. The remainder of this book seeks to provide answers to that question in relation to a variety of management activities and situations.

The figure given on the opposite page attempts to illustrate, for a few examples, the variety of intellectual constructs available. Each can be used to lead to the defensible statement that: using this particular (explicit) intellectual construct as a means of investigating a particular real-world phenomenon this particular

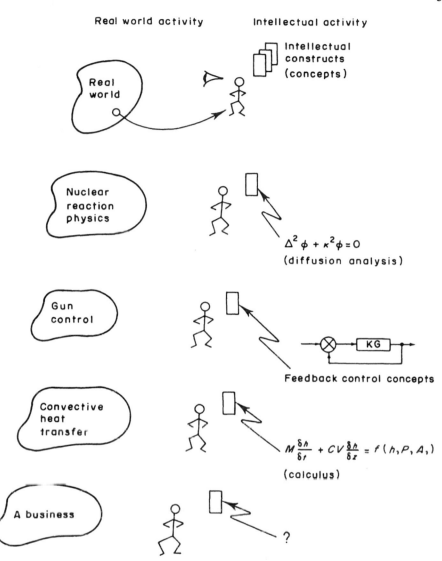

Real world activity   Intellectual activity

Intellectual
constructs
(concepts)

Real
world

Nuclear
reaction
physics

$$\Delta^2 \phi + \kappa^2 \phi = 0$$
(diffusion analysis)

Gun
control

KG

Feedback control concepts

Convective
heat
transfer

$$M\frac{\delta h}{\delta t} + CV\frac{\delta h}{\delta z} = f(h, P, A,)$$
(calculus)

A business

?

Examples of real-world/intellectual
world distinction

conclusion can be reached. It is equally the case that using some other construct in relation to the same situation can lead to other (but equally defensible) conclusions. At the level of complexity which is our concern (management problem-solving) the notion of the *correct* conclusion is one which is inappropriate. The best that we can achieve is to derive conclusions which are defensible (and hopefully appropriate to the situation). The defensibility can be argued on the basis of the intellectual constructs used; the appropriateness comes from the

6

selection of the intellectual constructs themselves. Thus, in any intervention, both the conclusions and the constructs need to be made explicit; hence the need for the real-world/intellectual distinction emphasized in this section.

Finally, the process of inquiry is one which takes the particular intellectual construct (or concept) and asks the qeustion: 'if we map this construct on to a particular bit of the real world what does it tell us about that bit of the real world?' The following diagram attempts to illustrate this process of mapping.

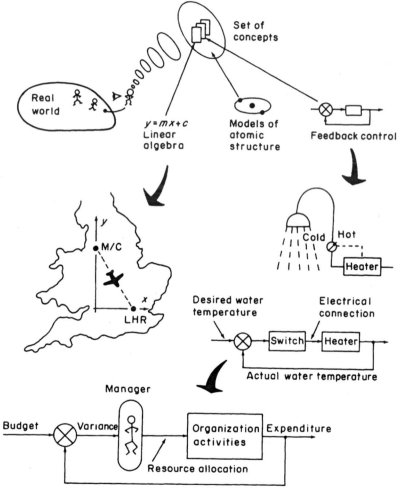

Mapping of constructs

The first example takes the constructs of linear algebra and addresses the process of flying from Manchester to London Heathrow. Using the simple equation shown (model), questions could be asked about flying time, direction, fuel usage etc., and some understanding of their interrelationships obtained.

The second example takes the constructs of feedback control (described later in Chapter 1) and applies it to a description of the working of a shower. Using this idea the relationship between heater power, switch settings and water temperature could be investigated. Also in this example, the same construct is shown applied to the area of budgeting control in which a manager is taken to be a controller. Here the relationships between expenditure and use of resources could be investigated as a regulatory process, given the input of a particular budget. The investigation would be valid and defensible on the basis of the construct used. It might not describe what actually happens in the real situation in which case some learning has been achieved. Modified or alternative constructs may then be selected and used.

The way that the constructs, or concepts, are assembled can be described as a model. More will be said about this in Chapter 1. The purpose of including this section here is to introduce the distinction between activity in the real world and intellectual activity. This is a distinction that needs to be retained while reading the remainder of the book. It must be remembered that every intervention is a potential source of learning; that potential can be realized by making good use of the distinction.

Finally, within this introduction the nature of problems and organization is worth a brief examination.

Problems and Organizations

Since the development process referred to above has been undertaken in outside organizations and has effectively been driven by the kinds of problems encountered within those organizations, it is useful briefly to examine the nature of such problems. This will help to explain the view taken later of the nature of the problems within the problem spectrum and the relationships of these problems to the particular methodologies used in their analysis.

A broad classification of problem types can be derived by taking the extremes of a spectrum which extends from 'hard' to 'soft' and by considering the distinction between questions which are concerned with *how* an activity should be undertaken as opposed to *what* the activity is. In this context, the well defined problem of the flat tyre, referred to previously, is a hard problem, whereas the situation in Northern Ireland is extremely soft.

A 'hard', or structured, problem is one which is exclusively concerned with a 'how' type of question. This can be exemplified by considering the problem confronting Brunel when faced with the need to span the Avon Gorge. There was no doubt in Brunel's mind as to what he wanted to do; the problem was how to do it. This kind of problem is the domain of the design engineer who seeks effective and economic answers to the 'how' type of question.

A 'soft', or unstructured, problem is one which is typified by being mixtures of both 'what' and 'how' questions. In the area of production, for example, a particular manager may be faced with the problem that production performance could be better. This statement of the problem gives no guide to *what* he should investigate to identify areas for potential improvement, or *how* he could then introduce change to realize that improvement. At the level of what he needs to do, he could (a) improve raw material to product conversion; (b) improve plant maintenance; (c) redesign production planning and scheduling methods; (d) improve marketing to production communications. Having decided on one or more of the above areas, he has then to determine *how* to bring about the improvement desired. This example is typical of 'management' problems and hence any methodology, which is aiming to help managers tackle the problems of management, needs to be capable of structuring the problems, i.e. of converting them from mixtures of 'what' and 'how' into problems only of 'how'. Thus a major output of the action research programme has been the change in emphasis towards the development of methodologies to structure problems and away from the development of techniques to 'solve' problems.

The concept of 'problem' is also one that has been found to be inappropriate. The notion that a problem can be defined suggests that a solution can be found which removes the problem. This is not unreasonable at the 'hard' end of the problem spectrum, but at the 'soft' end problems do not occur in a way which enables them to be readily isolated. It is more usual to find sets of problems which are highly interactive and it has been found to be more useful to examine, *not a problem, but a problem situation,* i.e. a situation in which there are perceived to be problems.

Adopting this stance enables a highly significant aspect of any management situation to be taken into account, namely the multiple perceptions of the various managers who may be involved in the particular area of concern. At the strategic level in an organization, for example, some managers may see the primary aim of the organization to be the satisfaction of a market need while others may see it as the need to make most efficient use of the productive resources. In reality, of course, some balance must be maintained between these two extremes. Exactly where that balance lies may be seen to be very different by different managers and hence any problem-solving methodology which is to be effective in an organizational context must be capable of accommodating the component of the problem situation due to such multiple perceptions. This aspect was found to be lacking in all of the engineering methodologies referred to earlier.

A second significant aspect of a problem situation is its relationship to the particular departmental role definitions in the organization concerned.

Any organization consists of a collection of resources: people, machines, materials, money, information, etc. In order to manage this total set, it is usual to subdivide the resources into a number of organizational units (i.e. departments) and to relate these units through some organizational structure. Although the departments so formed may be sensible groupings of resources, they are not unique. Two companies an exactly the same kind of business may choose different departmental boundaries and different structures. Thus, since there is nothing fundamental about this organization structure, problems occurring within the organization are unlikely to be structured in the same way. One may, therefore, expect 'management problems' to cut across organization boundaries and hence any description of such problem situations must be capable of identifying the essential activities in a way that is independent of the particular departmental structure.

The ways of describing the problem situations (modelling languages) need to be appropriate to the nature of the problem under investigation. Since the 'hard'/'soft' distinction refers to the extremes of a possible problem spectrum, the modelling languages can also be viewed in relation to these extremes.

Mathematics provides a general language which has been widely applied to 'hard' problems. The abstract constructs of calculus and statistics have been found to be useful when the elements of the situation can be assumed to behave in accordance with physical laws. When the elements of a 'soft' problem include such features as conflicting objectives, unclear or complex information flows, people with differing perceptions and attitudes, etc., it is difficult to see how a mathematically-based language can be appropriate.

Chapter 1
Models and Modelling

Introduction

Reference to Figure 1 shows that modelling, or the ability to describe the situation confronting the analyst, is crucial to the subsequent application of any methodology to analyse that situation. Modelling is something that we all do, whether it is a conscious or an unconscious activity. It precedes every decision in that some assessment will have been made of the likely outcome of that decision, however superficial that assessment may be. It preceeds the formulation of an opinion. The statement that the conduct of a meeting was poor is based upon an implicit model of what a good meeting is like. An analyst visiting a factory for the first time and observing that the process of production scheduling could be improved is likewise basing this opinion on some implicit model of production scheduling. In the latter case, before a decision is reached on what to do about it, an explicit model would probably be produced.

Modelling is a fascinating activity and this section attempts to survey the area in terms of the kinds of models produced, the language in which the models are constructed, and how they are produced. The discussion is necessarily brief as other, more detailed, coverage is available in the literature and it is not my intention to reproduce it here but to concentrate on those aspects which are particularly relevant to the actual cases used for illustration. Selected references will be provided where appropriate.

This section also issues a warning. Because modelling is so fascinating there is a great danger than it can become an end in itself. The measure of success in modelling is not that you can produce a model that is bigger and more sophisticated than anyone else's but that it adequate answers the original questions for which it was developed. The literature is full of descriptions of models that have never been used.

Although the situation has now improved, the word 'model' is frequently associated with a quantitative description of some process. The interpretation needs to be much broader than that. Chestnut (1965) attempts such an interpretation in his definition:

> A model is a qualitative or quantitative representation of a process or endeavour that shows the effects of those factors which are significant for the purposes being considered.

This is not unreasonable, except that what he appears to mean by qualitative is restricted to functional or procedural diagrams usually preceding a mathematical formulation. I would wish to take an even broader interpretation than this and would suggest the following definition:

Definition *A model is the explicit interpretation of one's understanding of a situation, or merely of one's ideas about that situation. It can be expressed in mathematics, symbols or words, but it is essentially a description of entities, processes or attributes and the relationships between them. It may be prescriptive or illustrative, but above all, it must be useful.*

This is extremely broad, but necessarily so, given the complexity of management situations – the context in which we wish to consider modelling. The fact that it is in terms of interpretations or ideas about a situation provides the freedom to model something believed to be *relevant* to the situation rather than a model *of* the situation itself. The last sentence of the definition is inserted in order to emphasize the warning given earlier, i.e. that the model is only part of a process of analysis and not the outcome.

In order to illustrate the kinds of model, modelling languages, and the purposes to which they are appropriate, a classification can be produced. Ackoff (1962) makes a distinction between three forms of model:

(a) *Iconic* This implies that the model is a miniature (though it is sometimes an enlarged) version of the real article and the relevant properties of the real article are represented by the properties themselves, but usually with a change of scale. Thus a replica is usually constructed which, with a fair degree of confidence, will be expected to reproduce the behaviour of the original. Examples are as follows: a pilot plant for a new chemical; a stress model of a bridge; an aircraft model for wind tunnel testing.

(b) *Analogic* A model of quite different physical appearance may be constructed which, nevertheless, is expected to reproduce representative behaviour. For example, water flow through small plastic tanks at room temperature is used to investigate the behaviour of molten glass in large furnaces at temperatures around 1000 °C. An electrical network can be used to represent the flow of water through pipes or the flow of heat between surfaces. In a more general way an analogue computer can be used to model a wide variety of situations, the values of the physical variables being obtained by measuring voltages at appropriate points in the analogue network.

(c) *Analytic* Mathematical or logical relationships can be developed which represent the physical laws which it is believed govern the behaviour of the situation under investigation. Such a development will usually precede an analogue model. A simple example would be the representation of the rate of cooling of a hot body. According to Newton the rate of fall of the temperature, θ, of a hot body is proportional to the excess of temperature over the temperature, θ_1, of the surrounding medium. Hence:

$$-\frac{d\theta}{dt} = k(\theta - \theta_1),$$

where k is a constant.

Although this classification provides a useful distinction, it only covers those models which are either physical in form or which can be formulated quantitatively. To these three forms I would wish to add the further category of *conceptual models*. These include those pictorial/symbolic models, referred to in the definition, which cover the qualitative aspects of the situation. Conceptual model building may, in fact, precede any of the other kinds of modelling as well as being a modelling form in its own right.

Iconic and, to a lesser extent, analogic models are relevant to a highly specialized kind of investigation, and although the use of an analogue computer may be an acceptable method of solution in relation to analytic models, our concern is more usefully directed towards the derivation of conceptual and analytic types of model and their use.

Conceptual Models

Although a major concern in this text is the development and use of models of human activity systems, which are a particular class of conceptual model, this is better left until Chapter 2, where they will be discussed in detail. This section is concerned with the use of qualitative models in general and may be illustrated by reference to four kinds of use: (a) as an aid to clarifying thinking about an area of concern; (b) as an illustration of a concept; (c) as an aid to defining structure and logic; (d) as a prerequisite to design.

Clarifying an area of concern

Prior to any study some appreciation of the situation will need to be acquired and issues such as the following will need to be considered:

    (a)   What is taken to be the boundary of the area under study?
    (b)   What interactions are assumed to exist in relation to this particular boundary?
    (c)   What kind of activities are likely to be present within this area?

I have always found it useful to construct a simple picture to illustrate the assumptions being made. This picture is itself a qualitative model and making our assumptions explicit through a picture is an efficient way of conveying the relationships and of sharpening up the thinking about the area of concern. The advantage of a pictorial display is that the information contained therein can be processed in parallel whereas information contained in prose can only be processed in series. Imagine the amount of prose required to convey adequately the information contained in that very familiar qualitative model, an Ordnance Survey map.

In an early project, I was concerned with the development of information systems to support the highly complex area of aircraft maintenance within British Airways. This is carried out along with other activities within the Engineering Department (BAED). To try to understand what went on within BAED, I needed to make some assumptions about its relationship to British Airways in total and to its outside environment. The simple picture at the top of Figure 2 makes clear what I am taking to be the major interactions. There will, of course, be many others, but my assumption is that these will be of secondary importance to those shown and to a large extent will be generated by the activities that have to exist within BAED to respond to the major interactions indicated.

The lower part of Figure 2 illustrates the distinction that I am making between (a) the activities required within BAED to maintain aircraft plus the other supporting activities (such as planning the development of BAED, planning the receipt of aircraft, keeping abreast of the developments in the business of aircraft

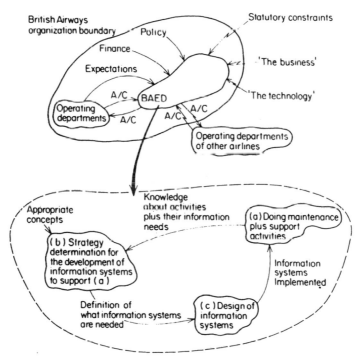

Figure 2. British Airways and information systems development activities. *Key:* BAED, British Airways Engineering Department; A/C, aircraft; ⟶, interaction internal to British Airways; ⟝⟋, interaction external to British Airways

maintenance and the technology, responding to the statutory constraints and the expectations of the operating departments, i.e. those activities required to respond to the interactions indicated in the upper picture); (b) the activities associated with deciding what information systems are needed to support the activities in (a); (c) the activities required to turn the requirements specified in (b) into actual operating systems.

Having made the distinction, it is clear that the expertise needed to maintain aircraft (engineering skill) is very different from the expertise needed to define an information network and the individual information systems within it (a conceptual analysis skill), which in turn is very different from the expertise needed to convert a statement of 'what' is required into the particular 'how' of providing it (a data processing skill). In reality these three kinds of activity are intricately mixed among a variety of people. It is very helpful (I believe essential) to know, at any time, whether you are describing (or doing) activities of type (a), type (b), or type (c).

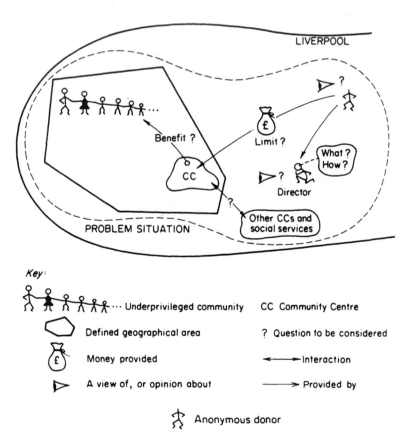

Figure 3. A community centre problem situation

A further example, which illustrates variety in the use of symbols, comes from an artificial exercise, though one based upon a real project undertaken by my colleague R. H. Anderton. The problem posed is briefly described by the following statement:

A local philanthropist wishes to donate a large sum of money for the establishment of a community centre for the benefit of an under-privileged sector of the community in a specific area of Liverpool. He wishes to remain anonymous and has appointed a director to manage its setting-up. How should he decide: (a) what to provide, and (b) how it should be managed?

The real project was concerned with providing guidance in relation to (a). The exercise is not concerned with answering the questions but only with deciding how to answer them, i.e. the methodology of analysis. This problem situation is illustrated by Figure 3.

Since the purpose of this kind of model is to help the analyst clarify his thinking about the situation, the range of symbols is only limited by his imagination. However, to be really useful the pictures need to be coherent and so, like an Ordnance Survey map, the symbols need to have a consistent meaning. A key helps to ensure this consistency and is essential if the pictures are used to convey an interpretation of the situation to a third party. The real benefit, however, from such a picture comes from its construction. In the case of Figure 3, assembling the picture helped me to arrive at the set of questions to be considered, such as: 'What is meant by benefit and who assesses it?' 'Do the director and philanthropist have preconceived views on the role of a community centre?' 'Is the money a limiting consideration?' These are obvious considerations in retrospect, but the construction also clarified the relationship between these considerations so that I knew what I would do with the answers to the questions.

An illustration of a concept

A number of concepts can be transferred from one discipline to another and in bringing about this transfer a clear definition of the concept is necessary. One such concept is the concept of feedback from control engineering. Figure 4 illustrates the mechanism and shows that by comparing the output from a process with some specification of desired output, an error can be formed. This error is then used to manipulate some variables in the process via a controller in order to reduce the error to zero. A common example of a device to apply

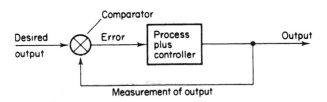

Figure 4. The concept of feedback

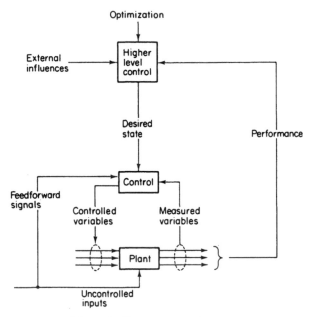

Figure 5. Process control system

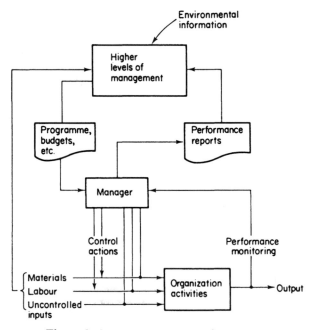

Figure 6. A management control system

this concept is a room thermostat (which is the measuring device plus controller) and the boiler in the central heating system which is controlled, by the thermostat, to supply heat or not depending on whether the room temperature is above or below its desired value. Figure 4 is a qualitative model illustrating this concept. A more sophisticated process control system is illustrated in Figure 5, in which the concept of feedback appears at two levels: one illustrating responsibility for the control of the plant and the second representing the higher level controller which acts on the lower one by manipulating the desired state. Also included in this model is a feedforward signal which enables the lower level controller to anticipate disturbances which cannot be controlled. This more sophisticated control concept can be translated into the area of management as a means of illustrating the processes relevant to a single manager within a management hierarchy. This is illustrated by Figure 6, in which the plant of Figure 5 has been replaced by the particular set of organization activities for which the manager has responsibility and the higher level controller is represented by the higher levels in the management hierarchy. The use of such a concept may assist in understanding the management process itself or help in defining information flows and responsibilities.

Defining structure and logic

Insight into a situation may frequently be enhanced by the development of a model which illustrates interactions in the form of cause–effect relationships. Systems dynamics provides a suitable modelling language in which to assemble this kind of model. It consists of relating rates of flow, levels, and states. For example, if the level of interest is population, then this will depend upon the interaction between the birth rate, the mortality rate, and the states of fertility and mortality:

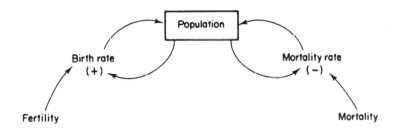

For a given level of population and a given fertility, a particular birth rate will tend to increase the level of population, thus indicating a positive feedback loop. Similarly, for a given level of population and a given mortality, a particular mortality rate will tend to decrease the level of population. This therefore indicates a negative feedback loop.

Models of the above form can be constructed to illustrate the relationships between a number of different levels and rates. Figure 7 is part of a model

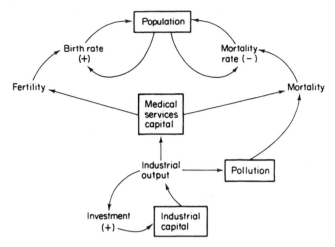

Figure 7. Relationship between industrial and medical activity and population

which adds to the above population model those interactions which come from industrial and medical activity. It makes the assumption that the use of industrial capital in producing industrial output will have an effect on pollution which in turn will affect mortality. Similarly industrial output will also be shared betwen investment and the level of medical services. Medical services in turn will affect both fertility and mortality. Models of this kind illustrating industrial dynamics can be found in Forrester (1961) and Coyle (1978) and, illustrating urban and world dynamics, in Forrester (1968, 1969) and Meadows *et al.* (1972).

A prerequisite to design

Design activity is concerned with defining *how* to achieve a particular purpose. Prior to design, however, comes the stage of deciding *what* is to be designed. Conceptual modelling in necessary in this stage. Take, for example, the design of a chemical plant. It is assumed that, as the result of some long- to medium-term planning activity, a decision has been reached to invest in a plant to produce a particular range of chemicals to satisfy a perceived market need. Thus, at the first broad level, the following transformation processes can be defined:

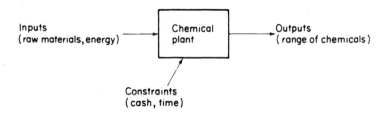

Dependent upon the defined range of chemicals to be produced and the particular raw materials used, the transformation can be expanded into, say, the following stages:

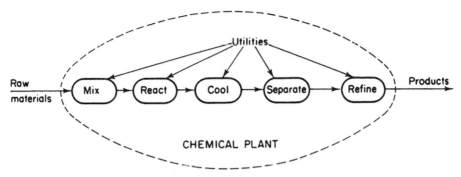

This model (still at a broad level) defines the process of converting the raw materials into the products required and defines what, in chemical engineering terms, are the unit processes. Before detailed design can proceed, this conceptual model needs to be expanded to include the particular structure which relates the various unit processes to one another (i.e. by including a particular 'how'). Figure 8 represents such a model. This is usually termed a schematic diagram but nevertheless it is still conceptual and remains so until the units themselves take on physical reality.

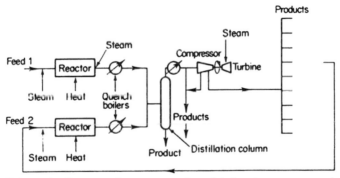

Figure 8. Schematic diagram of a conceptual model showing process and structure

The next stage in design is to size the various units and to define operating pressures, temperatures, flow rates, etc.

It is worth pointing out, in relation to this example, a general feature of the process of converting form a 'what' to a 'how', i.e. of converting from a statement of what is required to a definition of how it is to be provided. *There is not necessarily a one-to-one correspondence between the 'what' and its 'how'.* Thus, in this example, if what you want to do is to separate and refine (two activities), how it is done is with the one unit, the distillation column. Similarly the activities of mix and react are both done in the reactor.

In general, the process of implementation requires the conversion of 'what' to 'how'. So, if what you want to do is take exercise, you cannot do it directly without, first of all, deciding how to do it, i.e. by swimming, cycling, jogging, etc. This becomes vitally important later on when our concern is with problem-solving, in that observing real-world activity is always in terms of the particular 'hows' that have been implemented. Thus, in observing someone cycling, it is not apparent whether the 'what' underlying this 'how' is to take exercise or to make a journey from A to B. This reverse conversion from 'how' to 'what' becomes extremely difficult when that which is being observed is the activity of management in a particular problem situation.

This consideration is a diversion from our major concern in this section, general forms of conceptual models, but it is an important consideration and one that I will be returning to later on.

Analytic Models

A major effort has been devoted to the development of models and modelling languages which come under the general heading of analytic models. They are used mainly as a means of predicting the behaviour of some aspect of the real world and their development is usually associated with some form of validation. In order to survey this area of modelling it is useful to consider the four classes of model contained in the following matrix:

	Steady state	Dynamic
Deterministic {	Algebraic equations	Differential equations
Non-deterministic {	Statistical and probability relationships	Discrete-event simulation

The columns in the matrix make a distinction between those models which are time-dependent (dynamic) and those which are independent of time (steady-state or static). The rows differentiate between the events described in the models having a determined relationship – empirical or based upon physical laws (deterministic) – and those in which the parameters are uncertain and are based upon expected values from a statistical distribution (non-deterministic). The elements in the matrix represent the modelling language most commonly associated with the type of model.

It is not my aim, in presenting this overview of the kinds of modelling language appropriate to the 'hard' end of the problem spectrum, to describe each element of the above matrix in sufficient detail that the reader can become a practitioner in these methods. Rather it is my intention to survey this area so that the reader is aware that 'soft' systems modelling represents only one language, among others, when related to the total problem spectrum. A number of texts are already available which describe modelling techniques represented by the matrix elements as described above.

Appendix I provides a brief discussion of typical examples illustrating the simple use of these techniques together with selected references in which the reader can obtain more detailed descriptions.

Conclusion

The purpose of this section, taken together with Appendix I, has been to review a selection of modelling methods and languages that I personally have found useful. It has not been my intention to describe them in sufficient detail to enable a student to use them. The references have been included for that purpose. The activity of problem-solving covers the whole spectrum of problems and a variety of languages is available to cope with the variety of problems within that spectrum. Using the classification stated earlier in this chapter, examples have been included from each element within it. These examples have used modelling languages that have been pictorial, diagrammatical, and mathematical and provide the context in which conceptual model building needs to be discussed.

This chapter has provided a general survey of modelling languages and the process of modelling. Following a statement of a model classification, we briefly examined a range of conceptual models, developed for specific purposes. It is now my intention to return to conceptual models and conceptual model building and in particular to those considerations related to human activity systems. *But what is a human activity system?* Chapter 2 aims to answer that question.

Chapter 2

A Systems Language

Historical Perspective

Chapter 1, together with Appendix I, has briefly surveyed the activity of modelling and has presented a classification of modelling languages appropriate to problems within the spectrum from 'hard' to 'soft'. In this chapter we will concentrate on the development of models related to the first part of that classification, i.e. on those models that are conceptual in form. We will develop a language known as a 'systems language'.

It may be argued that systems languages of one kind or another have been in existence throughout the whole of recorded history. Ptolemy produced a model of the Universe that reproduced observed planetary motion. The Ancient Egyptians produced automatic control systems for opening and closing temple doors and hence must have had a conceptual understanding of the systems involved. However, it is the later developments in thinking, in the disciplines of engineering and biology, that are particularly relevant to the systems language that is our concern.

Following the Second World War, i.e. from the 1940s onwards, engineering design methodologies were being developed to cope with the need to design processes that were becoming more complex and costly. The market environments in which companies were operating were becoming more competitive and hence the processes used to manufacture products within these markets had to be designed and operated to be highly efficient. In parallel with this trend, computer developments were increasing the feasibility of carrying out the complex calculations required to achieve this increased efficiency. This situation gave rise to the emergence of systems engineering methodologies in which an attempt was made to integrate the effects of interactions between the process units themselves, between the process being designed and others on which it depended and between the interacting set of processes and the market environment being served. Thus, in developing these ideas the analysts were seeking to take in *more of the whole*, i.e. the boundary of the system being analysed was expanding.

At the turn of the century, and in parallel with the developments taking place in engineering, a group of biologists known as the 'organismic biologists' were concerned with studying the properties and behaviour of whole organisms. It was their thesis that living organisms had properties that were particular to their level of complexity and which were meaningless at lower levels. Thus the properties that make a man recognizable as the whole entity man (*Homo sapiens*) do not make sense when related to a foot or eyeball. Water has the property of 'wetness' which has no meaning when related to hydrogen and oxygen which are its constituent parts. These properties, known as emergent properties, enable complexity to be described in terms of a hierarchy of levels of organization in which each level is described in terms of its emergent properties. This idea of *wholeness* is the central concept in the general theory of systems proposed by L. von Bertalanffy in the mid-1940s. He argued that the organismic thinking of the biologists could be applied to organized complexity in general. (For a fuller discussion of this development see Checkland (1981).)

Also, since our concern is with the development and application of a systems language to complex problem situations involving people, some contribution from the activities of social science and sociology could be expected. However, these disciplines have not yet produced much in the way of methodologies that can actually be used in the analytic sense, though concepts related to perceptions

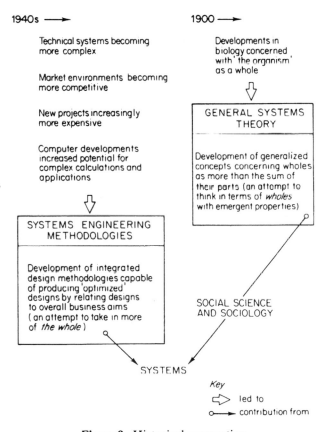

Figure 9. Historical perspective

and meaning, values, roles, and norms, have proved to be useful and have made a contribution to the particular language involving the use of human activity systems.

Figure 9 summarizes this development and leads us to the question, 'What are systems?'

A Systems Classification

The word 'system' has many interpretations depending upon the context in which it is used. It can mean, for example, a procedure, a process or its control, a network, or a computer-based data processing package. While these are all valid uses of the word, a definition is needed which will allow a particular interpretation to be placed in context. This is particularly necessary here if we are to make use of the concept, system, as a basis for a modelling language.

A useful starting point in arriving at a precise definition is to take a general definition that includes all of the interpretations mentioned above, i.e. the dictionary definition: 'a system is a structured set of objects and/or attributes together with the relationships between them'. This definition leads to the model given in Figure 10.

As Figure 10 illustrates, the system is first of all a set; i.e. it contains elements that have some reason for being taken together rather than some others. But it is more than just a set, it also includes the relationships that exist between the elements of that set. So that if the system is a computer package, the elements will be the instructions and the relationships will be defined by the particular program structure. If the system is a human being, the elements could be the heart, lungs, brain, etc., and the relationships determined by the particular functions of the nervous, hormonal, etc., systems.

This general definition, although providing an overall reference model on to which all interpretations can be mapped, is not good enough to be used as a modelling language.

The definition can be refined by first of all deriving a classification in terms of types of systems and then by developing a set of concepts appropriate to each type. The particular classification adopted at Lancaster (Checkland, 1971) may be summarized as follows:

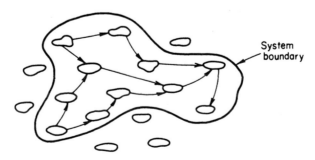

Figure 10. A model of a system

(a) *Natural systems* Physical systems which make up the Universe in a hierarchy from subatomic systems through the systems of ecology to galactic systems.

(b) *Designed systems* These can be both physical (tools, bridges, automated complexes) and abstract (mathematics, language, philosophy).

(c) *Human activity systems* Generally describing human beings undertaking purposeful activity such as man-machine systems, industrial activity, political systems, etc.

(d) *Social and cultural systems* Most human activity will exist within a social system where the elements will be human beings and the relationships will be interpersonal. This is different in nature to the other three classes in that it spans the interface between natural and human activity systems. Examples of social systems would be the family, a community, and the boy-scout movement, as well as the set of systems formed by groups of human beings getting together to perform some other purposeful activity, such as an industrial concern, a choral society, or a conference.

NB.   In the above classification the word 'system' is being used in its everyday sense. To be precise we should refer to 'those things or activities on which we can map the constructs of natural systems etc.'

A problem situation which can be described as 'hard', in the context referred to earlier, may be analysed as a designed system of the physical variety where the modelling language may be mathematics. A 'soft' problem situation can be analysed as a set of interacting human activity systems where the modelling language consists of activities (or verbs). Figures 11 and 12 illustrate a manufacturing enterprise described as a set of interacting human activity systems.

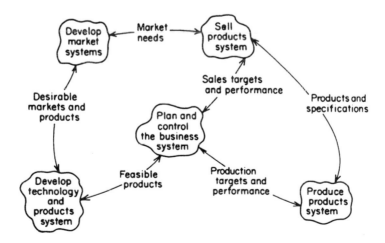

Figure 11. A systems description related to a manufacturing company

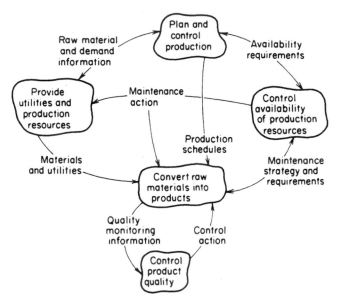

Figure 12. A 'produce products' system

Figure 11 illustrates a set of systems relevant to the total enterprise at a broad level and Figure 12 illustrates a set of subsystems relevant to the 'producing system' at a more detailed level. Sets of subsystems relevant to the level of detail of Figure 12 could be produced for the remaining systems in Figure 11 giving an expanded description relevant to the total enterprise. Figures 11 and 12 together illustrate the hierarchical nature of the modelling process, the relationships between the various systems and subsystems being maintained as the model develops to the level of detail appropriate to the nature of the analysis being undertaken. It must be emphasized that this is *not* a description of a set of departments within the organization but is a description which is independent of any internal organization structure. Neither is it an account of what ought to exist or some 'ideal'. It is an intellectual *construct*, a model, in terms of an interacting set of human activities, which may or may not turn out to be a useful way of analysing some problem situation within a manufacturing enterprise.

Within the above classification it is the human activity system type which has proved to be of value in the analysis of management problem situations and it is this type which will be further developed.

Human Activity Systems

As illustrated in Figures 11 and 12, a human activity system can be described as an interacting set of subsystems or an interacting set of activities. A subsystem is no different to a system except in terms of level of detail and hence a subsystem

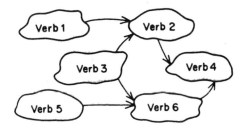

Figure 13. A basic human activity system

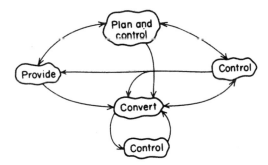

Figure 14. A 'produce products' system (cf. Figure 12)

can be redefined as a system and modelled as a set of activities. Thus the term 'system' and 'activity' can be used interchangeably. The word 'activity' implies action and hence the language in which human activity systems are modelled is in terms of *verbs*. A model of a human activity system (HAS) in its most basic form is illustrated by Figure 13. A model in such a basic form is not usually very useful. It is usual to add qualifying statements to the verbs; for example, see Figure 12. This would have been almost meaningless had the verbs only been used, as in Figure 14.

By using the above classification, our definition can become more precise. We can now identify whether what we are describing is a natural system, a designed system, a social system, or a human activity system. In terms of a human activity system the general systems model of Figure 10 can be replaced by the HAS model of Figure 13 (given the additional requirement of qualifying statements after the verbs).

An important relationship exists between the two latter classes of system. We can decompose a human activity system into two others systems: a system of activities and, if such activities can be said to exist in the real world, a social system whose boundary is coincident with the HAS boundary. Thus, *if and only if* a human activity system is mappable on to a particular organization, the system of activities can be used to define 'what' to change but it is the social system that defines 'how' that change may be implemented or whether or not

28

the change is acceptable. Although it is usually the case that a HAS is modelled as the system of activities, it must not be forgotten that one may also be defining an accompanying social system. To provide emphasis the following diagram illustrates a way of thinking about this relationship:

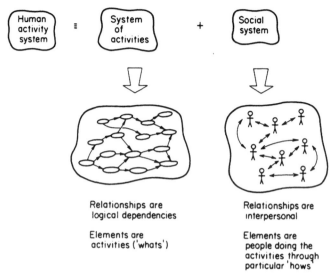

Relationships are logical dependencies

Elements are activities ('whats')

Relationships are interpersonal

Elements are people doing the activities through particular 'hows'

If we now concentrate on the HAS as that class of system most relevant to the investigation of the set of problems of management, we can examine the set of generalized concepts which have been found to be useful in practical application. Since these concepts are general, they represent a model to which any HAS can be related. There is no theoretical basis for this particular set, but they are derived from the experience of real-world problem-solving and are important aspects of real-world activity which, when omitted or inadequately represented, lead to the kind of problems encountered. Thus, if our HAS is to be a language which is relevant to real-world activity, our model of a HAS cannot be deficient in any of these features.

Generalized Concepts

The most basic concept related to a model of a human activity system is that it is a *transformation process*. This means that the set of activities contained in the model represent that interconnected set of actions necessary to transform some input(s) into some output(s):

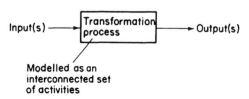

Input(s) → Transformation process → Output(s)

Modelled as an interconnected set of activities

Thus, as an example, take a designed physical system – e.g. the chemical process referred to in Chapter 1. The input could be raw material of some kind and the output the range of products. Here the activities are mix, react, etc. In this example the inputs and outputs are physical in nature but they do not need to be. The transformation can also be of an abstract kind. For example a HAS model of a manufacturing enterprise could be derived by taking it to be a system for transforming a perceived market need into a satisfaction of that need.

Following from the general dictionary definition of a system it can be argued that, for a HAS to be a system, a minimum degree of *connectivity* must exist between each entity (verb or activity). Within the model this connectivity is defined as logical dependence. For example if one activity within the model is 'convert raw material into products', it can be argued that this should be preceded by the activities 'decide what products to make' and 'obtain raw material'. Hence, 'convert raw material' can be said to be logically dependent upon the other two activities. A particular kind of connectivity is that associated with information flow and, of late, considerable attention has been given to problems related to information system design. We shall be giving consideration later to the development of particular kinds of models of HAS in which the connectivity is of the nature of information. However, since all models of human activity systems are initially developed by defining the logical connectivity, we will restrict our discussion here to this aspect.

Our concern in developing such models is with purposeful systems and hence that *purpose*, or *objective* needs to be made clear. The distinction between purpose, objective, goal, mission, etc., is not important here as what we wish to define is the '*raison d'être*'; clearly, in terms of the model, this is to achieve the particular transformation that has been defined.

It has been found useful to import ideas from control engineering and to define our model of a HAS as a controlled system. This implies that, if our system is to achieve a particular objective, some measure of that degree of achievement must be derived and activities included in the model that make use of the measure to take control action to improve that degree of achievement. This is defined as a *measure of performance*, and information collected according to that measure will be used by some *decision-taking procedure* to take control action through *control mechanisms*. Thus, if the system objective is defined as the satisfaction of a perceived market need, the measure of performance must be related to how well the particular sector of the market is satisfied, i.e. in terms of market share or customer complaints or some combination of the two. Based upon information collected in these categories action can be taken to improve the product or improve the market definition or selling activities. In reality, of course, a particular company may wish to maintain a balance between achieving market satisfaction and the cost incurred in so doing. If a cost constraint is regarded as important, this should be included in the definition of the original objective. The analyst can choose what to include, or what not to include, in the objective specified for the model and can hence explore the implications of the choice made.

The decision-taking procedure, defined above, can only take control action within a particular area of responsibility. This area defines the system *boundary*. If it is undecided whether or not a particular activity should be included in the system, the answer can be obtained by examining the nature of the decision-taking procedure and questioning whether or not the decision-taking procedure has authority over that particular activity. If the answer is yes, then, by definition, the activity is within the system boundary. If the answer is no, then, by definition, the activity is outside the system boundary.

In order to achieve objectives defined for the system, *resources* need to be available within the system boundary and need to be appropriate for the activities to be undertaken. It is these resources that are under the control of the decision-taking procedure and which can be acquired and deployed to achieve the objectives of the system. Hence activities must be included in the model of the HAS which perform that acquisition and deployment.

The concept of resolution level is crucial to the development of systems models. This is best described through the notion of *systems hierarchy*. This means that the boundary of the system chosen places the system at a particular level within a series of levels.

Thus a system is, at the same time, a subsystem of some wider system and is itself a wider system to its subsystems. What we define to be 'a system' is a choice of resolution level or the choice of level of detail at which we wish to describe the activities. It is a choice: there is no absolute definition of what is a system or what is a subsystem. Groups of activities only become subsystems when we have defined what we are taking to be 'the system'. The extremes of this hierarchy are defined by the omission of one or more of the above concepts, usually the decision-taking and control mechanisms. If the boundary of the system is widened to such an extent that, in principle, the system could not be engineered, i.e. a decision-taking procedure could not be conceived that could have control over such an area, this is termed 'an environment'. Similarly, if the boundary is reduced to such an extent that decision-taking is an inappropriate activity, systems at this level of detail are termed components.

The above set of concepts describes what Checkland (1981) refers to as a 'formal systems model', i.e. that set of concepts against which any model of a HAS can be validated. These basic concepts of the formal systems model may be listed as follows: (a) objectives, purpose, etc.; (b) connectivity; (c) measures of performance; (d) monitoring and control mechanisms; (e) decision-taking procedures; (f) boundary; (g) resources; (h) systems hierarchy.

Unlike the other categories of systems, defined in the systems classification, i.e. natural systems, designed systems, etc., human activity systems cannot be mapped on to reality with the same degree of agreement (between observers of that reality). Given equal technical competence, two observers of a motor car would produce the same description of it. Similarly a tree would be described in identical terms by equally qualified observers. Human activity, however, is not amenable to the same degree of certainty or precision in its description. A football match, for example, could be seen as a competitive system in which one team tries to place a ball in a certain location while the other team tries to stop them. It might also be described as a particular kind of entertainment-providing system or a gambling opportunity provision system. It is no use arguing about which is correct, they are all legitimate. It is also no use arguing that a football match is a mixture of all the possible interpretations. No group of observers would agree on what the mixture is; they may not even agree on the range of interpretations!

The characteristic which differentiates a human activity system from the other types within the classification can be further illustrated by examining the mode of description applied to a particular designed physical system (DPS) and then comparing this with the description of legitimate human activity systems which make use of the same DPS.

If we take the aircraft Concorde to be a designed physical system, this could be described in the following terms:

Shape

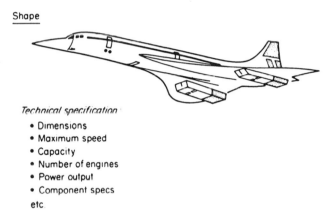

Technical specification

- Dimensions
- Maximum speed
- Capacity
- Number of engines
- Power output
- Component specs

etc.

The essence of this description is that it is factual and unambiguous. Given the necessary degree of detail in the specification and the same degree of competence in the observers of Concorde, complete agreement would be obtained on a description of it as a designed physical system.

If we now turn to the *use* of Concorde as a human activity system and ask the question, 'In what terms can we describe the operation of Concorde?' we may generate several answers, amongst which would be the following: (a) a transportation system for carrying passengers over intercontinental routes at speeds greater than that of sound; (b) a system to reduce the desirability of our environment by producing noise and other pollution; (c) a system to waste more of the taxpayers' money; (d) an obsolete political system for persuading the French to let Great Britain join the Common Market; (e) an aesthetically pleasing system which continually celebrates technical and engineering achievement; and so on. The essence of this description is that each definition is based upon a particular point of view. Although a given individual may disagree with a particular interpretation, he will not be able to argue that such an interpretation is 'wrong'. Thus whereas Concorde (as a designed physical system) is amenable to a single description, the operation of Concorde (seen as a human activity system) cannot be described in a way that ensures consensus. Thus a crucial characteristic of the mode of description of human activity systems is this aspect of the multiple perceptions that are possible and legitimate.

Having argued that the type of system, the human activity system, is the basis of a language in which to describe the human activity associated with management, it is necessary to accommodate this aspect of multiple perceptions within the set of concepts if they are to be applicable to real managers doing the real task of management. The interpretation of what that task is will also be subject to the multiple perceptions of the managers concerned. This concept is described by the German word *Weltanschauung*. Literally translated it means 'word-view', i.e. that view of the world which enables each observer to attribute meaning to what is observed. With reference to the following diagram it may be likened to a filter in the head of an observer which has been formed and is continually moulded by experience, personality, politics, society, and the situation:

It is usual, at Lancaster, to refer to this concept as 'W' rather than use of the whole word. Thus, an individual has a particular 'W', which is unquestioned, but which results in particular meaning being attached to events leading to particular actions based upon that interpretation.

Any management situation will contain a number of managers undertaking a number of tasks; taking action, or not, based upon their own, particular, Ws. Hence any analysis of that situation will have to be at the level of an analysis of Ws. It is not feasible to determine some mixture of Ws that will lead to the 'correct' description of the situation; hence it is not feasible to develop a HAS model *of* the situation. It is only possible to develop models which are *relevant to* (or about) the situation. Neither is it possible for an analyst (who is an observer in these terms) to ascribe a particular W to a particular manager. The interpretation which leads to that action will have been dependent upon the analyst's W.

Whereas, in a system engineering study, it might be reasonable to develop a model of a production process and validate it based upon data collected from the process, it is more useful, in the analysis of a management problem situation, to identify a range of Ws that are considered appropriate and to produce a number of models, each one representing a particular W (without attempting to ascribe that W to any one manager).

The concept, W, is incorporated into the model of a HAS through what is termed a *root definition*. This replaces the objective referred to earlier and attempts to capture the essence of the system being described. A root definition is hence more than a mere statement of the objectives of the system. It incorporates the point of view that makes the activities and performance of the system meaningful.

Any area of concern can be investigated in terms of a number of possible transformation processes, and it will also be possible to derive a number of root definitions. A manufacturing enterprise, for example, could be represented by the following transformation processes:

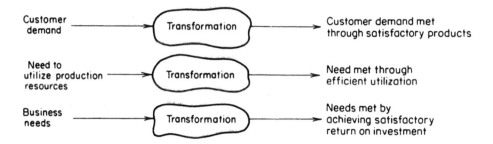

Root definitions, relevant to the first two, might be:

A system to meet the market needs for a particular range of products within the constraints of the production resources

A system to achieve the efficient utilization of production resources whilst maintaining secure employment and acceptable working conditions of employees

What the analyst is doing, in this kind of analysis, is choosing to view the situation in particular ways and, through the subsequent process of modelling each system, exploring the implications of the views taken. The first root definition above is based upon a market-oriented W, while the second is based upon a production-oriented W. Clearly each model derived from each root definition will contain different activities.

Model Development

It cannot be emphasized too strongly that what the analyst is doing, in developing a HAS model, is *not* trying to describe *what exists* but *is* modelling *a view of what exists.*Hence the model that is produced is no more than a model of the system described by the root definition. In Chapter 3 we will develop ways of using these models (methodologies), but first let us concentrate on the development of such models.

The model of a HAS is an intellectual construct. It is no different to a differential equation. That is also a model of a particular view of a situation. However, the latter is in terms of the language of mathematics, the former is in the language of systems. Systems models are termed *conceptual models* and are derived by defining the *minimum, necessary* set of activities at a particular level of detail (resolution level) for the system to be that described by the root definition. So that, whereas the root definition describes what the system *is*, the conceptual model describes the set of activities that the system must *do* to be the system so defined.

The constraint 'necessary' is added to ensure that the models are not deficient and 'minimum' is included so that what is aimed for is the simplest description. There are an infinite number of models that are not minimum; there should be

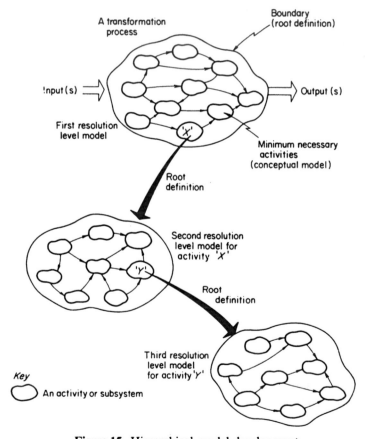

Figure 15. Hierarchical model development

a unique model that is. In practice this is seldom the case, since two analysts modelling the same root definition may choose slightly different resolution levels and may choose different words to describe the same activity. In developing HAS models it is necessary to live with the problem of semantics. Words are not as precise as mathematics but the richness and variety of description that is possible with words matches the complexity of management problem situations in a way that mathematics can not.

The first resolution level model, derived from the root definition, will usually contain between five and ten activities. Any attempt to develop models more detailed, and hence more complex, than this will make it more difficult to defend them in terms of the minimum and necessary conditions. Each activity in this first level model can, itself, be defined as a system and, through the mechanism of a root definition, be modelled. Thus, if at each stage seven activities are produced, two stages of model development will produce a model of 49 activities. This is still a model of the system defined by the first root definition but, whereas it would be impossible to defend 49 activities in terms of the above conditions, it is possible to do it in two stages. This process of hierarchical model development is illustrated by Figure 15.

As an example of this process consider an investigation that I undertook while in the role of Managing Director of ISCOL Limited, the company established by the Department of Systems in 1970. I was concerned about the allocation of managerial roles within the company and chose to view it in a particular way as a means of identifying those activities that needed to be managed. Since I was occupying the roles of both analyst and managing director, I took only one root definition which represented my W. In retrospect it might have been more enlightening to have considered several, but my main aim was to examine alternative ways to grouping activities into acceptable managerial roles rather than to initiate debate about what activities might, or might not be undertaken by the company. The root definition chosen was as follows:

> A bridging system between the Department and the outside world with the task of assisting the Department in teaching, researching, and promoting a systems approach to problem-solving through profitable involvement in activities in the outside world.

This particular root definition attempts to capture the view that ISCOL is concerned with the interface activities between an academic department and the world outside the University and also with satisfying particular departmental needs and doing so profitably (both intellectually and financially).

At the first resolution level, five activities only were modelled:

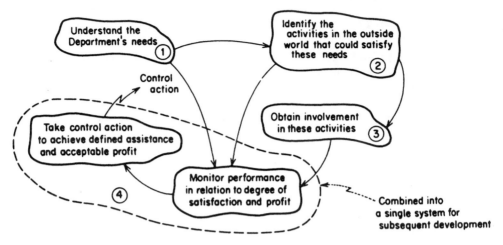

In this model the words 'Department's needs' are intended to cover the needs for assistance in teaching, researching, promoting, and profitable involvement.

At the next resolution level, the constraint was built into the root definitions that these activities would be carried out by ISCOL and that they would consist of two kinds of activities: projects and courses. Also at this stage the monitor and control activities were defined in a single root definition. These root definitions were as follows:

(1)

> A system to establish a continuing role for ISCOL which meets the needs of the systems department.

(2)

> A system to create, in the outside world, an awareness of the particular expertise within ISCOL (through courses and other means of promotion) so that appropriate outside activities can be identified.

(3)

> A system to ensure effective involvement in outside activities (through both MA and full-time projects) so that both intellectual and financial gain can be continuously achieved.

(4)

> A system to monitor and control the total activities within ISCOL so that it can achieve its desired role at an acceptable return on effort invested.

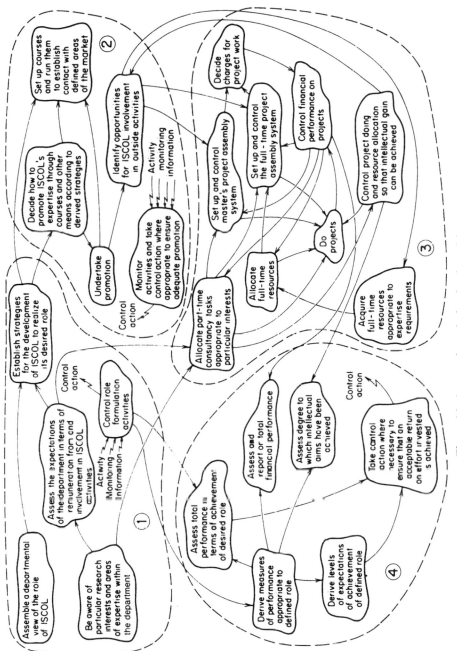

Figure 16. A second resolution level model

In the root definition for system 3 the project involvement is identified as two kinds: projects associated with the master's degree programme, and those undertaken by the full-time consultants employed by the company. The model derived after these two stages of development is illustrated by Figure 16.

Using the formal systems model as means of questioning the adequacy of the model in Figure 16 leads to the following:

(a) *Objectives* ⎱ Both are defined by the root definition and are clearly
(b) *Boundary* ⎰ represented by the set of activities derived.
(c) *Connectivity*   This was checked and could be defended.
(d) *Measure of performance and decision-taking procedures*   These are both defined in the monitor and control systems and the area of authority of each controller is defined by the appropriate boundaries.
(e) *Resources*   Activities are included which relate to their acquisition and deployment.
(f) *Hierarchy*   This concept is used in the model development and two levels in the hierarchy are represented in Figure 16. Further wider systems are not appropriate. Decision-taking and control mechanisms which encompass both the academic and the outside world in general are not capable of being engineered and hence anything wider can only be considered as an environment.

Further expansion of the model could be undertaken as a whole for selected activities; however, in this actual study it was used at this level to examine the significance (in management terms) of alternative groupings of activities.

The foregoing description has provided an example of the production of a conceptual model from a root definition. It does not indicate, however, *how* to produce a conceptual model. Unfortunately there is no algorithm or recipe which will ensure that, if followed, a conceptual model, in terms of minimum, necessary activities, will result. I can only provide guidelines which I find useful. The ability to do conceptual model building can only come from practice. Suppose that we take familiar scientific activity to be a HAS. It can be described by the following root definition:

A system to establish new, scientifically supported knowledge about area *X*.

Here the specific area of investigation has not been defined as I wish to remain general. However, since the words 'scientifically supported' have been included, the accepted method of science has been implied, which is to propose a hypothesis and then devise a repeatable experiment in which that hypothesis can be tested.

Figure 17 shows the transformation process contained in the root definition and the first level conceptual model derived from the root definition.

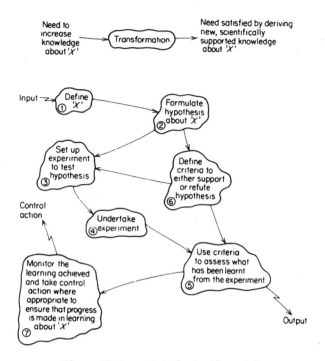

Figure 17. A model of scientific activity

Since a model of HAS is initially a model of a transformation process, it must be possible to trace that transformation through the activities. Thus, since 'X' has not been defined there must initially be an activity which defines 'X'. Based upon what X is and because the knowledge gained is to be scientifically supported knowledge, activities follow which are of the form, 'Formulate hypothesis about X', and 'Set up experiment to test hypothesis'. These must then be followed by activities concerned with *doing the experiment* and *learning from doing* if the output of the transformation is to be realized. Thus activities 1–5 represent an expansion of the transformation process. Activity number 6 has been included as a support activity, since I would argue that if you are going to learn from doing something it is necessary to include a statement of *how* you are going to assess what has been learnt. Activity number 7 is included in order that this model *is* a model of a HAS. Reference to the formal systems model requires the inclusion of a monitoring and controlling activity to ensure that the output is achieved. It could be argued that activities to do with resource acquisition and deployment have been omitted, but I would defend their omission at this level on the grounds that activities numbered 3 and 4 have been worded so that they are implied and would appear at the next level of expansion, i.e. you cannot set up an experiment without acquiring resources and, in activity 4, you cannot undertake an experiment without deploying them.

The question of resolution level is a difficult one when the modelling language is in terms of verbs. If in a model of a manufacturing enterprise, the activities 'Derive company policy' and 'Prepare invoice' appear, these are obviously at vastly different levels of detail and should not be in the same model (unless partial expansion of the model has been undertaken deliberately). However, the distinction is not usually that obvious. The words chosen in the root definition define what is important and, if it can be argued that, *because of the root definition*, activity A is equally as defensible as activity B, then they are both at the same level of resolution.

The question of 'where to start' when modelling the transformation process is also interesting. In the above example, concerned with scientific activity, I started at the input and this seems to be a reasonable choice, *if the root definition can mainly be expressed as a transformation process of the following kind*:

Take, for example, an earlier root definition;

A system to achieve the efficient utilization of production resources whilst maintaining secure employment and acceptable working conditions of employees.

The transformation process in this root definition is

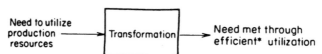

Hence a model of this transformation process would be

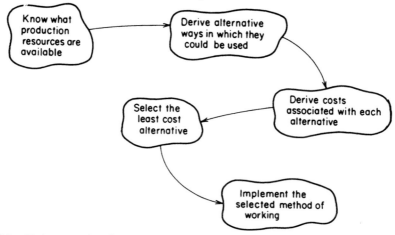

*'Efficient' is interpreted as 'least cost'.

To complete the model, however, the other words in the root definition need to be recognized and also monitoring and control activities need to be included so that the model is defensible as a model of a HAS. In the root definition a constraint has been stated: 'whilst maintaining secure employment and acceptable working conditions of employees'. This constraint may affect the choice of method of working and may cause the inclusion of other factors than just cost, in financial terms, into the assessment of each alternative. Hence the complete model would be that contained in Figure 18.

Since 'maintain secure employment' and 'acceptable working conditions' are not precise in meaning, an activity has been included which defines their interpretation. This interpretation is then used as a constraint on which the selection of alternatives is based. 'How to implement' is also taken to be subject to the same constraint. (Using external resources, for example, may not be acceptable.) Thus a dependency on this constraint is defensible for the 'implement' activity. Also any control action that is taken will be subject to the same constraint.

The above examples represent relatively simple root definitions and, as long as the form of transformation indicated above can be reasonably expressed, the method of derivation described seems useful. More complex, or sophisticated, root definitions lead to a more difficult-to-describe process of conceptual model development. A characteristic which is present in such models of that the set of activities associated with the transformation itself are a minor, rather than a major set. In the examples quoted above the expansion of the transformation led to almost all of the activities in the models.

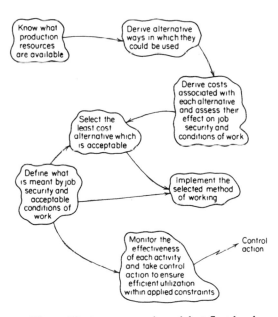

Figure 18. A conceptual model at first level

Given a more complex root definition, a useful first stage is to identify a small set of subsystems which are different in nature but which are arguably defensible from the root definition. Each of these subsystems can then be expressed initially in the transformation form (need → satisfaction of need) discussed above and the models then derived as described.

As an example of this approach let us return to a project mentioned in Chapter 1. This was a project concerned with information systems development but which required, first of all, a model of activities that were to be served by such information systems. The project was within the Engineering Department of British Airways (BAED). The picture of the situation appeared, in Chapter 1, as shown in Figure 2, but the relevant part is reproduced here for convenience:

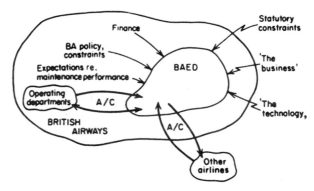

Figure 2. BAED and its environment (cf. Figure 2 in Chapter 1). *Key:* ——→, internal information flows; ⤳ information flows external to the system boundary; A/C, aircraft

This view of BAED and its environment led to the following root definition:

A British Airways owned system concerned with the continuously effective and efficient planned maintenance of aircraft belonging to BA and other contracted airlines, with a performance acceptable to operating departments but within statutory, local, and BA-applied constraints.

As a first stage in model development five subsystems were defined, each one having a distinct purpose. How these systems relate to one another and to the environment, depicted in Figure 2 above, is illustrated in Figure 19. Since the system is 'British Airways owned' and because it is concerned with 'continuously effective and efficient planned maintenance' a planning system is included which relates the development of the BAED system as a whole to the development of BA itself. Thus this is the system that links activity within BAED to the wider system of BA to which it belongs. Again, because 'effective and efficient' maintenance is required, a system that is concerned with developing the technology and other associated resources is included and it is this system that links the maintenance activity to the relevant part of the environment, i.e. that part concerned with the technology of aircraft maintenance and the airline business as a whole.

A system concerned with obtaining business is included because of the need to link with the operating departments of BA and those of other airlines. It is this system that negotiates maintenance contracts and agrees performance expectations with the 'customers' of the maintenance activity. The actual transformation process represented in the root definition, i.e. that of taking in aircraft to be maintained and transforming them into aircraft ready to go into operation, is represented by the BAED aircraft maintenance system.

Finally, a system concerned with the overall control of BAED activities is included in order that this should be a model of a HAS. It is this system that, as well as taking overall action, decides how to respond to the environmental constraints and ensures that any control action that is taken is within the total set of constraints.

Each one of these systems is different in nature and has a distinct purpose. However, each is defensible on the basis of the root definition. In this example, the activities resulting from the transformation of aircraft requiring maintenance into aircraft maintained are relevant to only one subsystem (of the BAED system) out of five.

The expansion in Figure 19 was felt to be a useful first stage in model

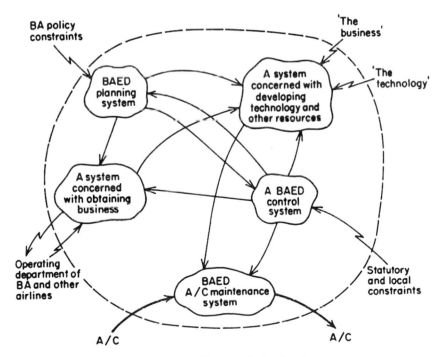

Figure 19. First resolution level

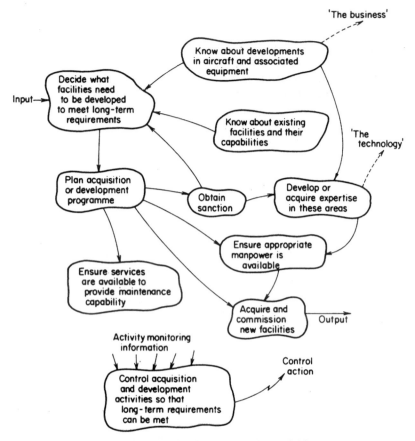

Figure 20. A resource development and acquisition system

development. Each of the five systems was then described by a root definition and the models derived through the process described earlier.

For the resource acquisition and development system the root definition used was

A BAED owned system for the acquisition and development of those maintenance resources needed to meet the requirements specified by the planning process so that best use is made of available knowledge and equipment.

The output of the BAED planning system was taken to be long-term resource requirements which then formed the input to this system. Seen as a transformation process it could be represented as

The model resulting from this expansion is illustrated by Figure 20, where the inputs and outputs are shown together with the interactions with the environment. Further use will be made of a development of this model later, but here it has served the purpose of illustrating the two stages involved in deriving a model from a complex root defining (i.e. one in which the activities resulting from the transformation process only, represented a minor set in the model as a whole).

Root Definitions

The set of concepts related to the formal systems model provides a way of checking if the conceptual model, once developed, is a defensible model of a HAS. What help can be given at the root definition (RD) stage to decide if it is a 'good' root definition or not? This question was faced by D. S. Smyth, a doctoral student within the Department. What he did was to examine a number of root definitions that had been used in past projects that we believed were successful, in the sense that the clients had agreed that useful outcomes from the projects had been achieved. He then asked the question, 'Do these root definitions contain elements that are common or represent identifiable characteristics that are translatable from root definition to root definition?' The result of this inquiry was a set of elements that can be used to test a root definition. They *do not* tell you if the root definition is good in the sense of useful, they *do* tell you if the root definition is good in the sense of *well formulated*. These elements are summarized by reference to a set of letters which can be remembered in the mnemonic CATWOE. The definitions of the elements are contained in the following table (from Smyth and Checkland, 1976):

Consideration	Amplification
(1) 'Ownership' (O)	Ownership of the system, control, concern or sponsorship; a wider system which may discourse *about* the system
(2) 'Actor(s)' (A)	The agents who carry out, or cause to be carried out, the transformation process(es) or activities of the system
(3) 'Transformation' (T)	The core of the RD; A transformation process carried out by the system; assumed to include the direct object of the main activity verb(s)
(4) 'Customer' (C)	Client (of the activity), beneficiary, or victim, the subsystem affected by the main activity(ies); the indirect object of the main activity verb(s)
(5) 'Environmental and wider system constraints' (E)	Environmental impositions; perhaps interactions with wider systems other than that included in (1) above, these wider systems being taken as given

To this list is added the sixth item which is, by nature, seldom if ever explicit in a root definition but is always implicit and always relevant:

(6) *'Weltanschauung'* (W)	The (often-unquestioned) outlook or taken-for-granted framework which *makes this particular RD a meaningful one*

The set of elements represents a useful checklist against which a root definition can be tested. Its usefulness lies, not in ensuring that each element is contained in the definition, but of ensuring that, if one is omitted, it is because a conscious decision has been made to do so. Since a basic concept of a HAS is that it is a transformation process, the element T must be present. Also, because the definition only makes sense from a particular point of view, it must be possible to identify the W which gives it meaning. All of the other elements can be included, or excluded, on the basis that they are either important, or not, according to the judgement of the analyst. The most frequent element to be omitted is A, as its inclusion usually constrains the system to be particular to a particular area of real-world activity. In our experience more insight into a situation can be gained, if, say, the boundary of a marketing system is not equated with the organizational boundary of a marketing department. There is a danger that such a constraint may be introduced unnecessarily if A is specified (in the above example) as the members of the marketing department.

Of course, it may be regarded as crucial to the analysis, to include in the root definition a specification of A. In a study undertaken by a postgraduate student in the Department, C. Y. Yuen, the subject of investigation was the operation of a particular voluntary hospital activity, that of visiting mental patients. The root definition taken in that study was

A regular, volunteer-student-manned, medically approved, mental hospital patient 'comforting' system.

In this root definition the 'customers' are clearly specified as the mental hospital patients. The 'actors' are the volunteer students, and the 'transformation' is to provide 'comfort'. Some environmental constraints are specified (i.e. those aspects of the system which the decision-taking process cannot change). The activity must be regular; the students have to be volunteers; and the whole operation must be medically approved. However, 'owner' is not specified and hence W is ambiguous. It may be stated as, 'comfort' can be provided through this kind of student activity. However, it can also be interpreted as, hospital visiting provides some form of self-satisfaction for the student volunteers; or as, free help is a justifiable way of reducing the workload of the medical practitioners. In this case it would have been helpful to remove the ambiguity by specifying owner.

This particular example illustrates that it is possible to construct three root definitions containing exactly the same transformation process but based upon three different Ws.

As well as being a useful test of a root definition, CATWOE is also a useful device which is helpful in their construction. Consideration can be given to the various elements prior to the formulation of the definition. However, once formulated, the root definition should be tested to ensure that the definition represents what was intended.

Pitfalls to be Avoided

It must be remembered that the systems language that has been described enables an analyst to make his thinking about some real-world situation explicit. It is not a way of describing what exists but is way of describing *an interpretation of what exists or some thinking relevant to what exists.* Common mistakes are therefore concerned with confusion associated with this distinction. In constructing root definitions it must be remembered that the definition relates to a system. In an exercise used within the course at Lancaster, the object of concern is the English pub. Root definitions are frequently produced which begin with the phrase: 'A pub is a building ...'. This immediately gives the definition the status of a description of what exists. *We cannot state what a pub is, we can only state what the system we are defining is.* Thus a root definition relevant to a pub could start as follows: A brewery-owned, landlord-operated systems for .... Of course this is only one example among many; however, in a root definition, we are defining *a system*, so call it a system.

The C of CATWOE refers to 'customer', the direct object of the transformation process. Frequently the customer specified in the root definition is several stages removed. For example, if our concern is with a system for controlling product quality, we might wish to define a system for setting specifications and quality standards. The output of this system, therefore would be the specifications and standards themselves, and the direct recipient of it might be the inspection activity. So whereas here the inspectors should be C, what is frequently specified as C are the actual customers for the product.

Root definitions are sometimes produced which are attempts at 'portmanteau' definitions. The thinking behind this is that, since the real-world situation is a mixture of Ws and transformation processes a definition needs to be produced that reflects this mixture. But bear in mind that the systems analysis that we are doing is an attempt to structure an unstructured situation through simplifying the complexity not by trying to mirror it. Hence a root definition should contain *one* transformation process and should have 100% commitment to a single W. Using multiple transformations in the root definition tends to build into it part of the conceptual model. Thus if the definition is of a system to do A, to do B, and to do C (i.e. three transformations), the conceptual model must contain the activities, do A, do B, do C, and control the doing. This has destroyed the fundamental relationship between 'being' and 'doing' that is the characteristic of the root definition ↔ conceptual model relationship. It must be emphasized that the root definition is a description of what the system *is* and the conceptual model is a description of what it must *do* to be that system.

The aim, therefore, should be for a single transformation process. In practice, however, it may be necessary to proceed with a root definition in which there is more than one T, for the simple reason that the analyst's linguistic ability may be such that, whereas two verbs describe what he wants to describe adequately, he is unable to find a single verb which captures the meaning precisely.

The whole CATWOE analysis process is concerned with identifying what is stated in the root definition, it is not a mechanism for unconsciously adding to the definition. It is not unusual to find elements described under the CATWOE headings that are in the mind of the analyst but which are not contained explicitly in the root definition. If the CATWOE analysis has prompted thought about certain elements, and they are judged to be important, then the root definition whould be modified to include them:

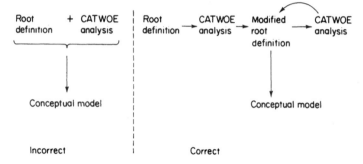

It is after all the root definition that is being modelled.

The systems language that has been described consists, essentially, of two linked ideas: the root definition and the conceptual model. It is the pair of ideas that is important. A common failing, particularly during the initial attempts to use this language, is in the production of sparse root definitions leading to sparse conceptual models. The pair are still logically defensible but not very useful. The use of CATWOE helps to overcome the failing but to emphasize the point, consider the following extreme root definition:

A system to do X

The conceptual model, at a broad level, derived from this definition will contain three activities: (a) decide how to do X; (b) do X: (c) monitor and control the doing. As a pair, the root definition and conceptual model have the necessary logical linking; however, they would probably not turn out to be very useful. Sparse root definitions of this kind are usually the result of superficial thinking about the situation. A thoughtful use of CATWOE will help to overcome this tendency.

Referring again to the importance of pairing the root definitions and the conceptual model leads to a further pitfall. The words used in the root definition need careful thought and each word should only be included if it is regarded as important. Because of this, activities should be included in the conceptual model which relate to those words. It often happens that phrases are included in a root definition in order to make the system so defined highly particular to the real-world situation being analysed. Yet when the conceptual model is derived, no activities exist which make reference to these phrases. If such phrases are *omitted* from the root definition and *it can be argued that the model is still logically defensible*, then the followng question needs to be considered: are the words actually necessary? If the answer is yes, then the model is deficient in terms of the ativities being the necessary set. If the answer is no, then the words should be omitted from the definition.

A pitfal which should be recognized is related to the modelling language itself. It has been stated that the language of activities is verbs. Models are often produced in which a mixture of verbs and nouns appear. Consistency of language needs to be maintained since, if a noun appears, this may mean that a real-world constraint (in terms of a particular 'how') has been introduced which may not be defensible on the basis of the root definition.

Take, as an example, the system concerned with control of quality mentioned earlier. A part of the model may be described as follows:

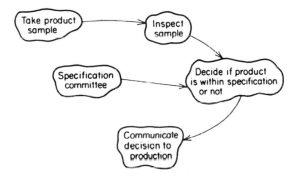

In this model, a part of the organization has been included as an activity, i.e. the specification committee. This is a particular 'how' of doing the activity 'set specifications'. For the model to be consistent the activity should be included as 'set specifications'. If this particular 'how' of doing the activity is a constraint, then the activity should not appear since, by definition, the model represents that area over which the decision-taking process has control. Constraints are outside the system boundary and in this case (where 'set sepcifications' it through the specification committee) the specifications would be an input to the system. Inconsistency in the use of modelling language is a common pitfall. It arises because of the use of words as the basic language. Such inconsistency does not arise in mathematics. No analyst would consider writing an equation of the following form:

$$M\frac{\mathrm{d}p}{\mathrm{d}t} + \mathrm{VALVE} = f(P_0).$$

Such inconsistency is not as obvious in a language based upon verbs as it is in a mathematically based language, but it is just as significant.

The distinction made at the beginning of the section, i.e. the distinction between activity of thinking about the real world and activity in the real world itself leads to a fairly common failing. Usually some picture of the situation is formed initially through interviewing or through acquired knowledge if the analyst is part of the situation. This picture is in terms of real-world activity and is the source of RDs.

The problem arises as a model is being developed from a particular RD. The analyst knows that activity goes on in the real world which is not being included in the model. This is not unreasonable since the RD represents only one view amongst the many which make up the complexity of the real world and so the mistake is to then include the activity even though it cannot be justified on the grounds of logic. This point is further illustrated in the following diagram:

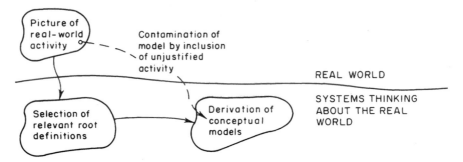

In the above diagram the full arrows indicate the legitimate flow of analysis. The broken arrow effectively eliminates the RD and hence weakens the model as a defensible representation of a particular view.

The choice of the words 'root definitions' as a technical term to describe the emergent property of a system was intended to convey the notion that this was a definition that contained the 'essence' of a particular view of the situation.

It is frequently the case that RDs are constructed which are so general that they could refer to a wide variety of situations as well as that to which they are intended to be relevant. So, for example, a system relevant to a library might be described as an employment system. This could be a legitimate view but it does not contain the essence of the situation which could be argued to be something to do with making stored information available to authorized borrowers. A useful test is to take the developed RD and to ask the question: 'can I infer the particular situation from the RD?'

It helps, in model building, if the structure of a RD is clearly understood. One possible structure could be to define the system as one to convert X into Y by doing Z. Activities would then be built into the model which are concerned with doing Z and other activities must be included which assess how well this converts X into Y. An alternative structure could be a system which does X to achieve Z through Y. Similar arguments can be mounted about activities to do

Y and an assessment of how well X is done and whether or not their joint achievement leads to Z. No one structure is necessarily better than another but understanding what the structure is, is what is important.

Qualifying statements are sometimes introduced into a RD which lead to the inclusion of activities in the resultant model outside the area of authority of the system decision-taker and hence outside the system boundary (thus they should not be included in the model). Returning to the English pub example, I have seen definitions of the form .... a landlord operated system to sell beer to customers who participate in social intercourse. The decision-taker of the system is the landlord and he can only control the sale of beer. The decision to participate in social intercourse or not is that of the customer and not the landlord. Thus activities cannot legitimately be introduced into the model which are concerned with making social intercourse happen. The phrase 'who participate in social intercourse' should therefore be deleted from the RD.

A common error in developing models is to include activities that are essentially not achievable. Since these are human activity systems models each activity, in principle, could be undertaken by an individual manager or operator and hence could represent their area of authority (job specification). In a model from the above definition of a system to self beer, activities of the form 'generate profit' appear. An individual could not be given this responsibility. Profit is the result of obtaining revenue and incurring cost and it is activities related to these latter two considerations that should be included. Taking the same example of the English pub, a customer oriented model may include activities of the form 'generate a pleasant social atmosphere'. This again is not achievable. Activities can be included which are of the form 'provide those amenities which it is believed will lead to a pleasant social atmosphere'. To this would need to be added 'monitor the atmosphere achieved' and 'control amenities provided to ensure that the atmosphere achieved is pleasant'. Thus these three activities would need to replace the original single activity. At a company level, a non-achievable activity which is frequently included is 'generate desirable company image'. Company image is something that results from the way in which a company does almost everything else. So, unless the company concerned is in the image building business rather than insurance, banking, manufacturing etc, this is an activity that should not appear. In these kinds of organizations company image can be monitored but it cannot be generated as an independent activity.

Because the formal systems model (p. 30) requires any model of HAS to contain monitor and control processes, they are frequently added, maybe as an afterthought, just to make the model complete. They appear at the bottom of a model rather like Tasmania off the coast of Australia; needed for a complete description but not really connected. Casual inclusion of this kind is not good enough. Thought needs to be given to what is being controlled and why. The inputs to the monitor are in the form of performance information flows (determined by examining the 3Es for each activity, see p. 92). Thus they are different kinds of arrows to the logical dependencies that describe the connectivity in the rest of the model so make them look different (either by using multiple arrows properly labelled or some other shape of arrow). The control action which represents the output of the control activity is also different to other activity outputs. It may be connected to any (or all) of the other activities depending upon the control action required. Thus it is of the form of 'occasional intervention' rather than continuous action aimed at every activity. I illustrate this selective linkage by a crooked arrow.

Thus, to summarise, always include qualifying statements after the verb monitor and control to identify what is being monitored and why control action might be taken. Because the input to the monitor and the output from the control are different to other inputs and outputs make them look different by using different kinds of arrows.

A final observation that it is worth making is in relation to the model-building process itself. The analyst cannot be sure at the start of the process that the RDs chosen will be relevant or represent a comprehensive or sufficient set. Comparing the models against the observed real world, assessing desirability etc., will result in the generation of new models and the abandonment of others. Thus the modelling process needs to be fairly rapid, though defensible, and this needs practice. If an analyst spends too much time expanding or elaborating a model he will be reluctant to abandon it, should in turn out to be irrelevant. He will have generated psychological ownership of the model and its retention may well distort the resultant analysis.

Summary of pitfalls

The pitfalls discussed in the preceding section refer to the following areas:

- Real world/intellectual world distinction
- Specification of customer in a RD.
- Portmanteau RDs.
- Loss of being–doing relationship in RD↔CM (conceptual model).
- Use of CATWOE analysis
- Sparseness of RD and CM.
- Overspecified RD or underdeveloped model
- Incoherent modelling language
- Inclusion of unjustified elements of the real world
- RDs lacking the essence of the situation
- Confusion over the structure of RDs
- Inclusion of qualifying statements in a RD leading to activities outside the decision – taking area of authority
- Inclusion of non-achievable activities
- Casual inclusion of monitoring and control
- Over-enthusiasm for, or excessive psychological ownership of CM.

Summary

Chapter 2 has been concerned with the specification and development of a particular modelling language based upon the concept of a human activity system (HAS). It is this language that has been developed via the operation of the 'action research' cycle described in Chapter 1. Hence the language has emerged from practical involvement in real-world problem situations and has been shown to be useful as a modelling language with application at the 'soft' end of the problem spectrum. Unlike 'hard' problems which are amenable to precise description, 'soft' problem situations are crucially dependent upon the perceptions of the various participants in the situation. Thus HAS models are models of perceptions relevant to the situation and are not models of the situations themselves. All models are only intellectual constructs and this is emphasized in the HAS model by making each one entirely dependent upon a particular *Weltanschauung*, or W.

The system chosen is initially defined by a *root definition*, whose formulation is tested through a series of questions based upon the mnemonic CATWOE.

The root definition defines the resolution level of what is taken to be 'the system' within the hierarchy of systems relevant to the area of concern. The root definition describes, therefore, the *emergent properties* appropriate to this level. Given this definition, the system can be modelled as a *whole entity* in terms of an interconnected set of activities in which the basic language is all the verbs available to the analyst. This *conceptual model* represents the minimum, necessary set of activities, at a particular level of detail, that the system must do to be the system defined by the root definition. Thus the conceptual model is a model of the root definition and is *not* a model of anything else.

The conceptual model can be assessed, in terms of its adequacy as a model of a HAS by reference to the *formal* systems model. It cannot be 'validated' as a model of an actual situation.

Because the model is expressed as a set of activities, it is in terms of *what* the system must do and not in terms of *how* it might do it.

The addition of this modelling language to the others available to an analyst (a survey of which was given in Chapter 1) provides a capability for the analysis of problems related to the whole of the problem spectrum from 'hard' to 'soft'. Our concern now must be with the *use* of these languages, i.e. with methodology.

Chapter 3
Systems Methodologies

Problem Content and Problem-solving

Our concern in Chapter 3 is with the use of systems languages (of whatever kind) in the solution of real-world problems (i.e. with methodology).

The word problem has been described earlier in terms of its relation to the extremes of a spectrum ranging from 'hard' to 'soft', where hard problems are concerned only with questions of 'how' and soft problems are complex mixtures of both 'what' and 'how' types of question. That distinction is still useful but, in discussing methodology, I wish to use a more general definition. I am taking 'a problem' to be any *expression of concern* about a situation. In this context methodology represents a structured set of guidelines which enable an analyst to derive ways of alleviating that concern. Methodology is not method (or technique). It needs to be more flexible than either of these in terms of its structure and its application if it is to be appropriate to the variety that exists in real-world concerns. A technique is characterized by guidelines which are precisely defined, so that if your concern is, say, how to solve a differential equation, techniques are available using either classical or Laplace operators which, if correctly applied, will lead to a solution which any competent analyst will agree is correct. Such a situation, however, relates to only a very small set of real-world concerns in general. The application of a methodology may involve the use of techniques, but it is the methodology which determines if a particular technique is appropriate or not.

It is the degree of precision in the guidelines within a technique that makes reliance on them, as a vehicle for real-world problem-solving, an inappropriate approach. To be 'technique oriented' is to introduce the danger that the problem situation will be distorted to fit the technique. It is a danger in the sense that, although a solution is guaranteed, the solution may not actually remove the initial concerns. A more successful approach is to be 'problem oriented' and to allow the situation to distort the way the analysis is being carried out. This orientation demands flexibility in the approach; hence the emphasis on methodology and not on technique.

The process of problem-solving can be viewed in terms described by Figure 21. In this figure, a distinction is made between activity which is in the real

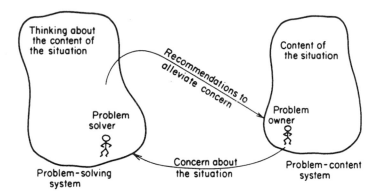

Figure 21. A view of problem-solving

world (the situation) and activity which is to do with thinking *about* the real world. The latter activity is to do with reacting to a concern expressed by some problem owner in the real world and is activity carried out by a problem solver.

It must be emphasized that this is an intellectual distinction and 'problem owner' and 'problem solver' represent the names of two *roles* and not of two people. A single individual can occupy both roles simultaneously. In the study described earlier to do with the management of ISCOL, when I was in the position of managing director, I was, at the same time, both the problem owner and the problem solver. In this example I was also in the role of client. The client may be defined as that role holder who causes the study to happen. The client may, or may not, be the problem owner. These are useful distinctions to make, as the client authorizes a particular study and will have some reasons for so doing. The problem owner may change as the study proceeds and the nature of the concern changes. Thus the identification of the problem owner provides some insight into whether or not the proposed 'solutions' will be acceptable and hence feasible. The problem solver may be a single individual or a group of individuals. Making the intellectual content of a study explicit (through the definition of the content of the problem-solving system) enables the approach to be defended, debated among the group and *consciously* modified if necessary. Making the thinking explicit is also important if the problem solver is a single individual in that the analyst will know, at any time, how what is being done is contributing to the overall analysis and can decide (again consciously) to modify the process of analysis, or not, based upon the progress being made.

The distinction made in Figure 21 enables us to view the intellectual activity of problem-solving itself as a human activity system. It is a system to transform the input (a concern about some situation) into the output (recommendations to alleviate that concern). Based upon a particular view of the value of problem-solving, a root definition can be derived as follows:

> A problem-oriented, problem-solving system to transform a statement of concern about some situation into recommendations to alleviate that concern that are acceptable to the problem owner.

In this definition, the owner and actors are not specified. The transformation is clearly stated and the customer is taken to be the problem owner. The W implied by the definition is that problem-oriented, problem-solving activity will lead to recommendations that are acceptable to the problem owner. This is a customer-oriented W, implied by the phrase 'acceptable to the problem owner', and it is implied that such activity is worthwhile Environmental constraints are included in the sense that the recommendations cannot be unacceptable and the problem-solving activity must be problem oriented.

A model of this system is given in Figure 22. In this model two control systems have been included. One is responsible for ensuring that the problem-solving process is adaptable and is, hence, capable of remaining problem oriented as an appreciation of the concern becomes deeper; the second is responsible for the control of the system as a whole. Although each activity must be carried out

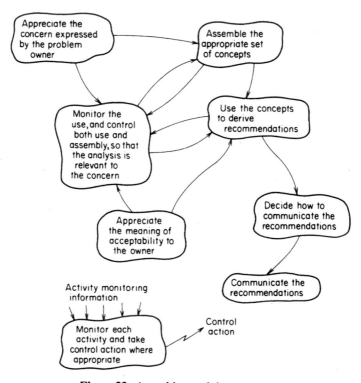

Figure 22. A problem-solving system

effectively if relevant recommendations are to be derived, the activity 'Assemble the appropriate set of concepts' is the most important with regard to our present discussion of methodology. We are interested in the use of systems concepts and *how they are assembled* represents the particular methodology which is regarded as appropriate to the nature of the concern. So that if the concern is about process optimization or plant design the concepts may be expressed in the language of mathematics and the particular assembly related to a 'hard' systems engineering methodology. On the other hand, if the concern is oriented towards the 'softer' end of the problem spectrum, our concepts will be those of human activity systems and the assembly will be a particular 'soft' systems methodology.

As the above activity indicates, and as a result of the general desire to be problem oriented, the methodology used in a study needs to be tailored to the actual situation under investigation. Chapters 4 and 5 contain a number of project descriptions which illustrate the flexible use of systems concepts and in which the resultant methodologies will be emphasized. However, work in the Department has led to a general soft-systems methodology, the Checkland methodology, which will be described later in this chapter. First of all let us briefly survey the systems engineering methodologies which led to the initial formulation of the 'hard' systems methodology which was used at the start of our action research programme.

Systems Engineering Methodologies

In Chapter 2 I gave a summary of the recent history in those areas of engineering and biology that had led to the set of concepts now used in our systems language. The motivation to undertake the development of system design methodologies arose because of the four characteristics that affected post-war industry:

(a) Technical systems were becoming more complex.
(b) Market environments were becoming highly competitive.
(c) New projects were increasingly more expensive.
(d) Computer developments made complex calculations more feasible.

These features gave rise to the need for integrated design methodologies that were capable of producing 'optimized' designs and the need to see design as part of business development planning. Thus there was a realization that technical system design was part of a wider environment which had to be accommodated in the design process.

The interactions are illustrated by Figure 23, in which a new plant is shown to have significant relationships with three aspects of its environment, i.e. the market to be served, the other production processes with which it may interact in terms of material flow (raw materials and products), and the business environment, in which the interaction will be in financial terms.

Perhaps the best illustration of an explicit design methodology which attempts

58

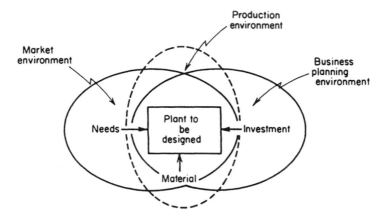

Figure 23. Plant design and its environment

to cater for these interactions is that developed by T. J. Williams while with the Monsanto Chemical Co. (Williams, 1961). This design process is represented by a hierarchical arrangement of feedback loops. The outer loop, illustrated by Figure 24, is what he describes as an 'economic feedback system', which seeks to define the potentially most profitable scheme for further detailed design. In

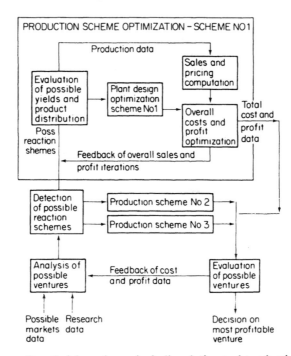

Figure 24. An overall optimizing scheme, including design and production optimization. (From *Systems Engineering for the Process Industries* by T. J. Williams. Copyright © 1961 McGraw-Hill Book Company. Used with permission of McGraw-Hill Book Company)

this loop data is assembled about the relevant market opportunities and research activities. This is then analysed to identify potential new ventures for the company. These are translated into possible reaction and production schemes and progressed to the level of plant design and further market analysis to determine the profitability of each postulated production scheme. Finally, the profit and cost data from all production schemes are compared to find out which would be the most profitable. The resulting data are then returned by a feedback loop to the initial analysis. There, the decision is made to proceed with more detailed design of the most profitable scheme or to seek another venture, based upon current market and research data.

Within this overall optimizing scheme are two other internal feedback loops (see Figure 25). Loop 1 contains repeated iterations on the possible sets of operating conditions to ensure that the most profitable set of products and byproducts is being used. Internal to loop 1 is an additional iteration on the possible design of the plant units themselves. This is included to ensure that the minimum production costs are obtained. The complete process of analysis is represented by the combination of Figure 24 with Figure 25.

This methodology illustrates the importance of adopting a hierarchical approach to investigation when the area of concern is highly complex and

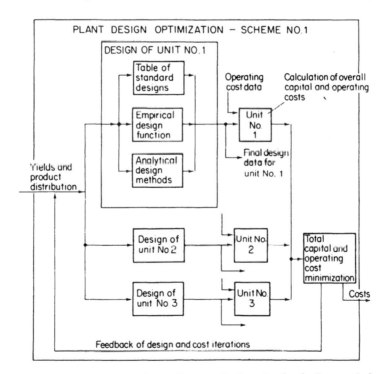

Figure 25. Detailed design loops. (From *Systems Engineering for the Process Industries* by T. J. Williams. Copyright © 1961 McGraw-Hill Book Company. Used with permission of McGraw-Hill Book Company)

contains high variety in the degree of detail to be considered. This is similar to the process of modelling described in Appendix I, where the detail of description is refined as the depth of analysis increases. In the approach described here the environmental interactions which are emphasized are those associated with the market and the business needs.

A similar approach to optimization was used in a design project that we were associated with in our first year of a activity (Swann, 1967). Here the major interactions were those related to the business and other product areas. This plant design project was undertaken in collaboration with an operating chemical company.

The plant was to manufacture an intermediate product which would be used as a feedstock for another chemical process. Hence the objective was to design a plant to produce the required amount of intermediate product at minimum cost over the lifetime of the plant. Since the plant only supplied the intermediate product to other plants within the one organization, the accepted accounting policy of that organization was adopted. This was to discount the cost of the plant at a fixed rate of return r. Hence the plant design was optimized with respect to the objective function

$$f = C + R/(1 + r) + R(1 + r)^2 + \cdots + R/(1 + r)^n,$$

where C is the gross capital cost of the plant, R is the annual running cost of the plant, and n is the plant lifetime in years. The function f, which is to be minimized, represents the total cost of the plant discounted over its lifetime to a net present value of cost.

To obtain the plant costs required for the calculation of the objective function, models of the individual units making up the system were generated. These local models were assembled into a global model of the whole process. Economic models were then added to determine capital and running charges. The plant parameters were then adjusted automatically by an optimization routine to determine the minimum value of f.

Since there were a large number of plant variables, the optimization was not carried out on all of them. Many variables were eliminated based on chemical engineering considerations since it could be argued that the objective function would be insensitive to them. Some variables were held at their top constraint and others were eliminated from practical engineering considerations. A final list of the ten most significant variables were selected for the optimizing calculations.

In the past, in this organization, plants had been designed by carrying out a few detailed design calculations and by selecting the one most economically attractive according to some specified criterion. The starting point for the optimization calculations was based on a design arrived at using this approach. The optimization calculations proceeded satisfactorily and on completion produced a design with a value of the objective function which was 16% less than the starting value – a considerable financial saving.

After locating the minimum, the response surface was investigated by perturbing each of the parameters affecting performance by $\pm 1\%$ and $\pm 2\%$ about their final values. (The response surface may be defined as the behaviour of the performance criterion in the region of its optimum value obtained by varying the plant parameters about their optimum values.) This exploration enabled two conclusions to be drawn. First, it showed that the 'optimum' point was very close to one of the plant constraints and, second, that the curvature of the surface in the region of the optimum was not severe. The curve of Figure 26 illustrates typical behaviour of the objective function with respect to variations in variable x_i about its optimum value $\bar{x}_i$.

The shallowness of this curve is extremely important since it shows that the plant design can be chosen some distance away from the constraint, at say $\hat{x}_i$, without incurring a large financial penalty. It also illustrates that the final design is not too sensitive to uncertainty in the individual variables x_i.

Finally, several starting conditions were tried to ensure that the best conditions found were an overall minimum and not just a local minimum. This project was felt to be particularly valuable in that it enabled a comparison to be made between the conventional design procedure and one based upon computer modelling and optimization, resulting in a saving of 16% in total plant costs (less, of course, the cost of the additional design effort required, but this was small).

In an environment which is highly competitive and well developed the above approach to systems design is essential. A further example is provided by electricity generation from nuclear fuel. To compete with the generation of electricity from fossil-fuelled stations, the nuclear industry has had to develop computer modelling methods. Stations such as Calder Hall, Chapelcross, and the early Magnox stations were built to demonstrate the feasibility of electricity generation on a commercial scale. Once this had been shown, it was necessary for survival to beome competitive with other fuel sources. In 1959 general

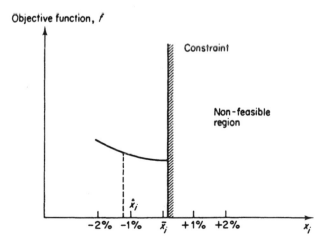

Figure 26. The objective function in the region of its optimum

purpose computer models were developed to aid the design and development of the new AGR (advanced gas-cooled reactor) stations. The computer programs provide a performance calculation procedure for the complete power station system, including the reactor core, pressure circuit, boilers, and power house, and are divided into a number of main sections as illustrated in the flow diagram shown in Figure 27. Because of the considerable complexity and field of coverage of the equations, many simplifications had to be made in the treatment of individual sections. The degree of detail included, however, is adequate for undertaking survey calculations during the development of a final specification.

To specify quantitatively, even in broad terms, a complete nuclear power station, values must be chosen for several hundred independent variables. Of these, about 20 are of prime importance in determining overall system

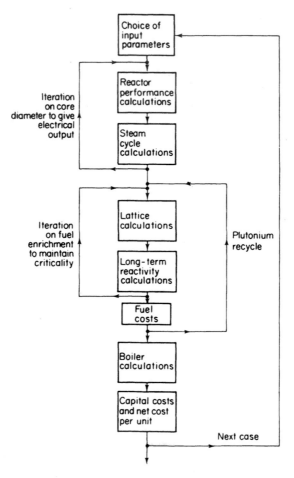

Figure 27. A flow diagram of a survey programme

performance and economics. These important variables include, for example, the fuel element and core dimensions, and the operating temperatures of the fuel, cans, reactor coolant, and steam. The parameter survey program provided a means of arriving at optimum values in each of these significant variables. The objective function in this case was related to the capital cost per unit.

The optimization of plant design using computer models is a great improvement upon the more conventional methods mentioned earlier. However, it is not the complete answer.

Although the importance of systems dynamics was often appreciated, it was seldom incorporated within the design procedure. Hence, systems were usually designed to meet a steady-state objective under the assumption that a control strategy could be devised to maintain operation under these conditions. In general, a more flexible design and operating system can be obtained by considering both the steady-state and dynamic aspects of the design together.

This approach is summarized in Figure 28, where it is indicated that, once a conceptual design has been established, both the steady-state and dynamic performance should be modelled and analysed simultaneously. The steady-state model provides a basis for optimizing the steady operating conditions within the constraints resulting from a consideration of the dynamic behaviour. It is essential at this stage that a complete dossier is obtained on likely loads and disturbances to which the system would be subjected under fault and normal operating conditions. It is also important to define system availability and the cost of non-availability since this can significantly affect the degree of redundancy and sophistication of control strategy required to overcome uncertainty in the estimation of subsystem performance and reliability. A system designed in this way allows feedback from the various operational considerations to the steady-state design to achieve a final design which is flexible in operation and of guaranteed performance. A project which made use of this design procedure will be described later in Chapter 4, but the design of a nuclear power station provides one example where this design philosophy is applied. It was only the changing duty of these stations that brought about a need to integrate dynamic

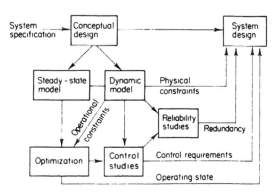

Figure 28. An overall design procedure

and steady-state considerations. All the first British gas-cooled reactors were designed as base load stations and hence were not expected to contribute to grid system stability. As the percentage of nuclear stations increased, however, it became necessary for them to respond to load changes and to operate in a dynamic mode. They have to be capable, for example, of responding satisfactorily to the kinds of disturbances illustrated in Figures A10 and A11 of Appendix I.

In summary, plant design methodologies (which are relevant to designed physical systems) have been created which take the system boundary to be the hardware (the plant together with its control system). These methodologies have, however, recognized that the system so defined resides within wider systems and environments represented by the market, the business, and production areas. They also recognize that an operating plant is seldom static and have, hence, incorporated dynamic considerations as an integral part of the design process.

As well as being concerned with design, systems engineering as a discipline was directed towards the development of methodologies for problem-solving in general. Notable examples of these are the methodologies described by Hall (1962), developed as a result of the experience within Bell Telephone Laboratories, and that of RAND systems analysis (Quade and Boucher, 1968).

Hall sees the systems engineering process to consist of five major stages, which may be summarized as follows: (1) program planning (in which the outcome is the focusing of attention on a specific project or problem area); (2) exploratory planning (leading to a decision that a specific development project will be undertaken. This may in turn lead to several development planning projects); (3) development planning (which produces a more detailed specification and action plan for the system to be developed); (4) system development (specifications produced are communicated to the factory and field); (5) system improvement (reacts to information returned from the factory and field which may lead to modifications or re-design).

The hierarchical approach to investigation is again built into this process of analysis, and the first three of the above stages involve iteration, at various levels of detail, of the following six steps:

1. Problem definition
   1.1 Define needs ('needs research').
   1.2 Search environment ('environmental research').
   1.3 List system inputs, outputs, and their relationships.
   1.4 Define system boundary and constraints.
2. Choosing objectives
   2.1 List objectives.
   2.2 'Optimize value system'.
3. System synthesis
   3.1 Collect alternatives.
   3.2 List system functions.
   3.3 Delineate subsystems.
   3.4 Use creativity!

4. Systems analysis
   4.1 Decide what to analyse.
   4.2 Select analytical tools – analyse.
   4.3 Deduce uncertain consequences.
   4.4 Compare system performance with objectives.
5. Selecting the 'optimum' system
   5.1 Define decision criteria.
   5.2 Evaluate consequences – rank systems.
   5.3 Document rejected alternatives.
6. Planning for action
   6.1 Write reports.
   6.2 Promote system plan.

The RAND systems analysis methodology is similarly described as a series of stages, summarized as shown below:

1. Formulation (the *conceptual* phase)
   1.1 Clearly, formulate, and limit the problem.
   1.2 Classify and select the objectives one hopes to attain with the system – and update when necessary.
   1.3 Select criteria for measuring achievement of objectives – and update continuously.
   1.4 State hypotheses (or possible solutions) in the light of statement of problem.
2. Search (the *research* phase)
   2.1 Establish facts and collect data on which analysis will be based – attach probabilities to those facts subject to uncertainty.
   2.2 Assess the cost of data collection.
   2.3 Generate alternative ways for achieving objectives.
3. Evaluation (the *analytical* phase)
   3.1 Carry out model building (conceptual or mathematical) to predict consequences of various alternatives – state approximations and assumptions on which such modelling is based.
   3.2 Carry out computation to explore consequences of model.
   3.3 Assess alternatives by weighing cost against effectiveness.
   3.4 Examine results of 3.3 from the point of view of sensitivity to changes in parameters and changes in assumptions.
4. Interpretation (the *judgemental* phase)
   4.1 Take account of non-quantifiable and incommensurable factors.
   4.2 Take account of 'real' unertainty as opposed to 'statistical' uncertainty.
   4.3 Present conclusions, distinguishing between what the analysis has shown and pooled judgement.
5. Verification (the *scientific* phase)
   5.1 If possible, test the conclusion by conducting experiments.

Both of these methodologies emphasize a *systematic* approach to problem investigation, though neither of them take the basic definition of a system to be more than the general definition, i.e. an interconnected set of entities. They both place considerable emphasis on the definition of the problem and on the need for consensus over objectives. The methodologies detail the stages involved in a complex analysis but give no guide as to how each stage should be undertaken. For example, in the RAND methodology stages 4.1 and 4.2 instruct the analyst to 'Take account of...', but there is no guidance on how to do the accounting. Similarly Hall's exhortation to 'use creativity' in stage 3.4 seems particularly unhelpful.

Significant claims are made in the various sources of the benefits from adopting the methodologies but, as the author has no direct experience of trying to use either methodology, the reader is directed to the examples of application quoted in the references. These illustrate, for particular problem situations, how the various stages were undertaken. They also illustrate that the methodologies do not use a coherent systems language in which to undertake the necessary debate at each stage. Because of this they cannot be said to be *systemic* in the sense discussed earlier in Chapter 2. They also make the assumption (even though the systems of concern are human activity systems) that the systems *exist* in reality.

The methodologies summarized briefly above, concerned with plant design and problem-solving in general, represent the area of systems engineering in which the Department at Lancaster started to operate.

The first professor of Systems Engineering at Lancaster was Gwilym Jenkins. He set out to test a particular approach to problem-solving, through involvement in real-world problems, by using mature postgraduate students in association with members of staff of the Department and the particular managers in the firms concerned. Thus the action research cycle, described earlier in Chapter 1, was initiated by using a particular 'hard' systems engineering methodology. It is not surprising, given the background of the members of staff at the time (Gwilym Jenkins, a statistician with experience in the chemical and paper industries; David Rippin, a chemical engineer with plant design experience; and myself, a control engineer with experience in the nuclear and electricity generation industry), that the projects that were undertaken were 'hard' in nature and concerned with plant operation and design. Hence the early experience tended to be confirmatory rather than to represent a test of the applicability of the methodology to 'soft' situations that were deliberately unsuitable. However, this was a necessary beginning to what has turned out to be a highly significant period of learning.

The Jenkins methodology was first made explicit in his inaugural lecture (Jenkins, 1967) and consisted of the following four major stages (elaborated later) and their associated substages:

1. Systems analysis
   1.1 Formulation of the problem
   1.2 Organization of the project

Within this methodology the concept 'system' had been defined as follows:

(a) A system is a complex grouping of men and machines.
(b) It can be broken down into subsystems which interact with each other.
(c) The system being studied will usually form part of a hierarchy of such systems.
(d) To function at all, a system must have an objective.
(e) To function efficiently, it must be designed in such a way that it is capable of achieving its overall objective.

In summary, Jenkins' methodology attempts to be both systematic and systemic, though the systems concepts are only a small subset of what is now known as the formal systems model (see Chapter 2). It still made the assumption that systems exist in the real world; i.e. the distinction had not been made between designed physical systems and human activity systems. Because of the inclusion of an analysis of both the system and its wider system and their respective objectives, there is an emphasis on the consistency of hierarchical objectives. The measure of 'performance' is expressed solely in economic terms. This stems from the above-stated concern that the systems should function efficiently. Finally, the methodology was based on the idea that engineering the system within which the problem lies would solve the problem.

The comments in the above summary are not meant to be critical as they arise from subsequent learning which has been derived from the application of the methodology itself, but it is useful to reflect on what was being taken as given at the time.

The methodology is equally applicable to the design of a new plant or process and to the improvement in the operation of an existing one. By way of illustration of its use, a case study which was concerned with the latter area of application was published in the *Journal of Systems Engineering*, Volume 1, Number 1 (Jenkins, 1969b), and this is reprinted as Appendix II in the present volume.

As the kind of problems encountered became 'softer', modification to the methodology were found to be necessary. For example, it was found to be useful to derive measures of performance that were not economic and to consider system boundaries that were not co-incidental with organizational boundaries. However, real difficulties were encountered when it was reaized that, in general, objectives could not be taken as given. There is usually no basis for assuming, at any level in an organization, that published objectives really represent what is being aimed for, or that there is anything like consensus about them. As a particular example, 'What are the objectives of management services activity within a company?' represents the kind of question to which it is extremely difficult to get a satisfactory answer. In Chapter 4 a project is described which explores this concern in particular. In general, questions about objectives, purpose, role, etc., usually form a significant element within a problem situation and need to be debated. A language in which such debate can be undertaken is provided by that described previously in Chapter 2, based upon the concept of the human activity system. The root definition provides a mechanism for describing any number of objectives viewed from many perceptions (i.e. with many Ws). Thus what is needed is a methodology within which an exploration can be accommodated of the significance or implications of adopting a particular stance in relation to the problem situation. None of the systems engineering methodologies contain this facility, and it was the recognition of this need that provided the major development from the Jenkins methodology.

Referring again to Figure 22 and, in particular, to the activity 'Assemble the appropriate set of concepts', the most general answer to the question 'What is this assembly?' is provided by what is now known as the Checkland methodology or soft systems methodology.

The Checkland Methodology

The Checkland methodology (Checkland, 1979) was derived experientially and represents the distillation of the learning achieved in a large number of 'action research' projects. It represents a major development in the history of the Department since, in retrospect, it can be seen as representing a paradigm shift. Systems engineering methodologies are based upon the paradigm 'optimization' whereas the Checkland methodology takes the paradigm to be one of 'learning'. This shift has been necessary given the increasing concern for ill-structured (soft) problems to which there are no such things as 'right', or optimized, answers.

In essence the methodology can be described as a seven-stage process of analysis which uses the concept of a human activity system as a means of getting from 'finding out' about the situation to 'taking action' to improve the situation. Figure 29 illustrates this process.

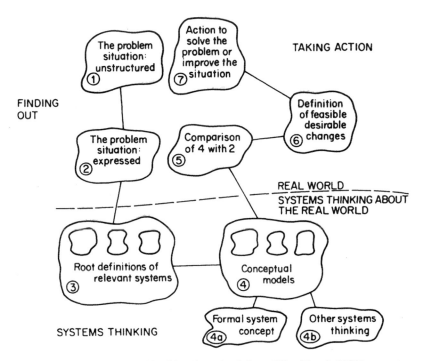

Figure 29. The Checkland methodology (Checkland, 1979)

The logical sequence illustrated by this figure is a useful way of describing the methodology but it does not necessarily represent the sequence in which it is used. In reality it represents a pattern of activities. The analyst may start with any activity, progress in any direction, and use significant iteration at any stage. The dotted line represents the boundary between activity which is in real world and activity related to the use of systems concepts to structure the thinking about the real world. Above the line the language of description will be in the everyday language of the particular situation while below the line it will be the systems language described in Chapter 2.

Stages 1 and 2: Finding out

The first two stages are concerned with finding out about the situation. The first is usually some statement about what makes the situation problematic and some basic facts about it. This will, of course, have been provided by some individual, or group of individuals, in the situation itself and will be seen to be important to the analyst (or not important) according to some Ws. Part of the finding out stage will be to try to identify what these Ws might be and to raise questions about what other Ws might also be relevant. This stage is particularly difficult in practice and the analyst must be careful that he does not impose his own W on the situation. It is attempting to be neutral that is difficult and to avoid tailoring the subsequent analysis to fit readily derived 'solutions' from the initial 'finding out'. In his book, *Systems Thinking, Systems Practice*, Checkland (1981, pp. 163–164) has the following to say about these stages:

> Stages 1 and 2 are an 'expression' phase during which an attempt is made to build up the richest possible picture, not of 'the problem' but of the *situation* in which there is perceived to be a problem. The most useful guideline here – in the interest of assembling a picture without, as far as possible, imposing a particular structure on it – has been found to be that this analysis should be done by recording elements of slow-to-change *structure* within the situation and elements of continuously-changing *process*, and forming a view of how structure and process relate to each other within the situation being investigated.

This relationship he calls *climate*. Elements of structure are defined as those features related to physical layout, power hierarchy, reporting structure, and the pattern of formal and informal communications. Process is related to the on-going activities of conversion of raw material into products, monitoring, decision-taking, and controlling. Although this approach provides useful guidelines, the analyst *cannot* be neutral. The questions asked (and the questions unasked) will be governed by some inherent W. An approach which I find useful (but which is accompanied by other dangers) is to use part of the methodology itself. The methodology is a learning system and hence can be made use of in doing these 'finding out' stages. This approach is based on the acceptance that the analyst cannot be neutral but it makes the non-neutrality apparent by making the assumptions explicit. I start with a very simple picture of the situation (see Figures 2 and 3 in Chapter 1 and the associated discussion) and develop a root definition (or root definitions) based upon that picture. The picture, and the subsequent broad level conceptual mode(s) then form the basis for a set of questions.

The danger in this approach is that the analyst will become committed to the initial choice of system and own, in a way which makes it difficult for him to abandon, the first conceptual model developed. However, it is the nature of the methodology that the choices of relevant systems, of stage 3, must be abandoned if the subsequent learning shows that they are less relevant than first thought. This danger can be minimized if, first of all, it is recognized and, secondly, if several models are chosen rather than a single model. The models chosen may be different because different transformations have been chosen and/or because different Ws have been thought legitimate.

An exmaple of question generation using this approach can be taken from a project for a paper mill which was at a particular geographical location within a group which owned other mills at other locations. For the sake of description we can call the group the Forest Group, and take the mill of interest to be located at Newtown. No specific problems have been identified, just a feeling that production performance could be improved.

The simple picture in Figure 30 represents an illustration of what is being taken as given at the start of the finding out stage. It is known that the Newtown paper mill exists with other mills within the Forest Group. It is assumed that

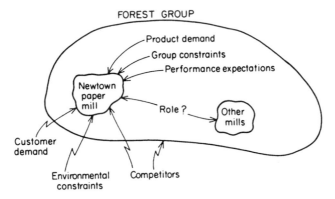

Figure 30. A problem situation expressed. *Key:* ⟶, influence internal to the group; , influence external to the group

this will give rise to a number of influences: mill to mill, mill to group, and mill and group to environment. The above assumptions about the nature of these influences leads to questions about the following features:

1. The nature of the product
   - Range
   - Standard
   - Specials
   - Quality
2. The nature of the demand
   - Customer characteristics (location, size, frequency)
   - Internal/external
3. Relationship with other mills
   (a) Specific role for Newtown
       - Static or changing
       - Constrained by group
       - Collaborative or competitive
   (b) Material interaction
   (c) Personnel interaction
4. Interactions with environment
   (a) Statutory constraints (effluent, pollution, etc.)
   (b) Local constraints
       - Labour availability/expertise
       - Attitude to mill
       - Use of contractors (transport, etc.)
   (c) Competitors
       - Products
       - Nature of competition
       - Market share

5. Group interactions
- Performance expectations
- Performance related to expectations
- Constraints (personnel, finance, etc.)

Given the apparent concern about production performance, a root definition of a relevant system emphasizing performance was taken:

A Forest Group owned system for the continuously effective and efficient conversion of raw materials into a range of paper products to meet customer demand while achieving the Group expectations for performance but within Group and environmental constraints.

The CATWOE analysis yield the following elements:

C   customers producing the demand
A   not specified
T   conversion of raw materials into a range of paper products
W   continuously effective and efficient conversion will enable group expectation to be met
O   Forest Group
E   Group and environmental constraints

An initial stage in the model development was to structure this system in terms of four subsystems: (a) a system to know about both the long- and the short-term demand (demand assessment); (b) a system to develop and maintain total resources so that demand can be met (Resource development and maintenance); (c) a system to convert raw materials into products (conversion); (d) a system to control the total generation to meet the group's expectation (total control).

A root definition was taken for each of these four systems that was consistent with the overall root definition and these were expanded into the model represented by Figure 31. Each activity in the model can now be used as a source of questions:

(a)   Does the activity exist?
(b)   How is the activity done at present?
(c)   Who is responsible for doing the activity?
(d)   Is the activity done well or badly? (What evidence is there to support this subjective view?)
(e)   Do the relationships exist?
(f)   In what form do they exist?
(g)   What are the relationships between the people doing the activities?

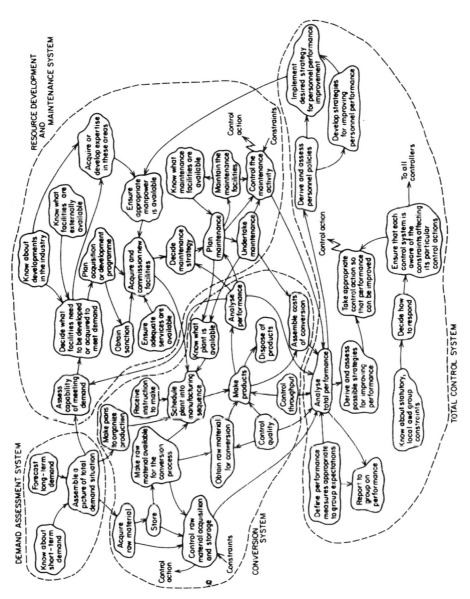

Figure 31. A conceptual model

These questions, together with those generated by the simple picture, form the basis for the finding out stage. A preliminary problem to be faced is to define who, in the organization, is the best individual (or the best group of individuals) to be the recipient of the questions. The organization chart may be the obvious place to start, but this may be refined by iteration, particularly as answers to questions (c) and (g) are obtained. During this interviewing phase it is useful not only to note the answers to the specific questions but also to record the information that is volunteered. What a manager feels is sufficiently important for him to tell you during an interview (other than in response to a direct question) tells you something about his W. This may be important for further root definition and model development.

This particular system was chosen because of the initial concern about production performance. Had the initial concern been about sales/production communications, a system emphasizing order processing activity may have been selected. Similarly, if in questioning the activity 'Receive instructions to make', it had become apparent that some concern existed about the reliability or timeliness of the input to it, a second choice of system could be made emphasizing this aspect. Thus the direction in which the model development proceeds is determined by the answers to the questions asked and is dictated by the situation itself.

In a less structured situation, say that represented by Figure 3 in Chapter 1, a number of models could be developed based upon an assumed community view, from the point of view of organized social services and/or from the point of view of the director or philanthropist. The set of models could then be the source of questions.

Models of the kind illustrated by Figure 31 may appear complex, but they can be produced quickly and need not be to this degree of detail to be useful. They provide the basis for a structured and coherent set of interviews. This is a much more efficient use of both the analyst's time and that of the individuals being questioned than a process of finding out based upon randomly generated questions. In practice managers are busy people and have limited amounts of time available. A set of well planned interviews helps in making the best use of this time and in establishing the analyst's credibility and rapport between the analyst and the managers. This is essential for later stages in a project if a readiness for change is to be cultivated so that recommendations can be accepted. In my experience the investment of the time needed to construct pictures of the situations and develop models as the source of questions is well justified. In the case of the model and related questions from Figures 30 and 31 this was about half of one day.

One of the most frequent questions asked by students in relation to this stage and one of the most difficult to answer is, 'How do you know when you have found out enough to move on to the next stage, i.e. to selecting a relevant system?' The only realistic answer, which does not appear to be very helpful, is that you don't. One never stops finding out throughout the duration of a project. The methodology itself is a learning system and hence the learning continues. It is a matter of judgement when to select systems relevant to the analysis, but since it is not a once and for all choice the analyst will learn from the choice made and be able to change the selection based upon what further learning has occurred. Thus in a sense it doesn't matter when the move is made from stage 2 to stage 3, but the good systems analyst is one whose selected systems turn out to be relevant. On the whole students spend too long in the finding our stage rather than doing the systems thinking about what has been found out and using it to define what else to find out. It is too easy to convince oneself that while you are interviewing you are making progress, but this is not necessarily the case.

Stage 3: Selection

In this stage the analyst is choosing to view the problem situation in ways that he believes will produce insight. He is now in a position to make use of the modelling language developed in Chapter 2 as a means of exploring the implications of the views chosen.

He is not attempting to define a system that ought to exist because, as argued earlier, 'what ought to exist' will be seen differently by the different individuals involved. Producing several root definitions will help to avoid any hoped for utopian analysis at this stage. The analyst is seeking root definitions of systems that are *relevant*, where this means, relevant to producing insight. Thus it *is* legitimate and might be useful to take a university department to be a system to stifle intellectual initiative as one view among others. Not many members of faculty would be likely to agree with this view, but it may nevertheless prove to be fruitful.

In making the choice of systems the analyst can only use his judgement that the choice will turn out to be useful. In the British Airways example quoted in Chapter 1, it had become apparent during initial stages that considerable cultural differences existed between those managers that had, before their merger, worked for British Overseas Airways Corporation (BOAC) and those that had worked for British European Airways (BEA). This difference seemed to be significant in all the departments encountered. Thus it would have been quite legitimate, and may have been very useful, for British Airways as a whole to have taken as a relevant system a system concerned with cultural integration within the airline. However, the expressed concern was about the development of information systems to support aircraft maintenance and it was my judgement that little progress would be made (in terms of any readiness, on the part of the particular managers, to take part in the resultant debate) by furthering the analysis based upon the above choice. In the event I believe that the analysis that was done was useful and, rather unexpectedly, made some contribution to the above-mentioned integration. This particular project, However, will be discussed in some detail in Chapter 6. The choice of relevant systems is helped by considering a range of possible input–output transformation processes and the possible points of view (Ws) from which such processes could be described.

Prime candidates for model development can then be selected and root definitions constructed using the CATWOE test as described in Chapter 2.

Stage 4: Model building

The root definition is a statement of what the system *is*. Stage 4 is concerned with the logical expansion of that definition into the activities that *the system* must *do* in order to be the system so defined. In addition these are the *minimum, necessary* activities at a particular resolution level and, as such, constitute the conceptual model. There must be one model for each root definition. Judgement must again be exercised at this stage in terms of the degree of detail contained in the conceptual model. The first resolution level model from a root definition should not contain more than about 12 activities, otherwise it becomes difficult to defend them as constituting a minimum, necessary set. It is frequently stated that the mind is only capable of retaining between five and nine entities at any time, and hence a single stage model expansion should only be of this order. This is probably true if the entities are independent, but since a conceptual model contains interdependent entities (i.e. activities and their connectivity) a slightly larger number than this is probably feasible. A first level model with this constraint on the number of activities may be at too broad a level for a useful comparison to be made. The detail can be increased by redefining each activity (or a selected set of activities) as the system and, through the mechanism of the root definition, produce a model at the second resolution level. Even accepting the lower of the two constraints mentioned above, the second level model can be of the order of 80 activities. I believe that it is useful to make a comparison at each stage of model building. An intuitive comparison has been made at the level of the root definitions and it is on the basis of this that the systems are said to be relevant. It is worth doing a superficial comparison at the first resolution level to support that belief. If comparison at this level suggests that alternative choices might usefully be made or that some modification of the original root definitions might be helpful, then the analyst will not have wasted his time elaborating the original models. The point was made in Chapter 1 that model building is a fascinating activity. Conceptual model building is no less fascinating and it is worth repeating the warning that the models are not an end in themselves. They are developed to serve a purpose and, although the analyst may find it comfortable to be temporarily insulated from the real world while doing this intellectual activity, continual immersion in the real world is necessary if the analysis is going to retain contact with it.

Because these models are conceptual, and because they are derived from root definitions in which 100% commitment to a single W has been made, they cannot be validated in the sense that an operations research model of a production process can be. However, since they are models of human activity systems, they can be checked against the formal systems model (see Chapter 2). This is the reason for the inclusion of stage 4a in Figure 29. Stage 4b is included so that help can be obtained from other systems thinking if it is felt to be appropriate. If the analyst likes the cybernetic models of Beer, for example (Beer, 1979), then he can use these to structure the conceptual models. A project is described in Chapter 4 (related to management services activity in the Central Electricity Generating Board) where specific use was made of control theory concepts at this stage. In this project they were crucial (to me) in overcoming a particularly difficult part of the analysis. This is not to say that other concepts might not have been even more helpful to another analyst who was not a control engineer.

Because of the tentative nature of the choices of relevant system made at stage 3, it is worth doing stages 3 and 4 quickly and moving on to stage 5, the comparison, in order to bring the models back into the real world to test the degree of relevance.

Stage 5: Comparison

Although this stage is called comparison, that is not strictly correct since one is not comparing like with like. The conceptual models are in terms of activities and hence describe a set of 'whats'. They define what must happen in a system described by the particular root definition. In the real world the analyst will observe a set of 'hows'. The distinction between 'what' and 'how' is a very powerful distinction; however, it can also be confusing. It is worth discussing this in more detail prior to a discussion of the ways in which comparison can be undertaken.

The 'what'/'how' distinction

A 'what' cannot be implemented without, first of all, deciding 'how' to do it. This implies a selection from a set of alternative 'hows'. For example, if 'what' you wish to do is 'invest spare cash', you cannot do it without selecting one, or more, 'hows' from the set following: {pay into a bank deposit account; buy savings certificates; purchase land; etc.}. in principle it is not possible to reverse this process and to decide, with any certainty, the nature of the 'what' underlying an observed 'how'. For example, if you observe someone buying a painting you cannot know if the 'what' is 'invest spare cash', 'add to collection', or 'increase stock for re-sale'. It is this aspect of the relationship that complicates the comparison stage. What is being observed in a problem situation is the resultant mixture of many selected 'hows'. The comparison is further complicated by the feature that there need not be a one to one correspondence between the 'whats' and the 'hows': this was illustrated in the chemical plant example in Chapter 1, where the activities 'mix' and 'react' were both done in the one vessel, the reactor.

These two aspects of the relationship give rise to complications, but the aspect that causes confusion is the fact that a 'how' is also a 'what'. So that 'buy paintings' (a 'how' of 'invest spare cash') is also a 'what' to which there will be a number of possible 'hows', such as 'visit the galleries', 'attend auctions', etc.

The key to understanding, and to making use of, the distinction is through the concept of resolution level.

Both the selection of an alternative 'how' and the expansion of an activity to a more detailed resolution level are achieved by responding to the question, 'How?' Figure 32 illustrates the two possibilities. The right-hand side of the figure represents the process of selection. Thus, if the answer to the 'how' question is A *or* B *or* C *or* D, the answer represents the 'how' of doing the activity *at the same resolution level*. If, as on the left-hand side of the figure, the answer is A *and* B *and* C *and* D *and* E, then this represents the set of 'whats' *at the next resolution level*, i.e. a more detailed expansion of the original activity. Thus relating an activity to the resolution level around which the discussion is centred will remove confusion and allow useful applications of the 'what'/'how' distinction.

In general, a conceptual model will consist of an interconnected set of 'whats' at the particular resolution level to be used for comparison. Specific 'hows' can only appear at the same level of resolution if they are introduced as a constraint through the root definition. So that, if our concern was for the development of an operational model for one of the major car manufacturers, our root definition might contain the following transformation: 'To assemble components into

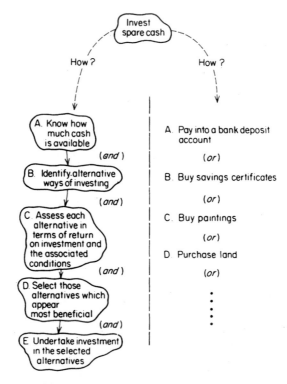

Figure 32. Responses to a 'how' question

finished vehicles by using mass production technology'. This transformation contains the 'what' (assemble components into finished vehicles) but it also constrains it to a particular 'how' (by using mass production technology). Having introduced this constraint into the root definition (i.e. as one of the elements of E in CATWOE), the conceptual model must now contain activities which are also constrained by this particular 'how'. It is not legitimate to introduce particular 'hows' into the conceptual model without, first of all, making them explicit in the root definition as above.

Constraints of this kind should not be included unless, according to the judgement of the analyst, they are necessary. The purpose of the models is to initiate debate about what exists and about how it exists, in order to introduce change. The introduction of constraints reduces the options available. It may, of course, be argued that in the above example this particular car manufacturer is not going to abandon mass production as the method of assembly and, hence, the constraint needs to be included if the choice of system can be said to be relevant. At the other extreme, if all the existing 'hows' are built into the root definitions, the models will end up describing what exists and there will be no basis for a debate about change.

Methods of comparison

Reflecting on the variety of projects that have been carried out, four methods of comparison emerge as the most frequently used. These are (a) general discussion; (b) question definition; (c) (historical) reconstruction; and (d) model overlay.

Irrespective of the method used, the analyst needs, first of all, to decide in what language the comparison is to be undertaken. It is at this point in this particular methodology that the transition is taking place between doing the analysis in a *systems language* and debating the outcome of that analysis in the *language of the real world*. Hence, unless the managers taking part in that debate are also familiar with the systems language, the analyst must translate the comparison into their language. In general, attempting to discuss root definitions and conceptual models is the quickest way to stifle debate. This is not unreasonable, nor is it a feature of this kind of analysis. Little progress would also be made if an analyst attempted to discuss the structure of a linear programming matrix with a production planning manager unless he was also familiar with linear programming. The same need exists irrespective of the nature of the intellectual constructs used by the analyst. It is, however, more important to be aware of the need here, since the modelling language is also in terms of the natural language of the analyst.

The first method of comparison is concerned with a general discussion about the nature of the models, and any organization implied by them, when related to the nature of what is believed to exist. Thus strategic issues tend to be raised during this kind of discussion in relation to role, and to the existence of certain activities rather than issues at a detailed procedural level. The most useful discussion of this type occurs when all the participants are familiar with the systems language. An interesting project, in which I was the manager, was undertaken for a particular works within a chemical company which was seeking a more effective organization structure and associated information systems. Since there were other works, at different geographical locations, manufacturing products of a similar nature, a preliminary part of the investigation was concerned with identifying the particular role for the works. Once this was established, more detailed analysis could proceed to determine the kind of organization and procedures needed to achieve this role. In this case the senior managers were interested in the way that the analysis was being done and wished to be fully involved. So a number of seminars were run in the works in order to provide some familiarity with the language and methodology. The first phase of the project consisted of a number of interviews, which I undertook. Each of the senior managers was interviewed separately and, although a number of questions that I wished to discuss had been previously circulated, I was more interested in what the managers wanted to tell me. This gave me an impression of what they thought was important and led to an identification of the range of Ws that might be relevant. The extremes of this range were used in the construction of two root definitions. Prior to the project this group of managers had spent some time deriving a set of objectives for the works and these were written down. A very explicit W appeared in these and so a third root definition was constructed based upon this set of objectives. The three root definitions were modelled at a broad level and the comparison took the form of a meeting, with all the senior managers present, at which the three pairs of concepts (the root definitions together with the respective models) were discussed. Initially the managers convinced themselves that the models were defensible as models of the root definitions. This involved some discussion of semantics in order that there could be a common understanding of the meanings behind the words used. Attention was then turned to the implications of the activities themselves in terms of whether or not they were desirable and feasible given the situation of this particular works, with its particular environment. This was an interesting discussion, with one significant outcome being the abandonment of the set of written objectives. This was due to the infeasibility of some of the activities and it was reasonably argued that, if what is implied by the objectives cannot be done, then the objectives themselves are infeasible. By way of illustration, the three root definitions, the CATWOE analyses, and the models used were as follows. To maintain confidentiality take the company to be the Cookwell Chemical Co. with the particular works situated at Kirkby.

Root definition 1 was stated as follows:

A Cookwell Chemical Company owned system, which responds to the company for the effective manufacture of a broad range of chemicals and which seeks to be major supplier to both the company and the market within the constraints applied by the company and the Kirkby environment and at a performance acceptable to the company.

CATWOE analysis then led to the following elements:

C   the company and the market
A   not specified
T   the manufacture of a broad range of chemicals
W   the system is reactive, but seeks to be a major supplier (hence competitive within the company)
O   the Cookwell Chemical Company
E   environmental and company constraints

Root definition 2 was:

> A Cookwell Chemical Company owned system which seeks to survive and establish its security by maximizing its contribution to the ongoing profitability of the company whilst having due regard to the interests of employees, shareholders, and the general public.

This resulted in the following elements:

C   the Cookwell Chemical Company
A   not specified
T   survive and establish security
W   maximizing contribution will ensure survival and security
O   the Cookwell Chemical Company
E   the constraints which come from having due regard for the interests of employees, shareholders, and the general public.

Note: This root definition was constructed from the set of objectives. The transformation process is solely in terms of maximizing contribution without any indication that this will be done by manufacturing chemicals. However, in modelling it was assumed that, as the works exist, the activities required to convert raw materials into proucts could be included as the particular 'how'. However, comparison with Kirkby Works at this resolution level raised questions about decision-taking authority with regard to how such contribution could be maximized. The company and not the works had the responsibility for marketing and selling some of the products from the works which were used as raw materials in some of the other works.

Finally, *Root definition 3* took the form

> A Cockwell Chemical Company owned system which seeks to fulfil its agreed role within the company by maintaining a viable manufacturing site, capable of meeting the continuing needs of the company by being highly professional and competitive in relation to its own areas of business whilst maintaining a responsible attitude towards statutory and environmental constraints.

Thus,

C   the company
A   not specified
T   fulfilment of agreed role
W   professional and competitive in relation to its own areas of business
O   the Cookwell Chemical Company
E   statutory and environmental constraints

The model which resulted from these three root definitions are illustrated by Figures 33, 34, and 35.

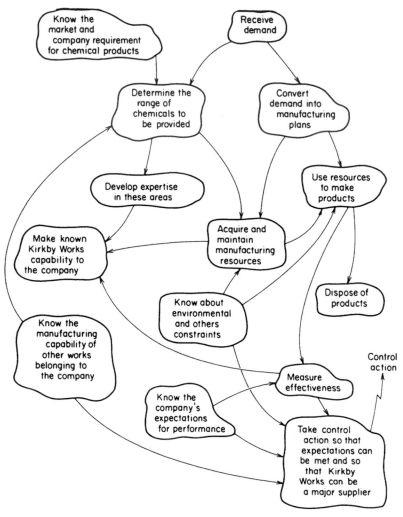

Figure 33. A conceptual model from root definition 1

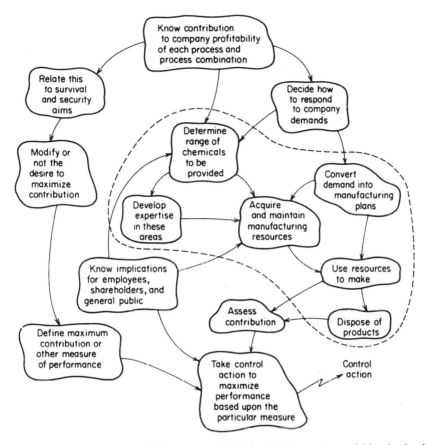

Figure 34. A conceptual model from root definition 2. The system within the broken boundary is that system concerned with the establishment and maintenance of effective manufacturing resources to meet a continuing demand

Discussion in relation to the model of Figure 33 centred around the desire to be more collaborative with other works within the company. It was agreed that all the activities were feasible since, in the past, they had collected information about other works so that decisions about the range of chemicals to be provided could be made in competition with them. However, the managers now felt that they were in a changing situation in which the implications of this model were less applicable.

In relation to the second model (Figure 34), it was quickly noted that in order to maximize contribution this system would need to have the degree of autonomy implied by the activity 'Decide how to respond to company demands'. Thus, if the company requested a particular product of low profitability, this system would be able to say, no. It was not quite like that in the real world. Also the activity 'Know contribution to Company profitability of each process and process combination', was not feasible in the Kirkby situation. This was because

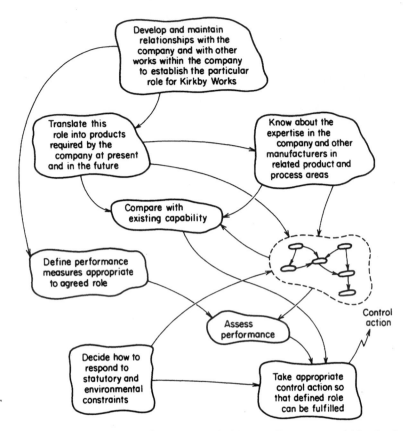

Figure 35. A conceptual model from root definition 3. The system within the broken boundary is as described in Figure 34

a number of the Kirkby products were used as raw materials for further processing at other works within the company. Thus Kirkby Works could have no control over the use of their products at other works and hence no control over the contribution that was achieved when their products were eventually sold in the market-place.

The model of Figure 35 seemed to be more relevant to the future desires of the senior managers and, after some discussion and modification, a similar root definition and model were taken for further development.

The second method of comparison, question generation, is the most commonly used. This has already been described earlier in this chapter in the discussion related to the finding out stage. However, at this stage, when the results of the comparison are going forward as a set of recommendations for change, I do the comparison more formally through a tabular display. Figure 36 represents one example of the kind of table used.

Columns 1, 2, and 3 represent the results of the questioning process referred to earlier. Since it is not possible to identify with certainty the 'whats' underlying the

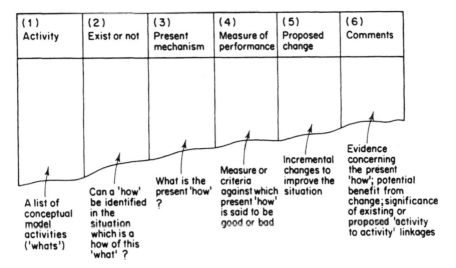

(1) Activity	(2) Exist or not	(3) Present mechanism	(4) Measure of performance	(5) Proposed change	(6) Comments
A list of conceptual model activities ('whats')	Can a 'how' be identified in the situation which is a how of this 'what' ?	What is the present 'how' ?	Measure or criteria against which present 'how' is said to be good or bad	Incremental changes to improve the situation	Evidence concerning the present 'how'; potential benefit from change; significance of existing or proposed 'activity to activity' linkages

Figure 36. Tabular display

real-word 'hows', the questions are asked the other way round, as in column 2, based upon a set of 'whats' from a coherent model. One method for identifying the 'hows' is to determine, first of all, what the output would be from the 'what' and to question whether or not that output exists. For example, if the 'what' is 'schedule production', the output would be a production schedule. If such a document can be found, then the activity can be said to exist irrespective of whether it is done well or not.

Columns 4, 5, and 6 represent the results of the assessment of what exists and the basis on which that assessment was made. This is all in the language of the situation and consists of the proposed changes together with the real-world evidence to support the recommendations. It is at this stage that the language translation, referred to earlier, takes place and it is the content of these columns that is communicated to the people in the problem situation.

The third method of comparison, historical reconstruction, is not one that I have had the opportunity of using. It consists essentially of reconstructing a sequence of events according to a conceptual model and then comparing this sequence with what actually happened. My colleague Peter Checkland describes two projects in which this method was used successfully (Checkland, 1981) but he issues the warning that it needs to be used with care. If it is seen as a means of recrimination about past performance, rather than a way of learning from it, it may produce antagonism from the participants. One of these studies concerned a project for a consultant who had had an unfortunate relationship, over one of his projects, with his client. He was anxious to learn from the experience so that he could avoid such outcomes in the future. In this case the problem situation consisted of the letters and other documentation about the projects held by the consultant and his client (to whom we had access). Thus it was possible to build up the history of the project from both sides and to compare this with a systems model of consultant–client interaction. This was a temporary system which had the following transformation process:

The resultant model and its comparison led to some significant learning both for the consultant and for ourselves in terms of the conscious engineering of a temporary problem-solving system. It was essentially this experience that led to the distinction (see Figure 21) between problem-solving and the content of the problem; a distinction that we have since found invaluable.

The final method of comparison, model overlay, consists of structuring the conceptual model in a manner which reflects as closely as possible the actual problem situation and, literally, overlaying one picture on top of the other (one being on transparent paper). The differences between the two are immediately apparent. I have used this method most successfully in studies concerned with organization structure where one is comparing decision-taking boundaries in a systems model with the areas of authority in an actual organization. A project using this method is described in some detail in Chapter 5 and is concerned with the reorganization of a company in the telecommunication business.

Stages 6 and 7: Recommendations for change and taking action

The comparisons of stage 5 will have yielded a set of recommendations for change which can be argued as desirable and evidence will have been collected about the relevant areas to support that view. It would be unrealistic, however, to expect that the complete set will be acceptable. A particular change may appear rational to the analyst, but to a manager who has lived through a particular history, who has to cope with particular relationships and internal politics (all of which help to mould his particular W), they may be anything but rational. In order to derive changes which can be implemented they need to meet two criteria. They must be desirable on the basis of the systems analysis and they must also be culturally feasible given the particular managers in their particular situation.

Although it is an intellectually separate activity from the comparison, an assessment of feasibility can be done at the same time as long as the relevant managers are participating. Comments about it can be assembled as an additional column in the table of Figure 36.

Once the set of acceptable changes has been assembled it becomes necessary to define how they are to be implemented, i.e. to determine those actions necessary to improve the problem situation. During the assessment of desirability and feasibility an examination of the alternative 'hows' will have been made.

The change may involve moving from one 'how' to another as well as change involving the implementation of a new 'what' through some selection of a 'how'. Action to introduce change must also be concerned with those structural and procedural changes which are necessary to accommodate the recommendations. Attitudinal changes may also be seen to be beneficial, but I don't believe that, in relation to changes of this kind, specific action can be defined which will produce an expected result. It is this aspect that differentiates human beings from automatons and, hence, changes should be defined as feasibe within existing attitudes. The need to implement can be seen as another problem within the same situation, to which the methodology can be applied as a design aid. At this stage a root definition and conceptual model can be developed of a temporary implementation system which will be in existence until it is agreed that the changes have been implemented. Since such a system does not exist, the comparison can only be carried out against the believed expectations. Each of the 'whats' in the model will need to be converted into a feasible and desirable 'how' before the implementation can proceed. Feasibility and desirability are based upon these expectations.

Although each root definition and model will have to be derived for the specific situation, a general model can be developed from the following root definition:

An organization owned and managed system for the cost-effective implementation of a set of agreed changes to existing structures and processes in a way which is acceptable to the organization and the appropriate unions and which minimizes the unintended disruption of the social situation.

A model resulting from this root definition is given as Figure 37. It must be emphasized that this is only included as a guide and *not* as a definitive statement of what an implementation system is.

It was pointed out earlier that the Checkland methodology would be described as a step by step process, proceeding from stage 1 through to stage 7. It can also be *used* in this way; however, it is more useful to regard it as a pattern of activities to be used in any order and at any starting point. The analyst using it needs to be clear at any time that what he is doing is finding out about the real world or doing some systems thinking or debating change through a comparison. In practice all of these activities may be mixed but, intellectually, the distinction needs to be made clear in order that the resulting analysis is coherent.

Irrespective of the starting point, significant iteration around stages 2, 3, 4, and 5 usually occurs and, to be done effectively, the analyst needs to be thinking forward. So that while a root definition is being formulated he can be thinking about the kinds of activities that will appear in the model and the likely outcomes

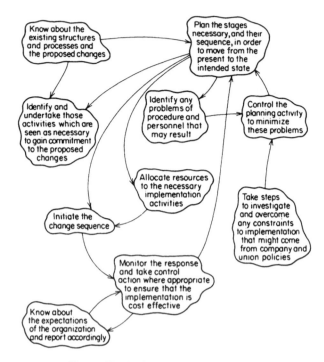

Figure 37. An implementation system

of a comparison. This helps in the production of root definitions that are relevent; not to match the real world but to lead to useful debate about the real world.

Referring back to Figure 22, what I have described here is the most general assembly of the set of systems concepts that has emerged from the action research programme. In any project it is worth, formally, assembling a picture of the situation (the problem-content system) and the related problem-solving system. In this way the analyst(s) can ask the question 'What is the appropriate assembly of systems concepts, within this problem-solving system, for the nature of the investigation to be undertaken?' In this way, the analyst can put together the set of concepts that, to him, seem reasonable, given his own expertise and that of others making up the problem-solving set of resources, and the kind of situation to which they will be applied. The systems language described in Chapter 2 is a very powerful language when one is confronted by a problem situation that is 'soft'. To be effective in its use and the use of other system languages, the analyst must remain problem oriented and take seriously the operation of the two control systems in the model in Figure 22. This demands flexibility in both the assembly and the use of systems concepts in total, over the range from 'hard' to 'soft'. Chapters 4, 5, 6, and 7 contain descriptions of a number of problem situations and the particular concepts and methodologies used. These are intended to demonstrate this flexibility and to act as a helpful guide. They are *not* prescriptions for methodologies for particular kinds of problem. Another analyst, in the same situations, may well have adopted different approaches. There is no way of knowing whether the outcomes would have been better or worse. All that can be asked is, 'Was the methodology appropriate for that analyst in that situation?' The answer is, yes, if individuals concerned in the situation agree that a useful outcome was achieved.

Prior to embarking on a discussion of these cases, I would like to return briefly to that part of the methodology, where a selection is being made of relevant systems. Although the choice is entirely up to the analyst, given the particular situation of interest it is possible to make a distinction between two types of analysis. In practice this is not a clear distinction, but it is helpful to consider if the nature of the concern being expressed tends to suggest one kind of analysis or the other. These are termed *issue-based* and *primary task* analyses (Checkland and Wilson, 1980).

Issue-based and Primary Task Analyses

In the project described earlier, for the Cookwell Chemical Company, an issue-based analysis was being undertaken—the issue being the nature of the role for Kirkby Works. The models illustrated, however, were all primary task models. This may initially appear confusing rather than helpful but we will encounter problem situations later in which the concern is the restructuring of organizations or information requirement analysis where this distinction is crucial to the analysis. Therefore it is worthwhile making the distinction clear.

In essence a primary task root definition is one in which the resultant system boundary could, in principle, map on to (and possibly coincide with) the organization boundary. Thus a primary task root definition could be a possible statement of the organization's mission or objective. In the Kirkby works example, the three root definitions are all plausible statements of its mission. Kirkby Works could be seen as an organization:

- To manufacture a broad range of chemicals.
- To maximize contributions to company profitability.
- To fulfil an agreed role within the company.

An issue-based analysis is one in which the above mapping is *unlikely* to be feasible. So that: A system to explore the role for Kirkby Works will not have a boundary co-incident with the organization boundary and hence will represent an issue-based definition. Its transformation,

'to explore the role for Kirkby Works'

is unlikely to be a plausible mission statement. No organization is likely to exist for the purpose of exploring its own role.

It is usually the case that, when embarking upon a primary task form of analysis one is choosing to model systems relevant to the situation which, it can be argued, are close to agreed perceptions of reality.

If an organization is concerned with manufacturing cars, for example, one can actually observe that components enter the organization and cars emerge. Hence it may be argued that a certain set of activities must be exist (irrespective of how they are done) in order that the transformation can take place. Thus, when carrying out a primary task analysis, the root definition chosen will lead to the model of a notional system which can be related very directly either to an organization as a whole or to a well established task carried out by a section, department, or division of the total organization.

In such studies a chosen relevant system is likely to be one which expresses the public or 'official' explicit task which is embodied in the organization or section or department. Such a choice takes as given that a certain explicit task is to be performed by or within an organization (without taking as given existing organizational boundaries). A root definition and systems model can then be derived which express this 'primary task'. In many studies located in manufacturing industry, for example, a chosen relevant system has been one which transforms raw materials into saleable products. In a major project referred to earlier, a primary task root definition described a notional system to carry out the task assigned to the engineering division of British Airways, namely to carry out planned maintenance on a fleet of aircraft effectively and efficiently under various constraints. It would be hard to argue that within an airline there might not be a manifestation of such a system, hence this is a good example of a root definition of a relevant system of the 'primary task' type. It expresses a primary task which must be manifest if the real-world organization is to be capable of fulfilling its public function.

Limiting the description to the public function of an organization, however, tends to produce root definitions which represent little more than the basic operations of the organization. Such definitions and models are likely to be sparse and are probably applicable to any organization which has such a public function.

Thus,

> 'a system to transport passengers by air over both national and international routes as a commercial operation while recognizing competition and constraints arriving from the CAA, FAA and other regulatory bodies'

is applicable to BA as it is to KLM or JAL.

To be useful a primary task analysis must attempt to be particular to the particular organization concerned. Each organization is unique and this uniqueness needs to be recognized. BA and JAL are different. Apart from cultural differences both organizations will have experienced different histories, they will have employed different generations of management and they will have achieved a different stage in their development.

A primary task model which just describes the public function is not rich enough to provide the required degree of insight to lead to an analysis which carries with it the acceptance of the particular managers in the particular organization.

Further illustration of an analyses of this kind will appear in later chapters but let us concentrate here on the production of a primary task model which attempts to recognize the uniqueness referred to above.

A significant feature of all organizations, (which contributes to uniqueness) is the *multiple perceptions* (of the managers concerned) of the role and nature of the organization that they are managing. Consider the following situation.

In 1983 we were asked to undertake a project for the Prison Department of the Home Office in the UK. The project may be summarized as follows:

The concern: Do we get value for money from our prison service?

The expectation: Information systems to answer this question on a continual basis.

Our problem: Costs can be accurately evaluated 'value' requires an answer to the question: *What is a prison for?*

The prison service consists of a variety of establishments, each operating a regime determined by the particular governor. Discussions with a number of governors and other prison officials indicated a variety of answers to the above question. The 'hard liners' believed that the inmates were there to settle a debt to Society while the more liberal minded saw it as a process of education and rehabilitation.

The situation is complicated by the fact that no one individual is ever 100% committed to a single perception. Thus the 'punishment'-oriented governor operates educational processes while the 'education'-oriented governor operates punishment routines. Hence each individual has a spectrum of perceptions related to the situation with which they are concerned. This rather extreme example illustrates the added complexity arising from a consideration of multiple perceptions and it is this feature which makes each situation unique because its effect is determined by the particular group of managers concerned.

The problem of developing a single concept for a company, while recognizing the multiple perception spectra of the managers concerned requires three stages:

(1) Decomposition of the spectra by assuming that specific perceptions can be isolated and incorporated into HAS definitions.
(2) Logical modelling of the HAS definitions.
(3) Re-composition of the perceptions through a process of accommodation resulting in a single HAS model, known as a consensus primary task model.

These stages are illustrated by Figure 38.

The notion of multiple perceptions is represented by the concept of 'Weltanschauung' (W), which, as mentioned previously, may be likened to a filter in the

92

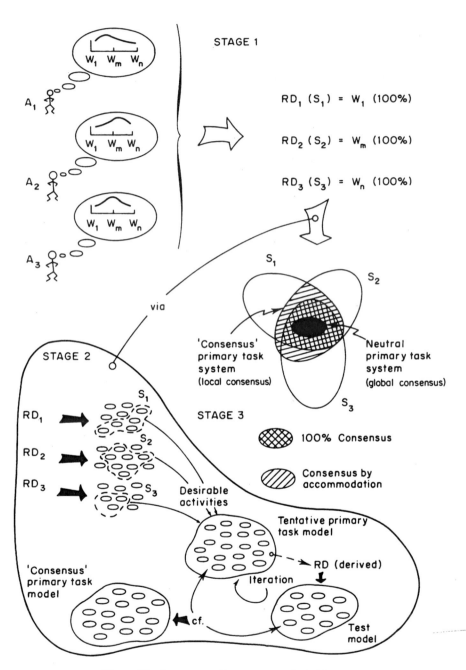

Figure 38. A process of systems condensation

head of an observer of certain events leading to a particular interpretation of those events. (Thus one observer of the prison service may interpret it as a punishment process while another observer may see it as a rehabilitation process.) In figure 38 three managers are shown with particular orientations towards various parts of the spectrum of Welthanschauungen. Here the legitimate range is from $W_1....W_n$ (where W_1 could be an orientation towards punishment and W_n could be an orientation towards rehabilitation). The decomposition stage is concerned with picking parts of the range (believed to correspond to major and significant Ws) and with developing root definitions corresponding to these chosen Ws. In practice there may be any number of root definitions (RDs), twelve being a usual upper limit, but only three are shown here for convenience.

Essentially the analyst is choosing to view the mission of the organization in a particular way; (as a human activity system) and, on the basis of a carefully produced root definition derives a logical model of what the system would have to *do* to be the one so defined. A second definition, based upon another perception, can be taken and modelled as a logical construct, and so on.

The second stage is concerned with the development of the conceptual models from the root definitions and the selection of activities from these models which are desirable in some way. Desirability may be expressed in terms of feasibility or in terms of necessity but it is defined and the choices are made by the group of managers to whom the range of Ws is relevant (in this case A_1, A_2, and A_3). Stage 2 in Figure 38 illustrates the derivation of a consensus primary task model by a route which seeks to obtain coherence in the resultant model.

This process results in the mapping illustrated as stage 3 in Figure 38. The neutral primary task system represents that set of activities that the organization must do to be that kind of organization, (i.e. a prison, to be a prison, must receive prisoners, store them for particular periods of time and release them). The larger shaded area represents the set of activities taken to be the single concept for the organization (the consensus primary task model) achieved through the accommodation (of Ws) process described above.

The process illustrated above requires a number of assumptions to be made and it is worth examining each stage in more detail.

The various perceptions (Ws) used at the start of the process are those that are taken to be *legitimate* for the particular company or organization as derived from discussions with the particular group of managers who the analyst takes as appropriate to the situation. While interviewing managers I try to be sensitive to illustrations of these Ws. For example, it is often the case that, as well as answering specific questions, the managers will volunteer information. Phrases such as: 'I think you will find the following document(s) of interest', or 'By the way you will need to know......', are examples of the introduction of such information. These are indicators that the manager finds them of interest or important. The question can then be asked 'what "W" must be implied by such an emphasis?' and answered by the analyst.

Significantly different Ws (but legitimate in the context described above) can then be chosen as the source of root definitions of primary task systems. In the project for the prison service mentioned above the following perceptions were taken as relevant and legitimate for the range of UK prisons considered. In fact about eight definitions were used; these three are used for illustration.

- A system for the control of interaction between offenders and the community.
- A system to instil society's norms and values.
- A system to enhance criminal activity.

The latter definition is legitimate although it can be argued that it represents an unintended mission. The comparison of the activities in the resultant model with what actually happens in a particular prison may cause certain activities or procedures to be abandoned or introduced on the basis of their desirability.

It is the case that the models derived from these definitions would contain different activities. Thus, while accepting that a complex changing mixture of perceptions can be simplified into a number of single perceptions the analyst now faces the difficulty of how to reduce these multiple models to a single model. The procedure described earlier for achieving this consolidation of models is based on the following assumptions.

The first assumption is that, no matter what kind of organization is being considered, there will be a description of it that will achieve *global* consensus, i.e. a *neutral* primary task description. So far a prison, a neutral primary task description would be – *a system* for the receipt, storage, and release of prisoners. Thus no matter if some governors believe that a prison is essentially about rehabilitation, and some governors believe that a prison is essentially about punishment they would agree that for a prison to be a prison it must, at least, take in prisoners, retain them for various periods and then release them. Similarly for a refrigerator manufacturer to be such a company it must, at least, assemble components into finished refrigerators; an insurance company must provide financial guarantees in respect of certain defined happenings in return for regular payments etc.

These definitions represent sparse descriptions of the basic operations of the organization under review, and hence the models derived from them are sparse representations of the activities undertaken by the organizations.

The second assumption is that less sparse models can be obtained in a way that is defensible through the addition of individual activities about which there is *local* consensus over their desirability. Figure 39 illustrates this idea.

The black central system (S_n) represents the 'neutral' primary task system, i.e. that set of activities that the particular organization must do to be that kind of organization. Systems S_1, S_2, S_3.... represent those systems corresponding to the range of 'legitimate' perceptions as defined earlier (potential descriptions of the primary task of the organization in that the system boundaries could map onto the organization boundary). Given this restriction on choice of root definition it is to be expected that the set of activities within S_n will be common

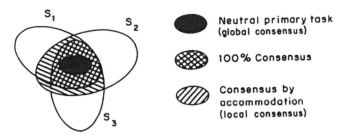

Figure 39. Illustration of 'consensus' primary task

to S_1, S_2, S_3.... etc., and can be taken to be part of the final primary task model.

Given the above two assumptions the procedure to be followed is in two stages. Each activity in systems S_1, S_2, S_3.... (excepting those in S_n) are questioned to determine their desirability. Desirability is determined in discussion with the group of managers or others who the analyst judges to be the group most concerned with the output of the particular analysis being undertaken.

Those activities about which there is a *local* consensus over desirability are added to those contained in S_n to form a 'tentative' primary task model. For those activities about which there is partial (rather than 100%) *local* consensus over desirability the analyst must use judgement on their retention or rejection. My criterion for retention is based on the resultant coherence of the model. If the activity is necessary (in a logical sense) in order to make the resultant model coherent then activity is included, otherwise not. An iterative procedure is used to explore coherence, illustrated by Figure 40.

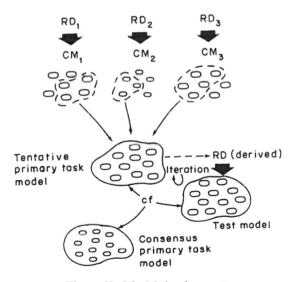

Figure 40. Model development

Here it is shown that the selection of 'desirable' activities from three models CM_1, CM_2, CM_3.... leads to the tentative primary task model which at this stage may be no more than an aggregate. Given the logical linking of root definition (RD) and conceptual model (CM) it should be possible to reverse the process and derive an RD from the tentative primary task model. This is a difficult step and so a 'test' model is derived from the RD so formed. It is known that this is a coherent model and hence it can be used to compare against the tentative primary task model derived and so on. The iteration is continued until the test model is derived which is an acceptable version of the tentative primary task model in that it represents the activities and is also coherent.

During this process it may have been necessary to add activities to the tentative primary task model (i.e. those about which there is not 100% consensus) but also it may have been necessary to add others. For example activities from CM_1, and CM_2, could in fact be conflicting if RD_1, and RD_2, represent conflicting perceptions. In this case an activity needs to be added which is concerned with resolving the conflict.

By adopting the above procedure any number of initial models CM_1.... CM_n can be reduced to a single model which, through the interaction with the relevant managers, can be taken to be a defensible 'consensus' primary task model relevant to the particular organization in its particular social context. The boundary of this model is represented by the shaded area in Figure 39.

The putting together of these stages results in the total process of deriving a primary task model illustrated by Figure 38. Since such a model is based upon an analysis of multiple perceptions (or Ws) it represents a taken-to-be or 'constructed reality'. It is similar to the 'selected target system' which is used as the initial stage in a conventional socio-technical systems approach (STS). The essential difference is that within STS, the target system is assumed to exist (i.e. to be reality).

An alternative approach to the process of re-composition which can be more rapid but less defensible is based upon the concept of a model of an enterprise (see Figure 72). As before a number of models will have been derived as a result of the analysis of Ws, and an indication of desirability/undesirability of activities obtained. Activities are then selected which are seen to correspond to the achievement of some transformation process (this will at least contain the neutral primary task activities). It is accepted as coherent if sufficient activities are included to enable the process to get from input to output. The analyst decides what additional activities are required (if deficient) and also what surplus activities to remove.

A similar assembly process is undertaken for these activities which are taken to represent support systems, linking systems and monitor and control systems.

The resultant assembly in total is then taken to be a coherent primary task model and it is also taken to represent consensus if there is general acceptance within the organization.

'Validation' of a Primary Task Model

It was mentioned earlier that, since a primary task model describes a public or a consensus view of some situations, the activities contained within the model should exist in that situation through a particular set of 'hows'. Thus if the model describes the process which transforms components into cars, then one would expect to find activities representing, for example, the storage and assembly of those components, in some form, otherwise the entity that is recognizable as a car could not be seen to emerge from the factory. The particular 'how' of assembly might be by hand or by using mass-production methods but all that is necessary, for 'validation', is to identify that such a 'how' does in fact exist. One is not concerned, at this stage, with whether or not the 'how' is an appropriate 'how', or whether it is done well or badly. Existence is the important criterion. One way of doing the 'validation' is merely to determine what the output of the activity might be and to look for the existence of that output. For example, if the activity is 'plan production' the expected output would be some form of document called a production plan. If such a document exists then activity producing it must also exist, irrespective of the production planning methods employed.

An alternative method which is useful, particularly if, as part of the analysis, it is also required to know who is responsible for doing the activities, is to identify the respective decision-takers. The assumption is that if a decision-taker can be identified with responsibility for an activity then it is legitimate to include the activity in the primary task model. This kind of analysis is usually associated with organizational restructuring or information systems analysis and a discussion of methodology appropriate to these concerns is contained later in Chapters 5 and 6. This method is known as 'organizational mapping'.

One of the most significant formal systems concepts referred to earlier is that the boundary of a system is defined as that which determines the area over which some decision-taking process has control. The purpose of organizational mapping is to determine those individual decision-takers who make up this process over the area represented by the primary task model. Thus one is converting an *organization-independent* model into an *organization-dependent* model by drawing boundaries around sets of activities representing the areas of authority of the relevant people. This is usually *not* done by asking individual managers which of the activities in the model they are responsible for. There are two main reasons for this.

Firstly, the language of the model is in terms of 'whats' whereas the managers are more familiar with the way in which the activities are executed through particular 'hows'. This inconsistency of language may lead to misinterpretation of what the activity covers both on the part of the manager and on the part of the analyst.

Secondly, the managers are human and it is not unusual for a manager to claim an area of responsibility which contains those activities for which he would like to be responsible as well as those for which he actually has responsibility, and to omit those activities for which he is responsible but would like not to be. This is more likely to be the case if some restructuring of responsibilities is a possible outcome of the project.

The method of organization mapping which I use, which tries to overcome these problems, is one which seeks to describe each manager in terms of his own primary task model and then to map each of these models onto the overall primary task model. Each model is derived by asking the following set of questions:

(a) What are the (primary) inputs to his area that cause activity to take place?
(b) What are the resultant activities?
(c) What outputs are generated by doing the activities?
(d) What additional activities are initiated internally and what are their outputs?
(e) What additional (secondary) inputs are initiated by doing the total set of activities?
(f) What are the constraints under which the activities are done?
(g) What are the sources of the inputs (primary and secondary)?
(h) Who receives the outputs?

There is no guarantee that this procedure will overcome the problems referred to earlier, but the questions are asked in the spirit of finding out what goes on and not explicitly as a means of defining responsibility.

The general model which results from the questioning procedure is illustrated by Figure 41. This figure also contains an actual example, by way of illustration, which comes from a project in Mexico where the concern was related to information systems for the control of the economic growth of a state in southern Mexico based upon an overall development plan produced by central government. These five activities represented the area of responsibility of the Policy and Economics Planning Manager within the state governing body. In relation to this model, the primary inputs were the relevant actions from the planning committee minutes, the federal plan, and the state development plan. Both of these plans came from the Office of the Governor and the planning committee consisted of the heads of the various planning departments, including the Policy and Economics Planning Manager. The constraints consisted of the economic policies produced as an output from this manager during the previous planning period but which were still relevant. A secondary input, which in this example was also a primary input to another activity, was the state development plan. The outputs were the investment priorities, which went to the various departmental directors (e.g. tourism, law enforcement, urban and rural development, etc.), performance reports, and modifications to the state development plan, both of which went to the Office of the Governor. Within the model, the activities 'Develop planning models relevant to the state' and 'Modify the state development plan' (the latter being dependent upon the former) were initiated internally rather than resulting from a primary input.

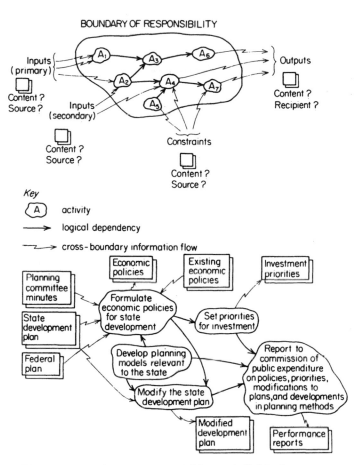

BOUNDARY OF RESPONSIBILITY

Figure 41. A primary task model for an individual manager

In mapping an individual model, such as this, on to the overall primary task model, it is likely to be the case that some of the activities will be at a different level of resolution. Thus, to do the mapping precisely, some differential expansion of the relevant activities will be necessary. The information inputs and outputs also serve as a useful check on the connectivity in the overall model since an information flow cannot exist without there also being a logical dependency.

'Validation' as described here is only relevant if some representation of *current* activity is required. In some circumstances primary task models may be developed to describe some future or intended company (department or individual) mission or activity. In producing an information plan or IT strategy for example, the business plan for an organization may be taken as the source of the consensus primary task model. 'Desirability' of an activity in this case may be in terms of its future significance rather than its current existence. 'Validation' must therefore be regarded as of limited concern and not a necessary part of the development of a primary task model.

At the start of a study the analyst, or group of analysts, need to start to do the activities in Figure 22. An appreciation of the area of concern is essential if concepts and their assembly are going to be appropriate. By 'appreciation' is meant a deep understanding, through some of the finding out activities referred to earlier, not just a casual acceptance of the statement of the problem. Chapters 1 and 2 have described some of the concepts available to the analyst, related to the whole of the problem spectrum from 'hard' to 'soft', with emphasis in Chapter 2 on the particular systems language based upon the concept of the human activity system.

Chapter 3 has surveyed a few of the systems engineering methodologies developed for process design and improvement as well as some of the earlier general problem-solving methodologies. These represented the starting point for the development of the 'soft' systems methodology which has been described in more detail in this chapter. In conclusion I wish to emphasize the need to be flexible in the use and assembly of concepts. This is essential if one wishes to remain problem oriented in the face of the variety that exists in the real world. Some examples of this flexibility are contained in the cases described in the remaining chapters, which illustrate a systems approach in action.

Chapter 4
Problem-solving and Methodology

Introduction

The problem spectrum, or more appropriately the spectrum of problem situations, becomes more complex as we move from the hard extreme to the soft extreme. By this I do not wish to imply that at the harder end of the spectrum the problems encountered are 'easy' or straightforward. This is certainly not the case, but the aspect of the situation that adds considerable complexity as we move towards the softer end of the spectrum is the *increasing uncertainty as to what the nature of the problem* is. It is this uncertainty (arising because of the multiple perceptions that exist in every 'soft' problem situation) that has led to the shift in emphasis towards methodologies to structure the situations and away from techniques to solve the problems. Management science has tended to make the assumption that problems recur and has largely been concerned with the development of techniques, usually based upon a mathematically oriented modelling language, to solve such recurring problems. My belief, however, is that every probelm situation is unique, since the multiple perceptions, that are a significant aspect of it, arise from Ws that have been formed by particular histories and particular social relationships. Thus if it is *not* reasonable to make the assumptions that problems or problem situations recur, then it is not reasonable either to work out in advance problem-related techniques that guarantee solutions. Such solutions are unlikely to turn out to be appropriate for the situations to which the technique has been applied. As mentioned previously, techniques based upon a mathematically oriented language are too limited and too rigid to cope with the richness and variety of real-world problem situations which can be described as soft.

A methodology, on the other hand, represents a structured set of guidelines within which the analyst can adapt, in a coherent way, the concepts being used. This enables him to remain problem oriented as the analysis progresses and the nature of the situation confronting him unfolds. Thus he stands more chance of producing results that will turn out to be appropriate for the particular, unique situation of concern.

Chapter 3 presented such a methodology, the Checkland (or soft systems) methodology. This represents the most general assembly of systems concepts (referring to Figure 22) and, by making use of the guidelines provided, an analyst can make progress in learning about the nature of the concerns within a 'soft' situation.

Experience from the action research programme suggests that other assemblies can be used with effect in relation to certain types of problem situations. It is the purpose of the remaining chapters of the book to present such a range of methodologies used in a variety of situations. I hope that this is not falling into the technique-orientation trap referred to above, since the methodologies are presented in terms of broadly based guidelines offering considerable flexibility in their use. The problem situations are also broadly specified, i.e. where the concern is about reorganization, or about information planning, or to do with undertaking a general analysis of an enterprise and its performance.

The projects discussed are all taken from my own involvement in the action research programme referred to earlier. It is my aim, in presenting this particular selection, to illustrate some of the variety that exists in real-world problem situations as well as the flexible use of systems ideas.

The first project relates to the harder end of the problem spectrum and makes use of ideas generated by the application of a systems engineering methodology. Later examples concentrate upon the use of the concept of a human activity system and, in some cases, illustrate the derivation and use of techniques and other systems thinking within 'soft' methodologies. The final section of this chapter concerns itself with a project to do with the analysis of a management service organization. From these and other projects, particular methodologies have been developed for use specifically in studies concerning information planning, organization redesign, and management control systems. These methodologies are presented (in Chapters 5, 6 and 7) and their use illustrated through the set of projects that led to their emergence. Where possible the actual organizations in which the studies were carried out are named, but in some cases the names have been changed to maintain confidentiality.

Design Assessment

The process of design is concerned with the 'what' $\longrightarrow$ 'how' conversion. 'What is needed' is not problematical, the difficulties are associated with determining *how* to provide what is needed. Thus the problems of design represent 'hard' or 'structured' problems as defined earlier. If there is any uncertainty about what is needed, the problem immediately becomes 'soft'. If, however, we can take it that there is no such uncertainty, then the methodologies of systems engineering become appropriate and the systems language used within the methodologies may take on a mathematical orientation.

Appendix II describes an application of the Jenkins' methodology to the performance improvement of a petrochemical plant in which a statistical description of the process was derived through experimentation. In this case the plant was already in existence and was operating. This project could therefore be seen as part of the re-appraisal of the initial design in order to achieve improved operation.

A project specifically concerned with design assessment was undertaken in 1971 for Glaxo Laboratories (now called Glaxochem Ltd) of Ulverston in the north-west of England. The project was undertaken by one of the students on the master's degree course together with a full-time consultant of ISCOL Ltd, with myself as project manager. The duration of the project was 16 weeks.

The works at Ulverston is concerned with the manufacture of antibiotics such as penicillin and streptomycin and the manufacturing process is one of fermentation. Essentially the culture grows at the interface between some carrier medium and air and, on a production scale, many such interfaces are produced by foaming the medium with sterile air at around 1.9 bar pressure. Prior to the start of fermentation the vessel containing the medium is sterilized by the use of a large quantity of low pressure steam. Other services required during the process are electricity, nitrogen, and cooling water.

Because of the relative isolation of the Ulverston works, the services used on the site, with the exception of electricity, had always been produced internally. When the expansion of one of the plants led to the need for further steam and air production capacity, a study, carried out by Glaxo into the total services requirement, showed that there was an economic argument for extending the production of services to include electricity generation. Thus a new services complex was designed to provide the following services: electricity; high pressure steam (HP steam); low pressure steam (LP steam); compressed air; nitrogen; cooling water.

Figure 42 shows a simplified version of the services complex together with the physical flows. In addition to satisfying process demands, the complex has to satisfy its own internal needs. For example, the steam generated derives the turbo-alternators and the air compressors prior to being used to satisfy process

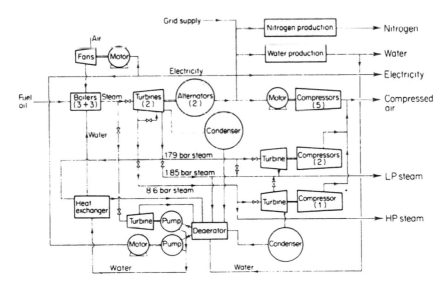

Figure 42. Services complex: flow diagram. Numbers in parentheses indicate numbers of units

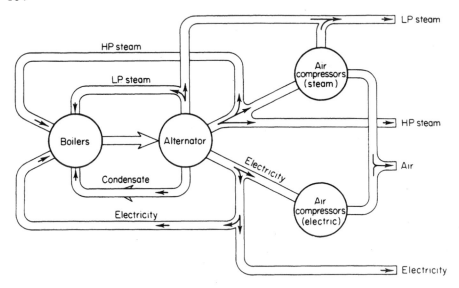

Figure 43. 'Service complex' interactions

demands. These internal flows make the items of plant operating within the complex highly interactive, and hence it is difficult to assess the total effects of a demand for a particular service on the supply of that and other services. Nitrogen and cooling water supply have only minor interactions and it was regarded as a reasonable assumption to treat these demands as an additional electrical demand. The major interactions within the complex are illustrated by Figure 43.

The system that was our concern was the services complex itself, i.e. a designed physical system, but in order to assess its performance it was necessary to relate it to its wider system, i.e. the services complex plus the production processes that it served. The characteristics of this wider system may be summarized as follows (Figure 44 shows the major process demands for a single unit over the fermentation cycle):

(a) The manufacturing processes use a complex biological reaction which is difficult to control.

(b) Because the processes are batch processes, and because there are a large number of units on site, the demands for services are constantly changing and are very difficult to predict.

(c) The raw material and conversion costs account for approximately equal proportions of the finished product cost.

(d) The services used in the manufacturing processes account for only a small proportion of the conversion cost.

(e) An adequate supply of services is critical for the successful operation of the manufacturing processes.

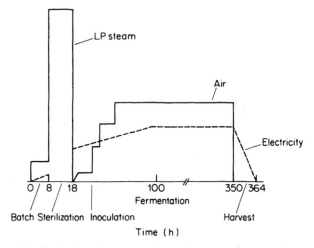

Figure 44. Variation of services demands over a fermentation cycle

Thus, although the cost of supplying services was worth reducing, the overriding performance criterion was that the probability of incurring production losses, through a failure of the supply of services, must be minimized. Hence the measure of performance for the services complex must in some way enable an assessment to be made of (a) the ability of the complex to meet the total current services demands and those in the future; (b) the costs of achieving (a); (c) the ability of the complex to respond dynamically to the demands in a satisfactory way.

In reaching the above conclusions we had effectively been carrying out stage 1 of the Jenkins methodology. However, the major part of the project was concerned with the assessment itself and help in this was derived from the design procedure contained in Figure 28 (see p. 63). Glaxo had already produced a design based upon manufactures' data for the individual units making up the

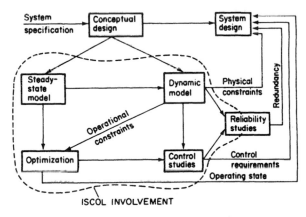

An overall design procedure (Figure 28)

106

complex (heat exchangers, compressors, etc.) A control system had also been designed based upon the controls required for the individual units, but no study had been done to investigate the effect of the internal interactions (i.e. the aspects which make it a system). Thus the conceptual design existed and our task was to undertake the developments and analyses highlighted in the modified form of Figure 28 reproduced here in order to convert it into a system design. Glaxo were already involved in some reliability studies based upon statistical data related to the individual units. Our contribution to these studies was through the dynamic behaviour and control studies. The actual plan for the project consisted of the activities shown in Figure 45.

The initial activity was to be concerned with the development of the two models representing the services complex and it was anticipated that a further model would be required of the processes themselves so that a prediction of present and future demands could be made. In the event this turned out to be over-ambitious and an alternative approach was adopted based upon extrapolation from historical data. In examining how the complex was to be controlled it was decided that control did not stop at the hardware but that an integral part of this activity was the management of the complex. If, as a result of the

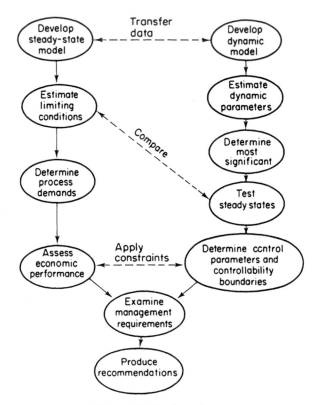

Figure 45. Project plan

economic analysis, several operating strategies were shown to be possible for any particular pattern of process demands, then decisions would be required as to which would be preferred related to equipment availability, predicted maintenance requirements and so on. Also, if the examination of controllability indicated that under certain demand situations there was a need to adopt particular actions to avoid undesirable transient conditions, then this information would need to be made available to a manager with the decision-taking responsibility. Thus any investigation of total performance must include a definition of how the plant is to be managed so that the relevant information needs can be determined and made available as part of the design activity.

The steady-state and dynamic areas were developed simultaneously in the project with a considerable amount of interaction between them. However, here they will be treated as distinct areas and described in turn.

Steady-state performance

Specification of the model

The first step which was taken in evolving a model to represent the operation of the services complex was the derivation of those relationships which represented the steady-state performance. The relationships derived were mass and energy balances for (a) individual plant items; (b) internal flows within the services complex between individual plant items; and (c) flows entering and leaving the services complex.

This procedure produced 33 equations and it was fairly obvious that, if a large range of operating states were to be examined, a digital computer model would provide a rapid and cheap method of solution. This decision raised two further questions:

(a) Which digital computer?
(b) What method of solution?

If the model was going to be available as an operational tool within Glaxo, then the answer to the first question would be constrained by the computing facilities available. An IBM 1130 computer at Glaxo's Ulverston works offered both a machine capable of handling the set of equations and the means for running the developed models. Thus the development was also undertaken on an IBM 1130.

The answer to the second question was not as straightforward and required an investigation of the possible methods of solution (i.e. dynamic programming, linear programming, or a FORTRAN model). Of these methods, linear programming was ultimately selected since

(a) It provided not only a method of solution but also a method of optimization.

(b) It was possible, simply and cheaply, to translate the set of equations into a form suitable for use in a linear program.

(c) It provided an economic and quick method of solution.

The completion of the model development required that the linear program be run and its results analysed for consistency and accuracy. This testing was done by introducing a number of sets of operating conditions into the model and manually checking the accuracy and consistency of the results.

Application of the linear programming model

To demonstrate the application of the model the extremities of operation were examined. this application was selected because a detailed knowledge of the actual demands was not required and it produced tangible results early in the project. In examining the combinations of demands that the complex could satisfy, an exhaustive survey would have meant that:

(a) Both maximum and minimum combinations of services demands should be investigated.

(b) All combinations of major items of services plant should be considered (i.e. boilers, turbo-alternators, and air compressors).

The time scale of the project did not allow an examination of all the possibilities in (a) and (b) above. It was necessary, therefore, to restrict the use of the model to an examination of the maximum demands that could be satisfied with all items of services plant operational and no import of electricity from the grid. By examining this particular case it was possible to establish the limitations on

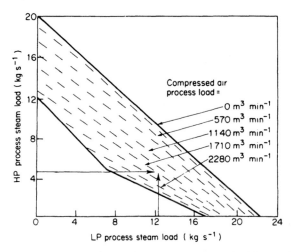

Figure 46. Operational envelope for a process electrical load of 13 MW

process demands which would exist during normal operations. The examination of other combinations of services plant was not neglected, but was used later as a means of familiarizing the Glaxo personnel with the use of the model once it has been transferred to the Glaxo machine.

To establish the limiting combinations of process demands for services, the linear programming model was run in its 'rating' mode. In this mode three of the four demands for services were specified and the model determined the maximum level of the fourth unspecified service. The results obtained from this analysis were summarized in a series of performance curves. One example from this series is illustrated in Figure 46. This figure shows the relationship between the low pressure (LP) process steam demand and the process compressed air demand at a process electrical demand of 13.0 MW. By selecting any point within this envelope, it is possible to establish a particular limiting combination of process loads. Thus if the LP steam and the high pressure (HP) steam process loads are 12.2 and $4.7 \, \mathrm{kg \, s^{-1}}$, respectively, then the compressed air load cannot exceed $1650 \, \mathrm{m^3 \, min^{-1}}$.

Estimation of process-demands

In order to derive the most economic way of operating the services complex it is necessary to know the process loads which the complex has to satisfy. Two methods of approach were possible to establish these loads:

Method A   To predict the process demands from a model of the production process.

Method B   To analyse historical data on demand conditions and to extrapolate these to produce predictions of future demands.

Method A had the advantage of (a) being able to simulate current and future operations, and (b) enabling changes in plant configuration and product mix to be examined. This method had the disadvantage of requiring a period of 6–8 man-weeks for its development. Method B had two disadvantages:

(a)   It relied on an analysis of historical data for particular plant configurations and product mixes.

(b)   It required extrapolation to examine future demands which necessitated the examination of plant configurations and product mixes not encountered in the historical analysis.

This method had the advantage of requiring only 1–2 man-weeks of effort for development. It was considered that the extrapolation of demands produced by Method B would be of sufficient accuracy to examine the economics of operating the services complex during the next 1–2 years, and, as Method B offered the shorter development period, it seemed the logical path to pursue in the light of the light time schedule. It was found useful in doing the extrapolation

110

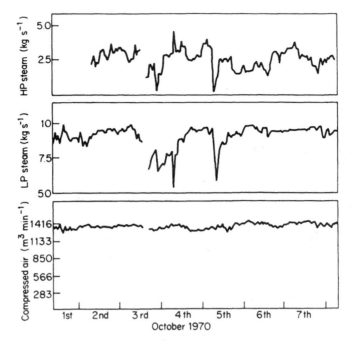

Figure 47. Process demands as a function of time

and, in using the data, to derive the process demands in three forms, namely
(a) as functions of time (Figure 47); (b) as histograms (Figure 48); and (c) in
terms of maxima, minima, and mean.

Economics of operation

The first part of the steady-state analysis had identified the limitations of the
complex in terms of the maximum levels of the services that could be supplied
for any pattern of demand. Within these limitations, choices existed with regard
to the pattern of plant units used to satisfy any of the above extrapolated sets
of demands. It was felt that significant economic benefit might be derived from
one choice rather than another. The operating strategy that could be adopted
at any point in time would depend upon the availability of the three major
categories of plant of which the complex is composed. These items of plant
would not always be available because (a) there will be occasions when routine
maintenance necessitates that items of plant be removed; (b) there will be
occasions when fault conditions prevent items of plant from being operated. It
was necessary in deriving the alternative strategies to consider the implications
of both (a) and (b) together, as they may well occur simultaneously.

Several operating strategies were examined using the linear programming
model and estimates of process demands and the most economic configuration
of services plant determined.

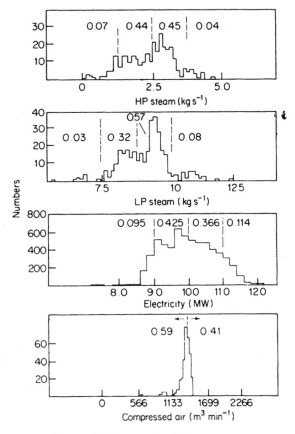

Figure 48. Process load histograms

Considerable cost savings were shown to be possible by running this most economic mix of plant rather than by running all the available plant.

Dynamic performance

Development of the model

The decision was made early on in the project to develop the dynamic model on an analogue computer. The reasons for this choice were based primarily on a need for direct user–machine interaction and secondly on the availability of visual display facilities and low running cost. A fully expanded PACE 231R with two consoles was available, giving a potential amplifier complement of 200 with extensive function generation capability. Thus there was unlikely to be a hardware limitation on the size of model produced. However, care was taken throughout the model development to limit the simulation to the minimum desirable level of complexity; the emphasis was placed on the analysis of the system based on an *acceptable* model rather than on the evolution of a highly sophisticated representation. There were two major reasons supporting this philosophy:

(a) Much of the available data was imprecise and many assumption had to be made.
(b) Large analogue computers are not commonly available. This study was intended, therefore, to produce useful results and conclusions, rather than a tool for analysis.

Inspection of the complete services complex (see Figure 42) showed that although involved irreversible interactions occur in the region of the turbo-alternators, some simplifications could be made at the downstream end. As mentioned previously, it was possible to consider the nitrogen and water supplies in terms of an equivalent electrical load since those items interact in a single direction without any closed loops.

The various steam supplies interact heavily, and care was required in making approximations in this area. Three distinct steam mains existed, operating at approximately 17.9 bar (260 p.s.i.), 8.6 bar (125 p.s.i.) and 1.9 bar (27 p.s.i.). Since the 8.6 bar steam was derived solely by means of degradation of the 17.9 bar steam, it was a valid approximation to transform all loads from the 8.6 bar main into equivalent loads on the 17.9 bar main. The pressure reducing valve motors at a constant rate, and hence, the transformation involved the alteration of all step changes in the 8.6 bar steam load into ramp changes in the 17.9 bar steam load.

The loop connecting the boiler feed to steam loads (e.g. de-aerator steam, feed-water pumps, etc.) was not completed because considerable uncertainty

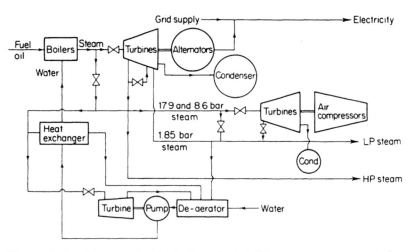

Figure 49. Services complex; basic flows included in computer representation

existed in this area. The time constants in this loop were likely to be long compared with the rest of the system and the load changes comparatively small and, hence, this approximation was considered reasonable and confirmed during the course of the study. It therefore followed that, in all the results, corrections had to be made to the quoted HP and LP steam load for steam used internally for the boiler feed-water in order to ascertain the quantities of steam available to the process.

By means of these approximations, considerable simplification of the system was achieved before the production of the model. The actual system is illustrated diagrammatically by Figure 42, whilst the simplified system modelled is shown in Figure 49.

The boiler model

The physical operation of a boiler with superheat under transient conditions is complex. Although the energy and mass balance equations to represent the system were derived, it was apparent that the resultant performance depended largely upon the thermal lags and transport delays within the system.

In view of the uncertainty concerning the precise operation of these boilers, the potentially complex model was rejected for the initial studies. It was intended that a simple model would be retained, unless more exact performance of the boilers was found to be of great significance.

The application of a simple model, and variation of the basic parameters over a wide range, confirmed the early opinion that the overall system performance was not significantly dependent on these units, and the simple representation was used throughout the studies.

The model used considered the boilers as consisting of three distinct physical features: (a) heat transfer to the water, and evaporation to form steam; (b) a volume reservoir of superheated steam; and (c) thermal and control loop time lags. These three features were assembled in conjuntion with a simple control loop relating heat release to steam pressure to form the model shown diagramatically in Figure 50.

The model was implemented in such a way that the parameters could be adjusted during the operation on the analogue computer to simulate the sudden loss of one unit.

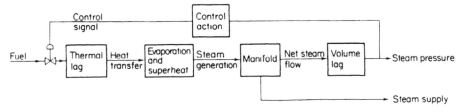

Figure 50. Boiler model assembly

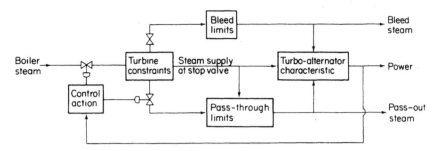

Figure 51. Turbo-alternator model assembly

The turbo-alternator model

The data used for the construction of the turbo-alternator model was derived from the characteristics quoted by the manufacturers in the proposal. Approximate equations were determined for the operating characteristics and the operating limits. These equations were then linearized and rearranged to form a set of defining characteristics. the equations corresponded to the quoted characteristics to within a few per cent for all conditions with a bleed extraction greater than zero.

The turbo-alternator control system was such that the turbine nozzles were brought into use sequentially in order to obtain optimum efficiency under all conditions. This sequential operation was not represented in the model; the main steam control system was assumed to have a linear characteristic.

Choking of the nozzles was assumed to exist at all loads, and on this basis the approximation that the total steam flow was proportional to the boiler steam pressure for a given control setting was employed.

The assembly of the model is shown in block form in Figure 51, from which it can be seen that non-linearities associated with the pass-out and bleed extraction constraints were also considered.

Performance index

In order to quantify the system performance, so that direct comparison between different modes of operation and control could be made, a performance index was required. Such a performance index should, in some manner, represent the important characteristics of each of the services provided. It was discovered after some experience had been gained in running the analogue model that an index involving only the HP process steam was an adequate representation of the complete system performance. This pressure was particularly difficult to control, and since it had a direct influence on the generation of electricity via the pass-through valve of the turbo-alternators, and on the LP process steam, it was considered the most important variable. Thus the performance index chosen was

$$E_f = \int |(\text{pass-out pressure deviation})| \, dt.$$

This performance index was used to identify the major parameters in the system which affected performance and as a result the optimum values for these parameters could be specified and incorporated into the design. Also the effect of uncertainty in other parameters could be examined by changing them from their expected value and observing the change in performance index. These studies confirmed that the boiler parameters, about which considerable uncertainty existed, played no significant part in the overall performance. The studies, however, identified the importance of the steam let-down valves.

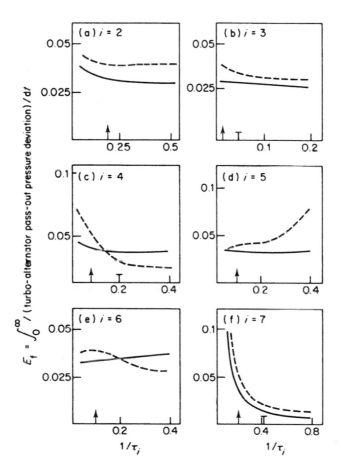

Figure 52. Variation of performance index as a function of system time constants τ_i. *Key:* ↑, standard values; T, revised values; ———, $\pm 5 \, \text{kg s}^{-1}$ LP steam; -----; $\pm 2 \, \text{kg s}^{-1}$ air compressor turbine

Figure 52 illustrates some of the effects of various system time constants on performance using this particular performance index as the basis of comparison. If the performance curve was flat in the region of the expected value of the parameter (inverse of time constant), then the uncertainty in that value would be of little significance (Figure 52(a)). On the other hand, if the performance curve had a high gradient in the region of the expected value (Figure 52(f)), then this value was critical and, if possible, the design was changed to move the particular time constant in the direction of reduced sensitivity. This was done in relation to certain control valve time constants and, although it produced a slight reduction in performance, it produced a more robust system. The curves shown in Figure 52 were produced for two particularly significant disturbances. These were positive and negative step changes about a particular operating level for low pressure steam flow rate and air flow rates.

Model testing

Once the assumptions had been tested, as far as possible, by examining the sensitivity of the performance index, it was necessary to obtain confidence in the model by comparing its behaviour with the linear programming model, i.e. it was necessary to determine that the steady-state and dynamic models were describing *the same services complex*. This was done by increasing the process demands on the dynamic model with it in 'operate' mode until the limits of supply were reached. Curves similar to those in Figure 46 were obtained and compared to those derived using the linear programming model. Fortunately, good agreement was obtained. Had this not been so further sophistication of the model would have been necessary with subsequent delays in the schelule.

Control and controllability

The overall control system had been assembled at the design stage by incorporating the separate control schemes appropriate to each unit. The model was now available to assess the suitability of this scheme.

The general philosophy used in the design of the control scheme was (a) to ensure that the interactions did not cause instability; (b) to match supply uncertainties with suitable redundancies; (c) to remove unnecessary interactions; (d) to utilize the plant generally in an efficient manner. At the same time the designed control scheme was adopted as far as possible.

In determining the recommended control strategy, the following requirements were also taken into account:

(a)   The services complex should remain under reasonable control following emergencies, particularly in the event of the failure of a turbo-alternator.
(b)   The control system should permit adequate control of the complex at both the initial load conditions and at the high load conditions which may be demanded in the future.

(c) Manual intervention should be kept to a minimum, and should only be necessary to effect an improvement in either the overall controllability or the efficiency of operation when this cannot be achieved adequately by the automatic control system used.

(d) The control system should maintain a reasonable operational efficiency consistent with the need for controllability over the required range.

In examining the ability of the complex to supply high loads it was discovered that the boundaries of controllability lay within those predicted by the steady-state model, particularly in the regions of high HP and low LP and high LP and low HP demands. This meant that the recommendations resulting from the linear programming model had to be modified to include the constraints resulting from the dynamic investigations. Operation in these regions could be obtained by manual intervention, and a procedure was derived for carrying this out.

This reduced area of operability due to dynamic instabilities is illustrated by Figure 53. An example of the system response (in terms of HP steam pressure deviations) as the LP steam flow is perturbed about a value within the shaded area and outside the shaded area is given in Figure 54.

Another aspect revealed by the study of system controllability was the significance of the capacity of the LP steam main between the boiler house and the rotary air compressors. As it materialized the actual capacity was within 20% of the optimum value, whereas a value of 50% less would have resulted in serious instabilities.

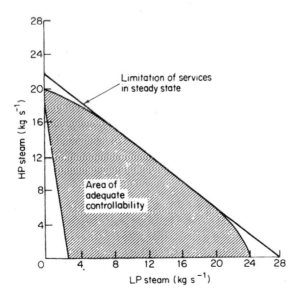

Figure 53. Operating limits. Electrical load 10 MW; compressed air pressure 566 m³ min⁻¹

118

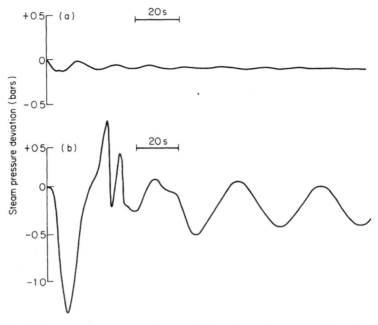

Figure 54. HP steam pressure deviations: (a) $21 \, \mathrm{kg \, s^{-1}}$ steady state, $+ 1 \, \mathrm{kg \, s^{-1}}$ step in LP steam; (b) $25 \, \mathrm{kg \, s^{-1}}$ steady state, $+ 1 \, \mathrm{kg \, s^{-1}}$ step in LP steam

A series of transient studies were carried out to assess the performance of the control scheme. These were initiated by large process demand changes and also by services plant failure. With the recommended parameters (derived by the sensitivity analysis) it was found that with minor modifications to the control scheme, performance was adequate.

Management of the services complex

Under normal conditions it had been shown that the services complex was a flexible and resilient system, capable of operating without major supervision or intervention. However, it was also clear from the work done on this project that this flexibility could only be maintained under certain conditions by making manual adjustments, and that the loss of any plant item could give rise to major operational difficulties.

If the loads on the complex were to remain reasonably constant for long periods, no problems should arise. However, considerable variations in demand were anticipated. To ensure acceptable performance at all times a supervisor or 'decision-taker' was required. The function of the decision-taker was to adapt the mode of operation of the complex to maintain adequate control and efficiency of operation. To do this he needed to assimilate data about the current and anticipated services demand, and data about the availability and operation of the services plant.

Thus, the decision-taker should (a) have a detailed understanding of the services complex and its interactions; (b) be kept informed of the current and likely changes in demand; (c) be aware of the precise operation of the services complex and of equipment availability; and (d) possess sufficient authority to be able to take whatever action is needed to minimize disruption in the event of a failure.

The services complex therefore required the installation of sufficient instrumentation to define the operation of the plant, and the implementation of an information system to support the decision-taker. It was disappointing in this particular instance that time did not permit a study of the human activity system (which was another wider system of the services complex), i.e. the services management system. However, specific recommendations were submitted to the company with regard to the broad requirements of the particular management information and control system needed.

Summary

This project had set out to examine the performance of a services supply system at the design stage and to show that actual and anticipated demands could be met in a well controlled way. The project managed to answer these questions and to show where modifications to the plant and control system were necessary in order that it might achieve this objective. This description has attempted to demonstrate that a design philosophy which includes the examination of both dynamic and steady-state aspects is the only effective way of ensuring that the expected performance can be achieved. Also it has tried to emphasize that control means total control—hardware plus management. To study one and omit the other is a procedure unlikely to result in maximum performance from the total system.

This case history has been an example of the analysis of a designed physical system where the elements were actual hardware (compressors, turbo-alternators, etc.) and the connectivity was the physical flows of steam, water, electricity, etc. The methodology adopted was essentially that due to Jenkins with some refinement in stage 2, related to model building and simulation. The modelling languages used were those associated with linear programming, differential equations and analogue representations. These were appropriate for the nature of the problem situation investigated. they would not have been wholly appropriate if the investigation could have been broadened to include the next higher level in the control hierarchy, i.e. that including the manager of the services complex.

Operational Systems

The term 'operational systems' is here meant to include that group of activities within an organization that can be said to exist and, although I am now referring to human activity systems, I am essentially restricting this group to what I have defined earlier as 'primary task' systems. Thus, if an organization manufactures a product which it sells in the market-place, then it must be accepted that systems exist which convert raw materials into products, which convert enquiries into orders which receive income, maintain production resources, etc.

The analysis of the performance of such systems requires the use of human activity systems concepts and related soft methodologies; however, in terms of position within the problem spectrum, they are towards the harder end.

Because of this, root definitions that are chosen to describe operational systems will contain transformation processes about which there will be consensus, since both the inputs and outputs will be observable in the real-world situations. The only aspect which may be debatable will be the performance expectations arising because of the particular W taken in the definition. If this turns out to be problematical, a number of Ws can be taken and implied performance examined for each one.

The Checkland methodology provides an appropriate assembly of systems concepts for undertaking this kind of performance evaluation and, in the main, is used sequentially. The 'harder' the nature of the analysis the less will be the iteration round the stages, 'select root definition' $\longrightarrow$ 'derive conceptual model' $\longrightarrow$ 'compare'.

The project chosen to illustrate this type of analysis was concerned with order processing and price determination within a company making low volume, custom-built rubber products.

The reason for making this choice is that, within this application of a 'soft' methodology, it was found necessary to derive a particular mathematically oriented technique. Hence this project description illustrates to some extent the relationship between methodology and technique. In this case the technique represented a particular 'how' of doing one of the 'whats' identified by the methodology.

The company, which we will call Lastric Ltd, was a small company employing about 1000 people in total. Of these, 400 were employed in the production of the low volume custom-built lines and the marketing of these products was dealt with by a General Trade Sales Office. In 1970 the company had become part of a larger group and, since the takeover, a rationalization of the high turnover product lines had taken place. However, a large proportion of the company's revenue originated from the custom-built products which were difficult to rationalize. The company also claimed, rightly, considerable expertise in this area and did not wish to see their activity reduced.

The total company product range included specialized rubber products for the medical profession (catheters, components for heart/kidney machines, gloves, etc.), rubber cones (for motorway hazard areas), rubber trays (for ice block manufacture and other freezer applications), specialized rubber mouldings, conveyor belting, rubber sheet, and a variety of rubber mouldings for the motor trade. The company as a whole was profitable but the custom-made product area was not.

ISCOL was invited to carry out a project in Lastric Ltd by the Marketing

Director. He believed that the custom-made product area could be made profitable. The group was applying pressure to close down this activity and he was resisting it. For certain reasons (which were never disclosed) our investigation was to be constrained within the General Trade Sales Office and our remit was to increase the profit from the custom-made lines within the following additional constraints: (a) no further rationalization of product lines; (b) no increase in staff or costs.

The staff in the General Trade Sales Office consisted of 15 sales correspondents who dealt with all customer communications (product enquiries, orders, complaints). They processed some 200 communications per day and were extremely busy people. Mainly due to an apparent lack of control procedures, they spent a considerable proportion of their time chasing the progress of enquiries or orders in response to customer queries. There was a mail clerk who applied a date stamp to all incoming mail and who was responsible for distributing mail around the office and for taking copies of orders to the production and accounts departments. There was a statistician who seemed to spend his time looking for data to plot, a filing clerk who spent her time looking for things that had been filed, and also a sales office controller.

Sales office procedure

Each morning a group consisting of the Marketing Director, the field sales controller, the accounts department manager and the production manager got together to go through the mail. The purpose of this meeting (known as the prayer meeting) was to identify any product enquiries that required special attention. For example, an enquiry might have been made for a moulded product which was totally different from anything that had previously been made. Thus there may have been production questions raised as well as the questions 'What price to charge?' and 'What delivery was possible?' and 'Should we make it?'

Once 'the specials' had been removed from the mail the remainder was distributed around the sales correspondents. The sales correspondents worked in groups. There were about five groups each concerned with a particular market area (e.g. medical, automotive, consumer, etc.). Their job was to process enquiries by quoting a price (related to quantity) and delivery, and to process orders by distributing copies to all who needed to know (like production and accounts). The method of deriving price was to take standard cost information and multiply this by a 'mark-up'. The mark-up for a particular product area was decided by the Marketing Director and was issued as a standard (periodically updated).

Enquiries (and complaints) were received by the sales correspondents by mail, by telex, and by telephone. Orders were usually by letter and these were checked against the original quote (if it could be found).

There were a number of immediate changes that could have been made to sales office procedure, based only upon observation, that would have been beneficial. However, unless these could have been related to an improvement in profit, which was our specific task, they would not have been acceptable.

Thus changes that were accepted and implemented came from a coherent analysis that was seen to be directed at maximizing profit.

The constraint that we should operate only within the sales office boundary was not accepted initially for the analysis, but it was necessary to observe the constraint when it came to recommending changes that were to be both desirable and feasible.

The relevant system that was selected came from the following root definition:

> A system for eliciting and receiving price and production enquiries and orders and for carrying out their processing at maximum profit.

This is a sparsely formulated root definition. The only elements of CATWOE specified are T and W. There are three transformation processes, represented by the verbs elicit, receive, and process. The W implied is that these activities should only be done as a means of maximizing profit. C, A, and O were omitted on the grounds that little advantage would be gained by specifying them, and

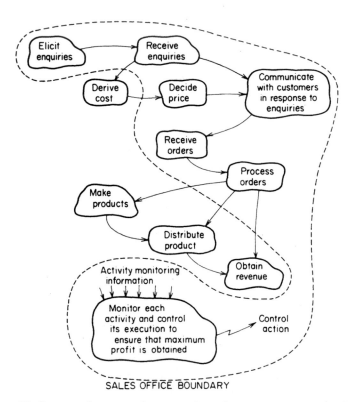

Figure 55. System relevant to the processing of customer communications

Table 4.1. Initial comparison

Activity	How performed	Relationship with profit
Elicit enquiries	Well	Little relationship with profit. Has more effect on level of business
Receive enquiries	Badly	More related to control than profit
Decide price	Uncertain	Major contributor to profit
Communicate with customer	Not very effectively	Little effect on profit; major relationship with control
Receive orders	Fairly well	Some aspects related to control
Process orders	Fairly well	Some aspects related to control
Monitor and control	Badly	Tenuous relation to profit but needs to be effective if maximum profit is to be realized

E, environmental constraints, were deliberately omitted so that the implications of the sales office boundary restriction could be assessed. The model resulting from the above definition is given in Figure 55. The set of activities within the sales office are shown within the dashed boundary. The outputs from these activities could be observed to exist within the sales office and so this was taken to be a primary task model relevant to the processing of customer communications.

In carrying out the comparison a subjective assessment was made of how well each activity was performed and the relationship of each activity to profit generation was examined in order to determine the direction of further detailed analysis. This initial comparison is summarized in Table 4.1.

As a result of this comparison, the activities selected for further analysis were those of 'monitoring and control' and 'decide price'. These two subsystems were each taken to be systems at the next resolution level and, through the mechanism of root definitions and conceptual models, were examined in turn. The control system was taken first since it was argued that unless effective control existed, profit maximization would not be possible. The root definition taken was as follows:

A system to ensure that all communications between customers and the company are dealt with correctly and in a timely manner.

The model resulting from this definition is given in Figure 56. The importance of timeliness is reflected in each activity by including the need to date each specific operation. Table 4.2 represents the summary of the comparison.

Before further analysis of 'a system to decide price' was undertaken, a clerical control system was designed to overcome the deficiencies indicated by the comparison. This entailed the design of a specific log for each sales correspondent

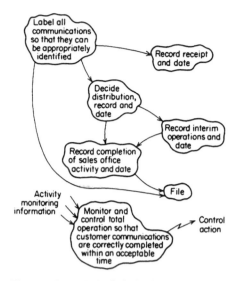

Figure 56. Model of clerical control system

so that they could each control their own activity and a general log used by the sales office controller so that he could control the activity in total. A procedure for identifying communications related to the nature of the content (enquiry, order, complaint, etc.) was designed in such a way that the filing activity could be related to the initial identification and to the sequence of progressing operations within the sales office.

The responsibilities of the various roles, in relation to control, are indicated by Figure 57. Some resistance to the formal recording of activity was apparent during the initial implementation as it was viewed as an extra load on sales correspondents who were already overworked. However, a large proportion of their normal work consisted of searching for orders or enquiries in order to

Table 4.2. Comparison of clerical control activities

Activity	How performed	Comments
Label	Ineffectively	Takes the form of a date stamp only; useless as a means of identification
Record receipt	Not done	There is thus no guarantee against loss
Decide distribution ⎫ Record interim operations ⎬ Record completion ⎭	Efficiently	The fact that receipt is not recorded removes their value as control activities
File	Inefficiently	Not related to any initial identification
Monitor and control	Inefficiently	No measures of performance and no means of tracing the progress of communications

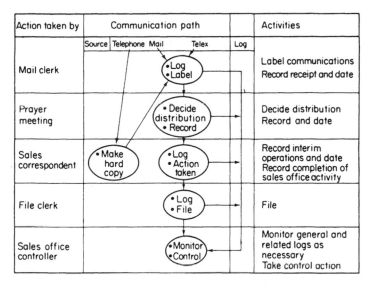

Action taken by	Communication path			Activities
	Source	Telephone Mail	Telex	Log	
Mail clerk		• Log • Label		Label communications Record receipt and date
Prayer meeting		• Decide distribution • Record		Decide distribution Record and date
Sales correspondent	• Make hard copy	• Log • Action taken		Record interim operations and date Record completion of sales office activity
File clerk		• Log • File		File
Sales office controller		• Monitor • Control		Monitor general and related logs as necessary Take control action

Figure 57. Clerical control procedure

respond to a progress query from a customer. The formal recording removed the need for this searching and hence the sales correspondents were able to spend their time more productively and, once the initial familiarization period was over, their workload was actually reduced.

In parallel with the implementation of the clerical control system, an analysis of the pricing activity was carried out. The 'prayer meeting' each morning divided the enquiries into 'specials' and normal production. The price derived for the specials was based upon intimate knowledge of the market and competitors. Thus considerable reliance was placed upon the 'feel' for the situation in terms of how badly the customer wanted the product, and also whether any known competitor would want to, or have the expertise to, manufacture the product. Hence price was determined more by an assessment of what the customer would be willing to pay than by the actual manufacturing costs.

In the case of normal production the price was calculated by taking the cost (derived outside the sales office) and multiplying it by a 'mark-up'. As mentioned earlier the product range was divided into five groups and a mark-up was determined for each group. Thus the sales correspondents would decide in which group a particular product enquiry should be located. They obtained the manufacturing cost from production and multiplied it by the standard mark-up for that group. It was the responsibility of the Marketing Director to set the mark-up and to keep it updated. In doing this he relied on a general knowledge of previous performance with respect to each group and also a 'feel' for the market. The main reason for our involvement was that he felt that this could be improved and also, because the process was so dependent upon his own personal knowledge, it was a process that could neither be quantified nor transmitted to anyone else. Thus a system for price determination that was capable of better performance and that was less dependent upon an individual was seen to be of considerable benefit. Our initial root definition for a system to improve the derivation of mark-up was as follows:

A system to 'home in' on the highest price that can be obtained for custom-built products.

We accepted that manufacturing cost, being determined outside the sales office, was to be a constraint, though this was not built into the root definition.

The major activities appearing in this model, together with comments about feasibility, were the following:

(a) Monitor performance with respect to pricing. (This was now possible with the improved clerical control system.)
(b) Define the factors affecting price. (This was also feasible but about 12 could be identified for each enquiry, though they were not always the same 12.)
(c) Assess the factors affecting price. (This was not possible, on the time scale necessary, due to the volume of enquiries received.)
(d) Combine the results of the assessment to determine mark-up. (An algorithm for doing this could, no doubt, be derived, but given the infeasibility of the previous activity it would be of no value.)

Based upon the assessment within this iteration, a further, more constrained, root definition was taken. The particular constraint introduced was due mainly to the infeasibility of activity (c) above. This root definition was

---

A system to 'home in' on the maximum realizable price for custom-built products based upon a statistical assessment of the factors affecting it.

The model derived from this root definition is given in Figure 58. In order to determine the feasibility of the activities within this model it is necessary to identify 'how' they might be done. Clearly each activity was done in the existing situation, though activities like 'monitor response', 'process data', and 'calculate mark-up' were all done intuitively. An explicit method was required to replace what existed as this was known to be inadequate. A mechanism already existed for doing the activities, 'Decide which enquiries are suitable for a statistical evaluation' and 'Decide mark-up for non-statistically processed enquiries', and this was the morning prayer meeting. The proposal was that this should continue, particularly as 'the specials' were a minor rather than a major proportion of

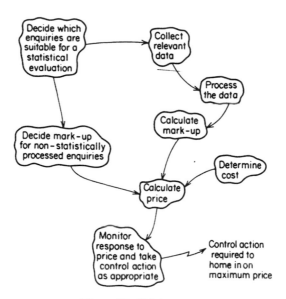

Figure 58. Pricing system

the total enquiries. 'Determine cost' was already done by requesting this data from production. We were constrained to leave that unchanged. 'Calculate price' was a simple multiplication and hence our major concern was with how to do the activities, 'calculate mark-up', 'process data', 'collect relevant data', and 'monitor response and take control action'.

A clue as to how to do these activities came from the use of the expression 'home in' in the root definition. This was used deliberately since the maximum price, in relation to this kind of business, was not fixed, or explicitly known, but was determined solely by what the customer would pay. Thus we likened this pricing system to a gun control system which homes in on a moving target by first of all locating it and then following it. In this case the target was the maximum realizable price, and the control system had to adjust mark-up rather than gun position. An assumption that we made was that the market response to a quotation was based entirely on price; delivery time and quality were not deciding factors. This was an assumption that in-house experience supported. Thus we could measure the acceptability of a price quotation on the ratio of orders placed to enquiries received. This we called the success ratio. In an ideal market, if all prices were acceptable, each quotation in response to an enquiry would result in an order being placed, i.e. success ratio would be unity. In reality this was not quite the case but it was reasonable to assume that the effect of considerations other than price could be neglected. Success ratio was therefore taken as the measure of performance which would be monitored to evaluate the response to price. The procedure adopted is now described.

Mark-up calculation

This section describes the general procedure for calculating mark-up. Each product group had a mark-up appropriate to its own area of business, but the procedure for arriving at it was as described here.

Consider two time periods T_1 and T_2. At the beginning of the first time period (T_1) a price is calculated based upon

$$\text{cost} \times \text{present mark-up} \times M_c,$$

where M_c is a mark-up multiplier and, within the time period T_1, is equal to unity.

During this period all enquiries receive the same pricing calculations and as orders materialize they are noted so that, for this initial period, the success ratio can be calculated as follows:

$$\frac{\text{orders received}}{\text{enquiries quoted}} = \left(\frac{O}{e}\right)_{T_1}.$$

(This ratio may be different for the different product groups.)

At the start of the second time period an arbitrary price increase is made by letting $M_c = 1.2$, say. During this period the same procedure is adopted so that a second ratio, $(O/e)_{T_2}$, can be calculated.

At the start of the next time period the price is calculated from

$$\text{cost} \times \text{current mark-up} \times \frac{(O/e)_{T_2}}{(O/e)_{T_1}} \times M_c,$$

where current mark-up is the value of mark-up used during period T_2 (i.e. initial mark-up $\times$ 1.2).

If the original price (at the start of T_1) was near the maximum (that the market could stand), then the success ratio $(O/e)_{T_2}$ would be less than $(O/e)_{T_1}$ and this would indicate that the arbitrary increase of 1.2 was too large. Thus the prices quoted at the start of the third time period would be multiplied by a factor

$$\left(\frac{O}{e}\right)_{T_2} \Big/ \left(\frac{O}{e}\right)_{T_1}$$

which would be less than unity, and M_c would be set at 1.0.

Alternatively, if the original pricing was far from maximum, then $(O/e)_{T_2} = (O/e)_{T_1}$ and M_c would be set at increased value of 1.2.

Thus if

$$\left(\frac{O}{e}\right)_{T_2} < \left(\frac{O}{e}\right)_{T_1}, \quad \text{then } M_c = 1.0.$$

Similarly, if

$$\left(\frac{O}{e}\right)_{T_2} = \left(\frac{O}{e}\right)_{T_1}, \quad \text{then } M_c = 1.2.$$

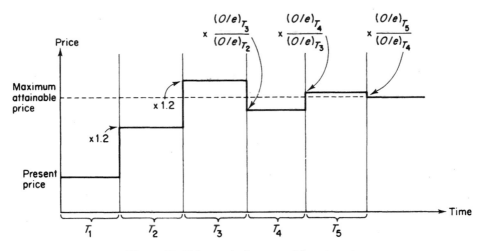

Figure 59. Price variations to achieve target

This price adjustment, via mark-up calculation, is an 'on-going' activity so that price can initially 'home in' on the maximum attainable and from then on will follow market variations.

The expected behaviour is illustrated in Figure 59.

Of course the maximum attainable price will not be a constant (as shown here for illustrative purposes), but will vary as market conditions change. This will not affect the principle of operation.

Calculation of time periods T_1, T_2, etc.

Once a quotation has been made, it is not known when the order will materialize and hence some iteration on the appropriate length of time period will be required. In Figure 60 a distribution of conversion delay is proposed where this delay is the time between a quotation being made and the order materializing.

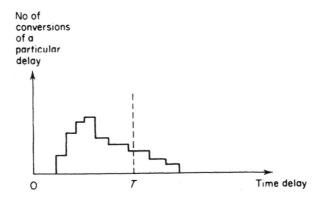

Figure 60. Conversion delay

The time T must be chosen so that the ratio (O/e) during the period T is a true reflection of the market response. It was found during the initial implementation that the period T was very much shorter than the period over which it was considered reasonable to make price changes.

Priority calculation

The calculation of mark-up has assumed that all enquiries are equally desirable and equally possible. In an actual production situation where resources are limited a priority calculation must be capable of differentiating between what can be made and what it is most profitable to make. These two features can be dealt with in turn.

Production constraints

At the enquiry stage it is desirable to know what resources are available and which production constraints are likely to be met. A constraint accumulation based upon expected enquiry-to-order conversion was suggested as a means of indicating this situation. A set of maximum production resources were identified (production man-hours for example), yielding the set of constraints, $C_1,...,C_n$. During each time period the production resources required for each enquiry are estimated and accumulated in the appropriate column (Figure 59). Although it is not known which enquiries will yield orders, an expected success ratio (O/e) is known and therefore each of the maximum resource levels can be increased by the reciprocal of this ratio. Thus if a success ratio of 50% is probable, the resource levels can be doubled and this new level used as the constraint on *enquiries*. This does not imply any changes in the actual resources available to meet *orders*. Thus, during a particular time period T_n, the resources used can be totalled as in Figure 61, until a constraint is reached. At this point decisions have to be made on whether or not quotations will be made on further enquiries and if so what deliveries should be quoted.

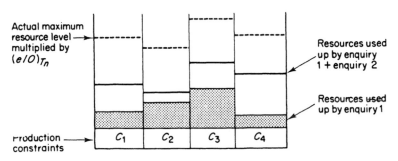

Figure 61. Constraint accumulator

Desirability calculation

In deciding 'what should be made' three aspects need to be examined:

(a) Orders which produce maximum contributions should get maximum priority.
(b) Orders which are 'easier' to make should get priority over those which are difficult.
(c) Orders which have low contribution but which are desirable in some other sense should get high priority (i.e. orders which will lead to high contribution orders, or orders which company policy dictates should be made).

A priority rating which accounts for aspect (a) can be assessed from the ratio

$$\frac{\text{contribution}}{\text{most scarce production resource}}.$$

Those orders with the highest number receive the highest priority. The other two aspects require some subjective assessment of priority but which, however, can be combined with the first to give an overall priority rating. This priority rating can then be combined with the previous mark-up calculation to give a differential mark-up based on priority.

For example, assume priority range 1–10 and multiply the previously calculated selling price by $10/P$. Thus if a particular enquiry has the highest priority (i.e. $P = 10$), then the selling price will be as determined by the previous mark-up calculation. If, however, the enquiry is less desirable, a priority; rating of 9.5 may be appropriate, yielding a selling price 5% greater than the standard. Similarly a potentially difficult enquiry may receive a rating, $P = 5$, resulting in a price double the standard. Therefore by applying this priority rating all enquiries become *equally desirable* and the market determines which eventually become converted into orders.

Summary

Figure 62 illustrates the interactions between the sales office activities and the other activities associated with pricing. In particular it shows the information flows that were necessary to establish the pricing and priority procedure proposed above.

The personnel involved in applying this procedure were (a) the sales correspondents, who assemble records of numbers of enquiries and orders, the times of receipt and who apply the derived mark-up to the quotations; (b) the statistician, who collects the data from the general and specific logs, processes it, obtains priority and constraint information, and supplies the sales correspondents with the derived mark-up; (c) The 'prayer meeting', which makes the decisions between 'special' and 'normal' enquiries and the appropriate

132

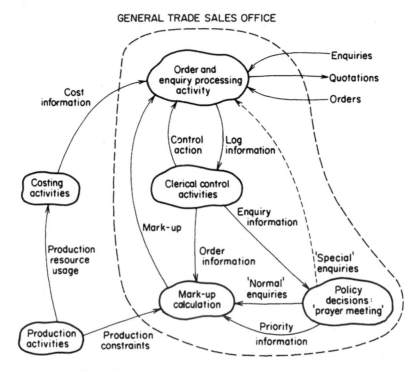

Figure 62. Total information flows relevant to pricing

members of that group who make the decisions with regard to priority; (d) production personnel, who provide information on critical constraints.

This project has illustrated the formal application of the Checkland methodology to low level operational activity within a small company. A number of deficiencies in the situation could have been identified and recommendations made based solely on the application of *ad hoc* 'common sense'. However, what is frequently called common sense is no more than an informal comparison of some implicit model of 'what ought to exist' in the head of the observers with 'what is observed to exist'. Using these particular systems concepts enables that implicit analysis to be made explicit. Thus the analysis can be seen to be coherent and any prejudices related to the view of what *ought* to exist are made apparent and can be challenged. Simple models are all that are required to achieve this and the time taken to produce them is minimal and is time well spent. The project has also illustrated the introduction of a technique and, although the decision on what the technique needs to be requires creative thought, the thought process is significantly aided by the initial systems analysis.

Testing an Hypothesis

Unlike traditional scientific research, in which the problem to be studied can be assembled in the laboratory and pursued via well defined and repeatable experiments, action research must be undertaken in the real world and hence must accept problem situations as they are presented. Because these situations and the environments in which they reside are in a dynamic state, the direction taken by the research cannot be predetermined. Given the above, it is difficult (and to some extent undesirable) to predict the expected outcomes. Again, unlike scientific research, it is difficult to test hypotheses. Even if an hypothesis can be formulated, it is unlikely that a real-world situation can be found which turns out to be just the vehicle for testing it.

Because of this, progress through action research tends to be less co-ordinated and less directed than the scientifically based mode of inquiry. However, there are certain instances where it is possible to formulate an hypothesis in relation to a specific situation so that learning can be achieved in relation to that situation rather than to the general body of knowledge with which the research is concerned.

The project briefly described here was, I believe, suitable for this kind of inquiry. The object of concern was the role of the headmaster in state-owned English secondary schools. The reason for our interest was that an organization in the north-west of England mounted short courses for headmasters with the intention of introducing them to the principles and practices of management. The view of this organization, and the motivation behind the courses, was that they believed that a *headmaster was more a manager of a set of resources than a head teacher*. We had had some discussion about project involvement with the principal of this organization and a project materialized when one of the headmasters, attending the short course and having been exposed to systems ideas, offered his school as a vehicle for undertaking an investigation of the actual management process.

The project was undertaken by two students from the master's degree programme, my colleague Ian Woodburn, and myself. No problems, or any particular concern, had been expressed by the headmaster and so the teams set out to answer two questions, 'How can a school be viewed as a human activity system?' and hence 'How can it be managed?' As the study developed, two different ways of looking at the management process were considered. One was to apply the Checkland methodology as a way of identifying any particular problems associated with management at this particular school. The investigation started at the stage of 'finding out' and proceeded to highlight areas of concern through the selection of root definitions relevant to the control options open to the headmaster.

The second approach took the above belief of the course-running organization as an hypothesis and, by expressing it as a root definition, developed a model of a resource management system with a headmaster as the decision-taker. The team undertaking the study was divided into two and the group adopting the second approach deliberately avoided any 'finding out' of the situation at the school. In terms of the Checkland methodology we were starting at the stage 'select root definitions of relevant systems'. In this case one definition was chosen and this was believed to be *relevant to the hypothesis* rather than relevant to the situation. The root definition led to a conceptual model which was then compared with activities in the school as a means of testing the hypothesis in that particular situation. The hypothesis was as follows:

Hypothesis *A headmaster is more a manager of resources than a head teacher.* The root definition of a system relevant to this view was taken to be

> A local authority owned system for the control and operation of those resources necessary for the formal education of 11–16 year old members of the community.

The conceptual model (at a first level of resolution) contained the following (minimum necessary) activities:

(a) Know the requirements of the local education) authority (LEA) and the community for the formal education of this group of 11–16 year olds.

(b) Assemble a set of activities* which will aim to achieve these requirements.

(c) Assess resources† required to carry out these activities.

(d) Acquire, allocate, and use resources.

(d) Monitor achievement of requirements and control all above activities to achieve requirements.

(f) Report on performance.

Each of the activities in the above conceptual model were further expanded to yield the more detailed model represented in Figure 63.

This expansion was achieved by defining, in turn, a system to realize each of the activities of the first level, and then by developing a conceptual model for each system.

The root definitions used for this further development were as follows:

(1) > A system to know about and appreciate the formal education requirements of the appropriate group of 11–16 year old members of the community through relationships with both LEA and the community.

---

*Statute 7 of the *Education Act 1944* suggests that these activities may be categorized in terms of the following: teach (academic subjects); train (vocational subjects); instil (social values); develop (physical needs); maintain (health).

†It could be argued that the pupils themselves represented one group of resources (i.e. the raw material), but that implied a more mechanistic view than was actually taken.

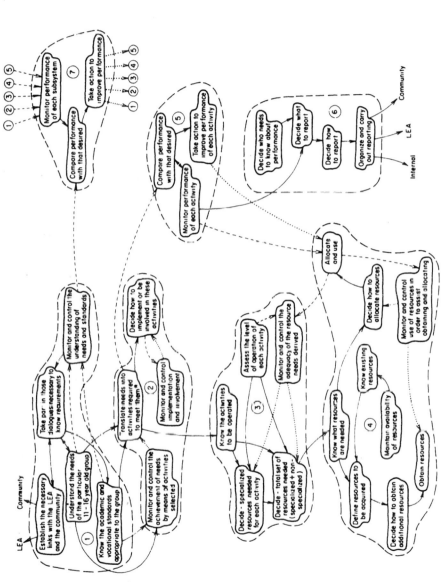

135

Figure 63. Conceptual model. System 1 is a system to establish needs and requirements; system 2 is a system to define activities to meet needs and requirements; system 3 is a system to establish resource needs; system 4 is a system to obtain and use the resources; system 5 is a system to control the use; system 6 is a performance reporting system; system 7 is a total system control system. *Key:* —→, major information links; ----→, performance monitoring information; ·····→, control action. Subsystem boundaries are enclosed by broken rules.

> (2) A system to assemble a set of activities relevant to the needs of this particular group which will aim to achieve the desired standards.

> (3) A system to assess the resources needed for the successful operation of the selected activities.

> (4) A system to acquire the necessary resources, plan their use, and use them to achieve satisfactory performance.

> (5) A system to monitor the achieved outcomes of the above activities and by comparing against desired outcomes take action to improve the overall system performance.

The activities present in this detailed model (see Figure 63) were compared with the actual situation at the school, yielding the following summary of observations. This comparison of the model with reality was intended to highlight the implications of taking the particular view described above. The differences identified were meant to initiate debate about particular areas and did not necessarily represent deficiencies which should be eliminated.

Communication with LEA and community (system 1)

LEA

Local authority advisers visited schools to discuss syllabus content. Headmasters and teachers saw this role as an advisory one, not as a control over school affairs. Thus although a mechanism existed for communication with the LEA it was not clear whether it was used for the purpose indicated by the model.

Parents

Parents had little influence over school affairs. A parents association existed but few parents took part in it. The Headmaster expressed regret for this non-participation but was reluctant to take a more leading role in the promotion of parent–teacher association (PTA) activities. Thus again a mechanism existed but it was debatable whether or not it was appropriate for the Headmaster to make it work. The model suggested that it was.

Employers

Most children on leaving school were employed locally but the Headmaster believed that the school should not be vocationally oriented. However, this link did have an impact on the curriculum. As the major decision-taker in the system we considered, the Headmaster had responsibility for the monitoring and control of the activities represented by the control subsystem. Thus the implication was that it was his responsibility to ensure that the requirements of the LEA and the community were fully appreciated in order to provide a suitable input to the next subsystem, i.e. the assembly of appropriate activities to meet these requirements. There was no doubt that, as far as the LEA was concerned, this was adequately carried out. It was not clear that sufficient understanding of community needs had been achieved but, in a situation where the responsibility for what happened between 9.00 a.m. and 4.00 p.m. had been so largely delegated to the school, could better be achieved? Certainly the mechanism existed for parents to communicate with the school through the PTA and parent–staff meetings, but this represented only a minor proportion of the community. Since most children remained in the local community after leaving school, how was their subsequent progress and behaviour related to their education between 11 and 16 years? the Headmaster was concerned about their needs while they were at school, and the particular measure of performance used was absenteeism. He argued that, if the assembly of activities provided at the school did not meet these needs, then absenteeism would be higher. In this area some negative indicators existed such as examination results, employers reports, police/social services reports, and contact with some parents and some past pupils, but there was no evidence of the existence of formal channels of communication.

In the school situation, the assessment of needs and requirements was mainly made as a result of the professionalism of the teachers themselves. Teachers developed their own appreciation of the aims of secondary education. These became part of 'a philosophy for the school' which then influenced many of the decisions about curriculum. Although formal procedures for this assessment of needs and, more importantly re-assessment, seemed non-existent, there were many examples of internal changes resulting from outside influences; for example, the decision to introduce a second language into the curriculum as a result of an expected change from secondary modern to comprehensive education, the decision to continue commercial courses (valued by parents but not by the Department of Education and Science), the decision to retain one formal examination per form per year plus a conventional school report to meet parents' wishes. Thus it was concluded that, while there was evidence that the activities represented by this area of the model were being carried out, the informality of the process made it difficult to assess how well they were being done.

Assembly of appropriate activities (system 2)

It was difficult to assess the degree to which needs were met by means of the activities selected. It was the Headmaster's belief that a broad range of activities provided the greatest likelihood of meeting these needs. For example, in addition to the usual academic and physical educational subjects, creative design was introduced as a necessary part of the educational development of the children. Although the Headmaster believed that the school should not be vocationally oriented, physics 'O' level was introduced into the curriculum as employers required it from children seeking engineering apprenticeships.

Assessment of resource needs (system 3)

The assessment, in terms of teaching resources, was done in consultation with the staff members concerned. At the start of the assessment year each head of department responsible for a particular activity was asked to assess the department's needs in terms of (a) those resources with which the department could just survive; (b) those resources necessary to maintain desired growth; and (c) those resources which would allow a specific new development to take place.

With this information and knowledge of the total budget available, the Headmaster decided on the appropriate allocation for that year. In terms of the total amount of teaching space, teaching staff, and materials, this was dictated by the LEA based upon expected number of pupils in each school year. Some discussion, however, did take place between the Headmaster and the LEA on the subject of space and staff needs.

Monitoring and control of the adequacy of the resources allocated was done internally by the Headmaster through consultation with the individual heads of departments, though it was unclear what performance measures were used in this assessment. With regard to the adequacy of the total resources provided, there was no evidence of any measures of performance, but in view of the LEA dictat these may have existed in the LEA itself.

Obtain and use resources (system 4)

The activities associated with obtaining resources were highly constrained in the school situation. The formal procedure for obtaining resources was through the LEA according to the rules laid down. Additional resources could only be obtained through informal fund-raising activities organized by the school or through parent associations.

Maintaining resources was an area in which some conflict appeared to have arisen. Maintenance as a result of school usage presented little problem, but some of the resources, particularly those associated with commercial or vocational activities, tended to be used by external organizations such as adult education, and this led to some confusion over responsibility.

The monitoring of the allocation and use of the total resources formed a significant input to the control activity of the previous subsystem and was discussed there.

Monitor and control (system 5)

In the detailed model represented by Figure 63, a distinction was made between the monitor and control of the individual activities undertaken in the school (system 5) and the overall control of the total system (system 7). Total system control was difficult to examine at this level of detail and a further systems analysis was undertaken of a control system particularly relevant to the school by the other members of the team based upon their appreciation of the actual control options.

The control of the 'assembly of activities' system has been referred to earlier. The absenteeism measure represented some degree of satisfaction with the assembly in total, but how was the performance measure for each activity derived and used? The responsibility for monitoring and control at this level resided with the individual staff members concerned, but it was the Headmaster's responsibility to see that it happened. The concern at this level, for example, was not '*What* should be taught?' but '*How* should it be taught?' Thus the model implied that the way in which a particular activity was carried out should be assessed and changes introduced to improve its effectiveness and interaction with other related areas. Although these types of change had happened to some extent (for example, a broad educational base had been introduced for first and second year pupils through team teaching of English, geography, and history under the label of 'humanities'), it was not clear how the assessment of previous experience had been undertaken which led to the change, or how the response to the new experience was being monitored and assessed.

Performance reporting (system 6)

The activity of 'reporting on performance' contained in the model concerned three areas relevant to (a) the LEA as 'owners' of the system; (b) the community as recipients of the output of the system; and (c) internally to the personnel responsible for operating the system.

No formal reporting mechanism seemed to exist with the LEA and it was assumed that such performance reporting as was seen to be necessary was carried out informally through the visits of the LEA advisers and possibly through the visits of the school inspectors. The links between the LEA and the Department of Education and Science were, however, outside the scope of the project and were not pursued.

Regular reports were submitted by the Headmaster to the governors and annual reports were submitted to the parents in respect of the individual pupils. Internally, reporting was carried out through various school meetings and through a staff newsletter. The latter, however, tended to be used more to advertise meetings and other events than to report on performance.

140

Conclusion

The particular model derived here served as a useful structuring device for questioning and hence learning about the actual situation in the school. There was no doubt that all of the activities contained in the model were being carried out in the school, but it was apparent that because, in the main, they were done informally, there was no way of assessing *how well* they were being done other than through a general 'feel for the situation'.

The statement that a headmaster is a manager of resources implies an ability to measure their allocation and use in such a way that control action can be taken to make that allocation more effective. This could only be done through a more explicit and formal set of activities than existed in the school at the time. As an important decision-taker in the system modelled here, the Headmaster's role was seen as one which undertook the monitor and control activities of systems 1 and 2 and the whole set of activities of systems 6 and 7. Although it could be argued that the Headmaster did fulfil this role, for him to be this manager with a concern for effectiveness, not only did he need to clarify the measures by which performance was assessed, but also the scope of the control action which was then open to him.

If the system of concern, as in this case, is an educational system, the effectiveness of the formal educational process can only be assessed some time after a pupil has left the system (unless one believes that examinations represent a true assessment of education). Of course, education continues and it is never possible to identify the particular contribution made by the initial formal process from the many other factors which eventually influence it. Thus, unlike a manufacturing process, in which product quality, yield, efficiency, etc., can all be measured, no measures of the effectiveness of the formal education process exist which a headmaster can use to derive control actions.

Similarly, unlike an industrial enterprise, formal communication and power structures are much more difficult to establish, making the control processes that much more complex.

With the exception of the Headmaster, all staff spent a considerable amount of time in the teaching situation, making communication difficult. Even outside the classroom, time was spent marking and preparing lessons. Time was therefore at a premium, making it difficult to make use of formal communication procedures. The main method of written communication was via the staff-room notice board and the weekly newsletter compiled by the Headmaster. Other communication was either completely informal, generally in the staff room at mid-morning and lunch breaks, or through the various committees that met outside school hours.

In addition, many staff had more than one role in the organization: 40 of the 65 teaching staff were also form teachers, which was a pastoral care activity. The Senior Master, who was a member of the Management Panel in the school, was also Head of Upper School (a pastoral care activity) and a teacher in the Mathematics Department. The Head of Middle School taught mathematics and therefore responded to the Head of Mathematics; the Head of Mathematics was a form teacher in the Middle School and, in this capacity responded to the Head of Middle School. This complex set of relationships, not uncommon in a school, are most unusual in an industrial or commercial enterprise.

Although it can be said that the hypothesis was tested, in the sense that the activities contained in the relevant conceptual model existed, how those activities were undertaken was characterized by a degree of informality very much higher than would exist in the kind of managerial situation envisaged by the course-giving organization when proposing that hypothesis.

In methodological terms, only three stages of the Checkland methodology had been used. These are illustrated by the following diagram:

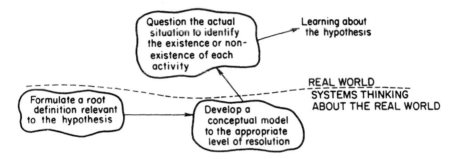

Here, the starting point was the root definition. Whereas it might be argued that the root definition was not, itself, a good representation of the hypothesis, and hence should have been changed, what was not feasible was a modification of the root definition based upon the results of the questioning process. Thus, iteration around this loop (a normal requirement of the Checkland methodology) is not permissible in this application. The belief is that, if the root definition is a reasonable representation of the hypothesis and a defensible model is derived from that root definition, then an identification of the existence of each activity in the model will indicate that the hypothesis is, itself, reasonable.

Service Systems

As companies grow they are faced with problems of management that are not merely concerned with the increase in personnel numbers. A company that undergoes the transition from small to medium-sized needs to contemplate the introduction of a more bureaucratic organization than had previously been necessary. A small company is not just a scaled-down version of a large company. One feature which differentiates between small and medium-sized companies is the existence of service functions. As companies become larger there tends to be a proliferation of such services. Examples are research and development departments, planning organizations, and management services. The analysis of problems related to service departments is particularly difficult as the very existence of these functions is a permanent issue. One approach, therefore, is to consider an issue-based analysis which seeks to identify a number of potential roles and develops root definitions of systems to realize these roles. The Checkland methodology can then be used as it stands as a way of exploring the implications of this variety of views of service systems within the particular company concerned and to identify procedural and/or organizational changes which can be definded as desirable and feasible. A modification to this approach, which was successfully used in one of our projects associated with the master's degree programme, made particular use of other systems thinking at the conceptual model development stage. The structuring of the conceptual model, which was achieved through the use of a particular concept from control engineering, was felt to be crucial to the progress of the project at this stage. Success in the project was assessed in two ways. Firstly, the project was seen as one way of introducing systems ideas and methodology to the personnel within the management services function of the company and, secondly, in terms of the degree and quality of debate about the nature, organization, and activities of the services function itself. Because of the first of these requirements, the project was preceded by a seminar to introduce systems ideas and the senior personnel took part in other seminars that were organized in parallel with the project. Since, through the seminars, the personnel became familiar with the particular systems language used, they were also actively involved in the analysis.

The project was undertaken for the Management Services Branch (MSB) of the South West Region of the Central Electricity Generating Board (CEGB) and the client was the head of the branch. We were asked to undertake a study of 'The management requirements of the Management Services Branch' both as a means of developing understanding of the nature of the branch and as a means through which the Lancaster systems approach might be demonstrated. The phrase 'management requirements' was deliberately vague in order to allow considerable scope for the project and flexibility in approach. The team with responsibility for the project consisted of two students from the master's degree programme, a senior manager from within the Branch, and myself, as project manager. It was recognized by the team that any discussion generated by the findings of the study would be constrained to personnel within the organizational boundary of the branch but that access to other parts of the CEGB Headquarters and the South West Region was permissible in order to assemble as complete a picture of the role and working of the branch as was feasible within the time-scale of the project (approximately 16 weeks). Given the nature of the 'problem-content system', an issue-based analysis was considered appropriate and the activities in the problem-solving system were as listed below. This effectively represented the intellectual development of the project and it was seen to consist of two phases. This approach was agreed with the branch management at the start of the project.

Phase I
(1) Assemble a picture of management service activity as viewed from within the branch and from outside (i.e. by their clients)

(2) Define 'a system' to realize management service needs within the Region.
(3) Select root definitions for this system which reflect the views expressed in (1).
(4) Develop conceptual models from these root definitions and examine the significance of any major differences which result.
(5) Decide to (a) accept a particular view; (b) accept a modified view; progress separately the models produced in (4) to identify the implications of the differences.

Phase II
(6) By comparing the conceptual model(s) derived in (4) with the actual situation, identify areas in which more detailed analysis would be beneficial. These may be in the context of defining 'management requirements' in specific areas as indicated in the previous phase.
(7) Set up a seminar with branch managers to discuss the areas of concern resulting from the investigation.

Activity (2) was the most difficult in that an analysis of 'what the branch does now' would take too much as given and might therefore have missed significant areas of study which could lead to what the branch *might* do. The approach taken here was to ask the question, 'What system does the Management Services Branch serve and hence of which is it a part?'

The system chosen was the South West Region electricity provision system of the CEGB. A broad level activity model was produced and those activities discarded which were seen to be executive decision-making activities and those concerned with operations management. What was left, it was argued, could be viewed as a management support system.

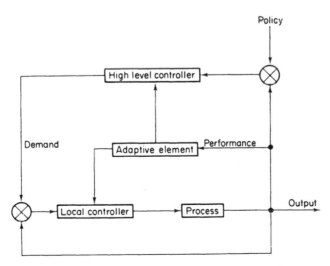

Figure 64. Simplified; adaptive feedback controller

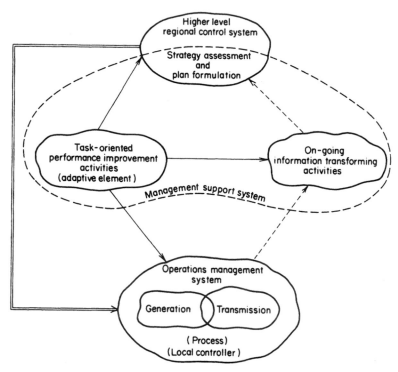

Figure 65. The Central Electricity Generating Board as an adaptive control system. *Key:*
--- →, performance information; ——→, actions to improve performance; ⟹, demand
and expectations for performance

The relevant system in this case was derived by reduction from a wider system.
The process of reduction was based upon making an analogy between the
CEGB electricity provision system and a general model of an adaptive control
system (Figure 64). Figure 65 represents the translation of this adaptive control
system into the structure of the wider CEGB system.

The root definition of this wider system was derived from the published set
of objectives for the south-western regional organization of the CEGB, and this
was argued as reasonable since the role of the CEGB was not being questioned,
only the role of the MSB within it.

Figure 66 represents a notional set of activities for the wider CEGB system.
Each activity was examined in turn to identify whether or not it could be
allocated to either the higher level regional control system or to the operations
management system. These are shown within the shaded areas. The remainder
could be viewed as management support activities. In addition to this process
of reduction further activities were discarded from this management support
set which, *within this particular situation,* could *not* be seen as MSB activities.

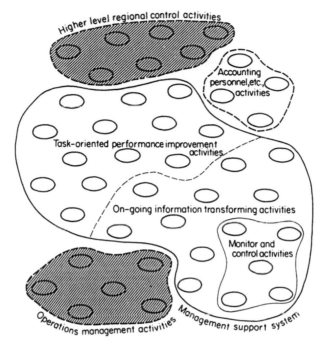

Figure 66. Management support system as part of a wider CEGB system

These were, for example, activities concerned with the accounting and personnel functions.

Thus, by this process of reduction, a set of activities remained which could be viewed as a set appropriate to the MSB.

However, when compared against the formal systems model, it was seen to be deficient as a model of a human activity system. It represented a set of activities, but none of these were concerned with the monitoring and control that was necessary if the set could also be said to be a human activity system. Thus control activities were added which were appropriate to the area of decision-taking represented by the boundary of the remaining support activities.

Returning to the activities of phase I, root definitions were also obtained which reflected the views of senior managers within the branch and from these root definitions further activity models were derived.

Apart from providing insight into the different interpretations of the branch's role, these models were used to 'validate' the derived model described above. By comparing the models from the manager's root definitions with the one derived by reduction, it could be established whether or not the derived model *at least* contained the others. Once the derived model had been validated it could then be used as a tool to compare against the actual branch situation to identify areas for improvement or for more detailed investigation.

Views were also sought from outside MSB, again with the intention of identifying possible activities missing from the derived model. In the event this proved less valuable as a means of validating the model but extremely valuable in assessing the effectiveness of some of the 'across boundary' MSB activities.

Having produced broad observations from a comparison of the total MSB model with the actual situation, the next stage was to concentrate on specific areas for a more detailed analysis. It was argued that the 'on-going' activities could readily be seen as necessary regional management support activities (see Figure 65) but that the task improvement activities represented a part of MSB role that was needed to be seen to be effective if it was to survive. Thus effort was concentrated on this particular area and, because the 'management' of the branch was crucial to effectiveness, the branch control activities were also examined. The necessary activities for a task-oriented improvement system and an MSB control system are given below.

Task-oriented system

(1) Establish a means for identifying areas for improvement.
(2) Ensure that skills are available to analyse and reach recommendations about improvements.
(3) Make known the existence of this capability.
(4) Take steps to obtain improvement tasks.
(5) Allocate the tasks.
(6) Carry out the tasks.
(7) Make recommendations on improvements and how to implement.
(8) Monitor implementation and review recommendations.
(9) Control the system to ensure that the specific tasks, their identification, and their execution are done effectively and contribute to the improvement in performance of the region.

MSB control system

(1) Decide on measures of performance and expectations relevant to each subsystem.
(2) Monitor performance of subsystems.
(3) Identify areas for improving the management of subsystems.
(4) Know and be competent in the application of modern management techniques and approaches so that subsystem management can be improved.
(5) Select from (4) and apply where appropriate to improve total system performance.

A listing of activities, such as the above, does not constitute a model. The connectivity must also be included. However, the listing is given here to illustrate the nature of the activities being considered.

Comparing these activities with the picture of the branch derived from both internal and external sources led to the identification of areas of concern within which specific recommendations could be made.

A summary of the comparison at this level of detail together with generalized comments about the activities is included to illustrate the kind of areas highlighted for debate.

Task-oriented system

Activities (1), (3), and (4) were all concerned with projecting to present and potential clients the capability of the branch as a problem-solving resource. Particularly as a result of outside questioning it appeared that activity (3) was not well done. Activities (1) and (4) were related to the decision on where the branch effort should be directed in seeking and obtaining problem-solving assignments. It was apparent that this decision was not taken corporately in the light of a disciplined survey of branch capability and future desired direction.

It was suggested that a useful aid to carrying out corporate strategy evaluation of this kind was a device that we called an 'experience matrix'. This was a particular recommendation that emerged from the comparison and will be discussed later.

Activity (2) suggested a need to assemble the appropriate skills once the 'marketing' decisions referred to above had been taken. A particular deficiency was seen to exist in that skills were assembled more on a technique orientation rather than on a problem-solving orientation. Activity (5) was seen to be done on a limited scale. Because of the technique orientation, problems tended to be allocated according to the technique that was likely to be of use and, hence, the resources used tended to be restricted to that particular expertise. Allocation also tended to be partial, i.e. the project manager and 'doer' were not always given full access to the problem situation.

Activity (6) seemed to be adequately performed except (as mentioned above) for the need to be involved in the problem situation as much as possible if the resultant 'solution' was going to be appropriate.

Activity (7) was deficient in the sense that 'how to implement' was rarely seen as another problem to which a problem-solving expertise should be developed. It was apparent that the reason for some on-going activities living on in the MSB was due to a failure to implement, the distinction here being the difference between designing and then operating a procedure.

Activity (9), which was concerned with the control of activities (1)–(8), is complex in the sense that a mixture of resolution levels of control is implied. Activities (1)–(4) were related to group effectiveness and hence should be the concern of the overall branch controller whereas activities (5)–(8) were related to within-group individual task effectiveness and hence should be the concern of section managers. There appeared to be a mixing of these levels within the then current branch control with, particularly, the involvement of the Head of Management Services in the lower level task control.

MSB control system

Activities (1)–(3) were concerned with the identification of the different kinds of activities going on within the branch and the development of measures of performance that were appropriate to them. This was not done in any *explicit* way and hence it was not possible to state whether or not it was done well. It was decided to make recommendations on the structure and components of measures of performance so that, if necessary, modifications to any existing measures could be made.

Activities (4) and (5) were concerned with keeping abreast of modern developments in problem-solving and other management activities and selecting, and developing if necessary, those particularly pertinent to the branch and the region's needs. The only deficiency to be noted here was that the process was rather *ad hoc*. There was no formal survey of what was available or under development in outside establishments and hence no rigorous selection of those worth monitoring.

As a result of comparison of these, and more detailed models for the 'task-oriented' and control systems, the following areas of concern were established within which detailed recommendations were formulated: (a) understanding, by

	Information system design	Maintenance organization			Productivity monitoring	Manpower planning	Wayleave acquiring systems
		2, 4			2	5	
		3, 5			2, 7		
		2, 4, 4				2	
	2						
	3					4	
						2, 5	

Figure 67. Experience matrix. Directions of marketing effort are as follows: (a) extend contacts in areas where prior experience (and extent of outcome) can be demonstrated and referred to; (b) decide to approach areas and clients where no previous activity exists and develop skills in these areas

MSB staff, of the role and capabilities of MSB; (b) understanding by clients and other functional areas of the role and capabilities of MSB; (c) development of resource skills within the branch; (d) development of external linkages between the branch and headquarters and the region; (e) development of MSB marketing policy; (f) organization structure of the branch; (g) measures of performance for control.

It is not really appropriate to discuss the detailed recommendations here, but one that may have more general applicability is the experience matrix.

Derivation of an experience matrix

As an aid to creating a greater appreciation of the extent of the branch's capability, as well as providing a vehicle by which conscious decisions could be made on the allocation of marketing effort, it was suggested that the following matrix be constructed. This showed, in the one picture, the areas of work undertaken by the branch as a whole and also the degree of effectiveness achieved. This is reproduced as Figure 67. The vertical scale was a listing of existing and potential clients and the horizontal scale was a listing of task areas. The elements of the matrix could be filled by a symbol for each project (the density of such symbols in any one element indicating the degree of experience in that area) or by a number (on a scale from 1 to 7, say) indicating the outcome of the project. Such a scale could be

      1 = enhanced understanding
      2 = improved organization of information
      3 = new organizational structure and/or processes
      4 = improved level of service or quality
      5 = better utilization of resources
      6 = reduced operating costs
      7 = increased productivity

Although one matrix only was considered, it is clearly possible to develop others which separately show level of activity in specific projects in particular client areas, or outcome in relation to particular clients, or outcome in relation to particular kinds of projects. The analysis of these matrices could provide information on how effective the branch was in relation to certain clients or in relation to particular kinds of projects.

The total picture could demonstrate those areas (both client and project) in which the branch had experience and those in which it had not. Thus decisions could be made on where, and how, to direct marketing effort and on what skills were required in order to provide the total problem-solving/advisory capability.

Such a matrix would have to be constructed for a particular historical period and subsequently updated. In this case the previous three years was suggested as a suitable period over which to survey existing branch capability and type of activity.

150

Reflections on the project

The methodology derived at the start of the project, listed previously as a two-phase analysis, was consciously modified as it was being used. Phase I, in particular, was changed as difficulty was experienced in using the issue-based models derived from the variety of views that existed about the role and activities of the MSB identified from the initial interviewing of branch staff and their clients. It was at this stage that the control analogy was used to extract a relevant model from a primary task model of the wider system related to electricity provision for the region as a whole. The resulting methodology that represented phase I is illustrated by Figure 68.

A similar approach was adopted in a project, undertaken recently, for a mining company in Mexico. This was a medium-sized company concerned with raw material acquisition and processing for the glass industry. The managing director of this company had identified a need for a new technical support function. Expertise existed within the company but it was distributed around the various departments. As part of a company-wide reorganization he had decided to centralize this expertise and hence form a new service department. Before arriving

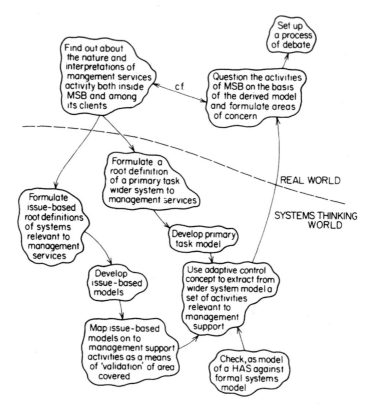

Figure 68. A methodology for management services analysis

at a structure for this department it was necessary to identify the kind and range of services to be provided and the activities that could be located appropriately within the departmental boundary.

Again, a primary task model of the wider system, of which the service system found part, was derived. This was a model of the company as a whole, but as this project was part of a company-wide study, an issue-based analysis was carried out first, using three root definitions and iteration, to decide what to take to be a primary task root definition and model.

From this, support activities could be extracted, in a way similar to that adopted in the previous report. However, it was not known which of these support activities might form the required set. This was not to be a management services type of department, as in the previous case, but which management support activities should be included, as well as technical support activities, was not clear.

A view of what might be included was obtained by interviewing personnel in the 'departments to be served' and identifying a set of expectations for service. From these expectations a model was developed of a system to meet these expectations. It was accepted that the activities in this model might be neither a coherent nor a desirable set; however, it formed a basis for comparison against the extracted model of support activities.

This comparison led to iteration of two kinds: firstly, in terms of the boundary of the new support function within the company model (since support activities that were not contained within this function had to be included elsewhere); secondly, in terms of the nature of the activities that were included. On completion of this iteration a model emerged that represented an acceptable primary task model for the new department. The next stage was to define the organization structure through which these activities could be carried out. Sub-groups of activities were formed based upon expertise required and these defined a tentative set of roles. These were assessed in terms of job feasibility (involving some iteration) and also in terms of the capabilities of the personnel who were available to fill these roles (involving further iteration). The two stages of iteration produced a set of roles (defined by the sub-groups of activities) that were agreed to be feasible. The connectivity between the sub-groups represented the relationships between the roles and hence defined the organization structure. Amplification of the activities within the sub-groups led to the detailed job specifications of the personnel within the department and also provided the basis for operating procedures and the necessary information systems.

This process of analysis is summarized in Figure 69. The company in which the project was undertaken was Materias Primas, part of the Vitro group of companies in Monterrey and the service department designed was called Process Control. This is a literal translation from the Spanish and as such is a misnomer. However, as a technical support function, its purpose was to assist in the control of the operating processes of the company.

I have mentioned earlier that modelling in words introduces semantic problems. These are compounded when the language used is not your own

152

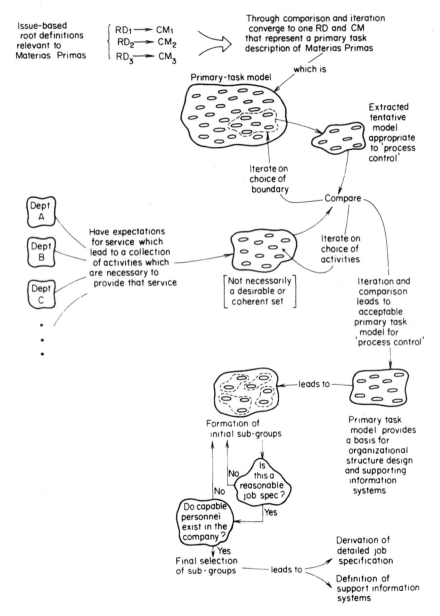

Figure 69. Process of new organizational design. *Key:* RD, root definition; CM, conceptual model

native tongue – as the above illustration of the department's title indicates. The necessity for translation introduced the discipline of precise definitions of meaning in the words used. This is not a bad idea even when the language used in your own. The project described was another example of work undertaken as part of the master's degree programme. In this case the team consisted of one student (a Mexican), the manager designate, and myself as project manager; the analysis, referred to above, occupied about 2 months of the total project period of 5 months.

These two examples illustrate the use of particular methodologies related to service systems. In one case the related department was in existence and in the second case a new department was to be designed. Both can be seen in retrospect as modifications to the Checkland methodology. Both were derived by asking the question, 'What assembly of systems concepts appears to be appropriate for the nature of the perceived problem-content system?' If help in answering that question can be derived by considering how to modify an existing explicit methodology, then do so. What is important is that the resulting methodology is appropriate for the particular analysts in their particular situation rather than the way it was derived.

One feature of service systems analysis has emerged from the many examples that we have been associated with. This is particularly significant if the service system under consideration is a management control system of any kind. It may even be stated as a law. *A useful model of a service system cannot be derived without, first of all, deriving a model of the system to be served.*

The four 'areas of concern' discussed in this chapter provide some illustration of the flexible use of systems ideas and methodology. A major area of work within the action research programme has been concerned with management control systems and information systems analysis. The concepts and methodologies that have emerged from this work are brought together in Chapters 5, 6, and 7.

Chapter 5
Management Control

Introduction

Considerable similarities exist between the process of control as applied to the regulation of production plants and the process of management applied to the control of organizations. The concepts of feedback, in particular, provides a powerful intellectual tool to relate topics in control theory, cybernetics, and management. These similarities have stimulated a number of authors to contribute to the literature on the subject of control, control languages, and their relation to various aspects of management (Beer, 1966; Pitt, 1968; Mallen, 1970; Gifford, 1970; Anderton, 1970; Schoderbek *et al.*, 1975; Bishop, 1979). These contributions are very valuable and completely valid when the analogies to control are used as a means of understanding the behaviour of a management system, but they are of less value when the concern is how to design or improve the management system itself. The major thrust in this intellectual endeavour occurred in the late 1960s and early 1970s and little progress has been reported since. The reason for this, I believe, is twofold:

(a) The effort was singularly concerned with exploiting the similarities.
(b) A clear distinction was not made between the use of control-based models *as aids* to the decision processes in the management system, and models *of* the management system itself. Most applications of system dynamics, for example, fall into the former category.

My concern in this part is with the latter category in (b) above, i.e. the use of models of management systems themselves, in order to improve their behaviour and eventually for their design.

Control and Management

As mentioned above, the simple concept of feedback provides a powerful intellectual tool for understanding the structure and behaviour of the management process. Figures 5 and 6 in Chapter 1 illustrate a hierarchical process control system and an equivalent management control system. It is this kind

154

Table 5.1. Common analogies

Process control	Management
Set point	Organization aims or plans
Measures of performance	Management information
Controller	Manager
Feedback control	Decision-making and corrective action
Dead band	Span of control
Noise	Personal motives
Friction and backlash	Bureaucratic procedures
Transport delay	Information delay
Sampled data	Periodic reports
Proportional band	Ratio of cause to effect
Hunting	Trade cycling
Cascade control	Management hierarchy
Manual override	Reorganization
On–off control	Go–no-go policies

of similarity, together with the list of common analogies in Table 5.1, which has led to the belief that a management system is really no different to a control system, though rather more complex, and is hence amenable to the same kind of analysis. Control, unfortunately, has emotive connotations when considered in the context of management. This is perhaps best illustrated by the cartoon in Figure 70. Because of this interpretation, crucial feedback information from lower levels of management to the higher levels tends to be in a form which is designed to be protective (to the lower levels) rather than in a form which enables control action to be taken where it can be most effectively applied; an aspect which does not feature in automatic control systems.

THE MANAGER

Figure 70. Control – a management concept

The philosopher Karl Popper (1963) advanced the argument that progress in scientific research can only be made by the establishment of experiments to refute hypotheses and not by the design of experiments for their support. I believe that this argument is equally applicable here and can be applied by exploring the differences between control and management and not by continually attempting to build upon the similarities. This philosophy will be adopted as we consider the various levels in the control hierarchy.

The Control Hierarchy

Although the argument mounted is completely general, let us take an industrial setting as a means of highlighting the distinction being made. The control hierarchy referred to here includes the basic production process and its automatic control systems, together with the higher levels of control described by the general term 'management control systems'.

A basic distinction is made between the lower levels of control and the higher levels of management. Thus:

(a) A process control system can be seen as a *designed physical system* (containing only inert physical elements).

(b) A management control system can be seen as a *human activity system* (containing autonomous human beings).

This distinction leads to the view that the higher levels in the control hierarchy, termed management, are *not* simply more complex control systems of the same kind. Adopting the Popper philosophy and exploring the differences rather than the similarities leads to an identification of several which are significant. The major differences are as follows:

(a) A real-world management control system is highly adaptive, but not necessarily to a consistent and unchanging objective function.

(b) It is capable of self-organization, hence the controller structure and mode of operation may change temporarily in response to particular disturbances.

(c) The elements of a real-world management control system are a set of roles and, although *what* each role-holder is responsible for can be pre-determined, *how* that responsibility is executed can not. In a process controller, both the 'what' and the 'how' are predetermined by the design and can only be changed by external intervention.

Figure 71 illustrates one context in which 'management control' needs to be considered. A designed physical system is shown as a process, plus its control system, which converts raw materials into products. The management control system contains this conversion system and operates to achieve the expectations of some business control system, according to its policy but within applied

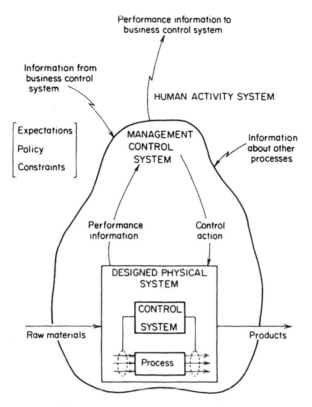

Figure 71. Management control system

constraints. It needs information about other processes with which it may interact and supplies information about performance back to the business control system so that that system, in turn, can take action to achieve expectations. Of course, this simple picture represents only one possible view of a management control system; the reader may wish to form an alternative view. This process of selection of a point of view represents an illustration of the special nature of human activity systems referred to earlier. It is unlikely that there would be much disagreement about the definition of the boundary containing the hardware (i.e. the designed physical system), but there may be considerable debate about what constitutes the management control system. For example, the management control system could be seen as 'a contribution-maximizing system' or 'a flexible manufacturing capability-maintaining system' or 'a market-satisfying system'. Whatever definition is used (and, as has already been emphasized, it is worthwhile exploring the implications of several), it needs to be made explicit as it is against this definition that the systems design, or the improvements recommended, can be logically argued.

Once this system is defined, questions such as the following need to be answered: ♥

(a) Within this system boundary, what is the *minimum necessary* set of activities for the realization of the particular view, i.e. what is taken to be the primary task model?

(b) What roles are appropriate, in terms of decision-taking responsibility, for what subsets of activities?

(c) What are the relationships between these roles and, hence, what is the structure through which they could operate?

(d) What measures of performance are appropriate given the particular relationships with activities outside the system boundary (such as the business control system and other management control systems)?

(e) What information systems are needed to support the activities undertaken and hence what role-to-role information flows are essential?

This set of questions represents the five broad areas of concern which are relevant to the design of management control systems. The system concept described in Chapter 2 represents an appropriate language in which to undertake such analysis and methodologies have been derived from the action research programme (Bowen and Wilson, 1971; Wilson, 1979, 1980(a)) specifically to answer the above questions.

The development of primary task descriptions of an enterprise, as a whole, or of specific sections within an enterprise, has been discussed earlier in Chapter 3. The remaining questions, which relate to management control systems and associated organization structures together with the supporting information systems, will be dealt with later. Let us, first of all, examine those processes which are currently embraced by the general label 'management'.

Management – A View of Current Practices

My aim in this brief introduction is to present a way of looking at the processes of management within any enterprise and to define the meaning and interpretation of some of the major concepts used. It is not my aim to cover the many functional activities and their management (such as marketing and marketing management) but to examine concepts which can be usefully applied to the management of any of these functions.

Initially, at a very general level, we can analyse the role of an enterprise and that of its sub-functions in terms of a simple model, following Anderton and Checkland, as shown in Figure 72.

In this model T represents the transformation process, L represents a set of linking activities that connect the enterprise to its environment, S represents the service functions required to support the transformation process and other activities within the enterprise, and PMC represents the planning, monitoring, and controlling activities necessary to steer the enterprise in the direction of

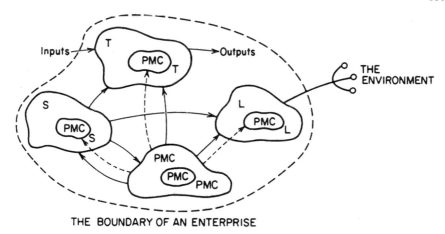

THE BOUNDARY OF AN ENTERPRISE

Figure 72. A model of an enterprise

attainment of its goals. Each of these four basic activities contain their own PMC subsystems. Although this is a simple model, it forms a useful basis on which to question the role of any enterprise. Suppose that the enterprise in question is a manufacturing company. A *production oriented view* might take the transformation process to be the conversion of raw material into products, in which the linking activities might be seen as those of marketing, selling, and distribution (though marketing might also be seen as a service to production and to corporate planning and control). A *market-oriented view,* on the other hand, may take the transformation process to be one of converting a perceived market need into the satisfaction of that need. Here production might appropriately be taken to be a service to marketing in enabling it to achieve that purpose. There is no one answer to the question, 'What is each of these elements when related to a particular enterprise?' However, an examination of a mapping of the total model on to a company will provide a useful and coherent debate about the nature of that company and the nature of the resulting relationships between the various organizational units within it.

We may use it here to define that level of management for which we can define some important characteristics. If we take the transformation process to represent the basic operation of an enterprise – converting raw material into products in a manufacturing company, providing cover in an insurance company, or providing health care in an area health authority – we can define 'operations management' in terms of $(PMC)_T$, i.e. those specialized planning, monitoring, and controlling activities that are directly related to (and defined by) the specific nature of the operations. If we take the levels of planning, monitoring, and controlling at a higher level than this, i.e. as the PMC element in Figure 72, the processes are less dependent upon the nature of the operations and can be discussed in general terms. This level is termed variously as stategic planning and control, management control, or corporate planning and control.

160

The key words in this process, irrespective of the particular label used to describe it, are *planning, monitoring,* and *controlling.*

Planning

Planning is a continuous process. It is concerned with defining ends and courses of action. The ends may be precisely defined, in terms of, say, the achievement of a desired market share, or they may be less precisely defined, in terms of a range of possible scenarios. At a strategic level, planning is the process by which a company, or enterprise, solves the problem of determining its relationships with its environment and the necessary internal structure and processes required to maintain these relationships. Because the environment and these relationships are continually changing, the planning process itself needs to be continuous. It may be viewed as a cyclical process illustrated by Figure 73.

This particular view of the planning process is not inconsistent with that adopted elsewhere. Anthony and Dearden (1976), for example, define it as follows:

> strategic planning is the process of deciding on goals of the organization, on changes in these goals, on the resources used to attain these goals, and on the policies that are to govern the acquisition, use and disposition of these resources.

Dermer (1977) has the following to say:

> A strategy is generally less fixed and more specific than a policy. But regardless of how one chooses to define these terms it is clear that both strategies and policies are the outputs of a process of information collection

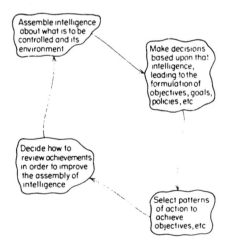

Figure 73. The planning process

and decision making. And so strategy formulation, policy development and similar terms and simply labels to describe certain planning processes.

Dermer makes no attempt in this passage to define what he means by the words 'strategy' and 'policy'. These words are frequently confused and hence I need to make by interpretation clear.

Definition *A* strategy *is is particular pattern of actions intended to attain certain desired ends (e.g. that set of actions required by a company to increase its share of a particular market or that set of actions required by a university department to ensure a flow of funds to maintain a particular research programme).*

Definition *A* policy *is a static guideline for repeated decisions or a preplanned decision waiting to be activated by the occurrence of the situation for which it was intended (e.g. companies may have policies to offer credit facilities when requested or employ minority groups whenever possible or make funds available to support university research).*

The planning process described is frequently related to different time horizons, so that planning may be referred to as long-, medium-, and short-term. As the time horizon reduces so the output of the process becomes more precise. However, the same cycle needs to operate irrespective of the time scale. It must be added that long-term or long-range planning and strategic planning are not necessarily the same.

Whereas a long-range plan deals mainly with predicting the consequences of decisions taken in the short term about current operations, a strategic plan is concerned with defining the basic orientations of the company, its key moves, and its key areas of competence for the future.

The activity 'decide to review achievements in order to improve the assembly of intelligence' implies making decisions about 'measures of performance' This is a crucial first stage in the monitoring process in that the definition of these measures determines what it is appropriate to monitor.

Monitoring

Historically, measure of performance has been interpreted in financial terms. This is understandable since money, and the often-used cost-accounting terminology, provides a common language in which to compare, and aggregate, the disparate activities of a company or business. However useful this might be for performance reporting, it is inadequate as a source of information from which to derive control action. Moreover the kind of information supplied to chief executives tends to concentrate upon short-term performance, rather than upon long-term issues.

In a survey carried out over a large number of comparable companies in the UK, the Federal Republic of Germany, and France, Horovitz (1979) found that the content of monthly reports supplied to the chief executives was heavily oriented (90%) towards detailed short-term costs, personnel numbers, and quantities produced. Information related to the outlook for the future (in terms of qualitative reports and/or projected figures) was present in less than 40% of the comapnies in France and Germany and of the order of 70% of the companies in the UK.

In defining measures of performance it must be recognized that these measures represent the link between strategic planning and the necessary strategic control. Planning without control is worthless and hence a critical examination of those features, which are regarded as key, in a strategic sense, for the particular company, represents an essential process in the development of measures of performance which can be said to be appropriate. It is then the performance, with respect to these key features, that must be monitored if the management of a company at corporate level is to be effective.

The features referred to above could be seen to lie in the following three areas: (a) assumptions related to the evolution of the environment and of the internal resources of the company; (b) areas of particular competence; (c) performance results and priorities. For each of these areas, norms, or standards, need to be set related to expected achievements. Measurement of degree of achievement would not normally come from the usual internal accounting information. For example, if a company wishes to exist in a competitive market environment by being a leader in the particular technology, standards could be set in terms of new product initiatives, presence in known-to-be technologically advanced customers, presence and status in professional meetings, staff development programmes, etc. It is information related to these considerations that should be acquired by top management and which should from the bulk of performance reports related to strategic control.

A similar approach is adopted by the Centre for Information Systems Research in the Sloan School of Management at the Massachusetts Institute of Technology (Rockart, 1979), though they refer to 'critical success factors' as the source of strategic information.

Controlling

It is not surprising, given the emphasis on financial measures referred to above, that control, when it is exercised, is also through financially oriented procedures. Budgetary control is the most common. A budget is frequently regarded as both a planning and a control tool. Thus the planning process establishes (usually through iteration around a number of management levels) annual budgets for a number of areas of responsibility. Performance reporting consists of statements of the budget breakdown for that area, actual costs or profits achieved, and an analysis of variances. The later analyses, it is hoped indicate the specific areas where control action is required. What that action is, is a function of many variables, including the definition of area of responsibility and degree of delegation within the prticular management hierarchy.

Exercising control in this way requires the definition of areas of responsibility, usually termed responsibility centres.

The essence of a responsibility centre is that it uses inputs, which are physical quantities of materials, hours of various types of labour, and a variety of services. It works with these resources and usually requires working capital, equipment, and other assets to do the work. As a result of this work, the responsibility centre produces outputs, which are classified either as goods or services. The goods and services produced by a responsibility centre may form the inputs to other responsibility centres or form the outputs of the organization itself.

Four categories of responsibility centre may be defined:

(a) *Cost centre*   This is the smallest organizational unit for which costs are accumulated and over which an individual has responsibility. Although every unit classified as a cost centre produces useful output, it is usually neither feasible nor desirable to measure it in monetary terms. Research and development and management services may be seen in these terms.

(b) *Revenue centre*   When only the output of a unit is measured in monetary terms, that unit lends itself to operation as a revenue centre. It will use resources in providing the output but usually no formal attempt is made to measure them. Areas of marketing and selling activity are frequently defined as such and control may be, for example, on sales variances.

(c) *Profit centre*   When performance can be measured in terms of both revenue earned and costs incurred, the unit can be controlled as a profit centre.

(d) *Investment centre*   This represents the most comprehensive forms of financial control. It is where measures are taken, not only of profit, but also of the capital employed to generate profit. Return on investment is usually taken to be measure of performance.

Irrespective of how the responsibility centres are defined, budgets can be an inadequate basis for evaluating managerial performance for several reasons. In the case of profit budgets, for example, it is impossible to provide an exact answer to the question 'What should be the annual profit objective for a particular profit centre?' Subjective judgement is required in setting such objectives and, in the usual review process, time does not allow for a thorough analysis. Often the most persuasive manager ends up with the easiest profit objective. In general, the profit objectives of the various profit centres are likely to vary considerably in relation to the ease or difficulty of achieving them.

In aiming at a profit objective, it is necessary to predict the conditions that will exist during the year. Most of these will probably be outside the area of control of the particular manager and hence will represent a set of likely-to-change constraints. The most significant are the economic climate, its effect on the particular environment of the company, and the competitive situation.

This situation is particularly unfortunate for the profit centre manager in that the uncertainty surrounding predictions of revenue is usually much greater than the uncertainty about costs. Thus the variance between actual and budgeted profit is often affected more by the ability to forecast than by the ability to manage.

A final major inadequacy in using budgetary assessment as a means of control is that the time span that elapses before many important decisions are reflected in financial performance is greater than the period covered by the budget.

The above difficulties have led to an alternative form of control based upon periodic financial evaluation. This kind of analysis is based upon two principles:

(a) The period covered by the evaluation should be long enough to make the evaluation valid.
(b) The evaluation should be made solely on the basis of what has actually been achieved, i.e. the effect of constraints needs to be recognized.

This type of evaluation is beneficial for two reasons. Firstly, the uncertainty surrounding the environmental conditions is removed and, secondly, it discourages the taking of short-term control actions to improve profits but which may have deleterious effects in the long term. Such evaluation need not be restricted to financial analysis but should, as mentioned earlier, be extended to cover key feature assessment.

An attempt to broaden the planning and control process and to specify formal procedures for its implementation emerged around the 1960s, known as "management by objectives'. It effectively consisted of five stages:

(1) The manager discusses with the subordinate, the subordinate's description of his own job.
(2) The manager and the subordinate agree to short-term performance targets.
(3) The manager and the subordinate discuss periodically the progress made towards meeting the targets.
(4) The manager and the subordinate agree to a series of checkpoints that will be used to measure progress.
(5) At the end of a defined period (usually one year), the manager discusses with his subordinate his assessment of the results of the subordinate's efforts.

The procedure was not restricted to any level of management but was intended to encompass the whole management structure. Mixed claims were made for the success, or non-success, of this procedure. Achievement of financial objectives still tended to overshadow the achievement of other objectives and, at best, it motivated managers to think in broader terms than purely financial. At worst, it produced incoherent and possibly conflicting objectives for the company as a whole.

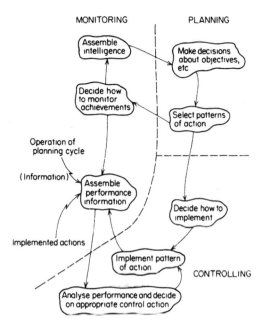

Figure 74. The total planning, monitoring, and control cycle

Control, of whatever kind, is the process of implementing the patterns of action produced by the planning process (see Figure 73) and of ensuring that the cycle is maintained. Additional activities, which represent the process of monitoring and control, are added to those of planning and the total cycle is represented by Figure 74.

At a strategic level, this cycle may be operated through what is known as scenario analysis. Using this approach, specific objectives are not set but instead a number of possible scenarios are derived, which represent the relationships between the company and its environment at some time in the future. These may be undesirable as well as desirable scenarios. To do this an intimate knowledge of current capabilities is required and, through a number of assumptions about environmental movements, critical resource availability, and costs, the set of scenarios can be derived. Once this set is available key moves can be defined which, it is believed, will take the company from its current state towards desirable scenarios and away from undesirable scenarios. These then become the 'patterns of action' which are the output of the planning process of Figure 74. Strategic measures of performance will need to be defined which will enable the monitoring process to collect that information which will show whether or not that belief was justified. Strategic control is then concerned with implementing those key moves. Information is then collected according to the defined measures of performance and, through the interpretation and assessment of this information, a new set of updated capabilities can be derived and the cycle initiated again. This process needs to be regular, but how regular it is, is determined by the response time of the company and its immediate environment to the key moves that have been implemented.

Plans are implemented, and control is exercised, by people through people. Thus, as well as considering the procedures involved, consideration must be given to (a) commitment to and ownership of plans, (b) management style, and (c) motivation and reward.

Commitment and ownership usually means involvement in the planning process from managers in line positions and requires the producers of plans to have credibility and respect within the organization. These features need to be recognized in the design of planning and control systems both in terms of the design of the procedures themselves and in the allocation of roles (and role definition).

In this latter part of the design process, the allocation of roles, attention needs to be given to the particular style and characteristics of the managers available. This is particularly important in relation to the control activities and may be viewed in terms of power base and style. Misshauk (1979) summarizes these features as follows:

Power base	Source	Comment
Reward	Ability to provide something of value in exchange for desired behaviour	Must identify what is perceived as reward and be able to convice that it can be delivered
Coercive	Ability to punish undesired behaviour	Can be effective but may eventually be de-motivating and lead to loss of morale
Referent	Charismatic personality	Highly effective but cannot be acquired
Expert	Acknowledged expertise in a given area	Highly effective but limited to area of expertise
Legitimate	Formal position or title	Based on internalized values, this power base is becoming less effective

A way of classifying managers has been produced by Blake and Mouton (1978), known as the Blake grid. This is based upon a simple model of managerial behaviour and consists of a matrix whose axes (scaled from 1 to 9) represent 'orientation to people' and 'orientation to the task'. The characteristics of particular managers, relative to elements within the grid, are specified so that a company can map on to it the particular managers under consideration. Although this model is completely inadequate, as a model of a manager, it is a device which enables an unemotional discussion to be held about relative styles. Although such features as management style, motivation, etc., represent crucial aspects of behaviour within companies and determine the effectiveness of control procedures within the company, models which attempt to describe them are simplistic and totally inadequate as devices to predict behaviour.

McGregor, (1960), suggested that it was useful to view style in terms of two extreme views. This is the 'theory X, theory Y' approach to management and may be summarized as follows:

	Theory X	Theory Y
People	are lazy will avoid work will take advantage	enjoy work will take initiative will seek responsibility
Hence managers	apply tight control offer little freedom supervise closely	apply loose control offer freedom and opportunity to use initiative

In an attempt to analyse behaviour related to job enrichment, Herzberg *et al.* (1959) proposed an approach known widely as the 'two-factor' theory. The two factors are 'motivating factors' and 'hygiene factors', defined as follows:

Motivating factors

(a) *Achievement* in doing something useful
(b) *Recognition* of such achievement
(c) *Meaningfulness* of the work
(d) *Responsibility* for making decisions
(e) *Opporunity* to grow and progress

Hygiene factors

(a) *Physical* (safe place to work and eat)
(b) *Social* (the need to be with other people)
(c) *Status* (role in the company)
(d) *Orientation* (function relative to the company)
(e) *Security* (job continuity)
(f) *Economic* (salaries and bonus rewards)

An examination of these factors provides an intuitively attractive approach to job design or role specification in that as a minimum, the hygiene factors must be satisfied before consideration can be given to those aspects of the job that can provide the required degree of motivation.

A alternative approach to the analysis of motivation is contained within Maslow's hierarchy of needs (Maslow, 1954). His basic premise is that a satisfied need no longer motivates behaviour: it is only the need that a person is striving to satisfy that motivates a behaviour pattern. The hierarchy ranges from the basic physiological and security needs, through social and affiliation needs, ego and esteem needs to self-actualization, i.e. to reaching one's potential. Again this appears intuitively helpful, but it is difficult to apply. Fortunately (though it makes prediction difficult), individuals are individual and, even if a need can be identified, it is not usually possible to identify how a given individual will seek satisfaction.

It is behavioural considerations such as those mentioned above that gave rise to the emergent discipline of organization development (OD). This is defined by Huse (1980) as

> a process by which behavioral science principles and practises are used in an ongoing organization in a planned and systematic way to attain such goals as developing greater organizational competence, bringing about organization improvement, improving the quality of work life and improving organizational effectiveness.

Initially, OD was concerned with the process of intervention and the development of behavioural techniques that could facilitate the introduction of change. This emphasis provided a focus on the internal workings of the organization and the way in which both management and workers could operate together more effectively. Thus its scope was organization-wide but, because of the emphasis on behavioural aspects only, its practices were limited to those that were believed to increase the readiness for change. It was not, at the same time, concerned with developing methodologies for identifying what these changes might be. In reviewing the current trends in OD, Huse observes,

> As change becomes more rapid, and the environment more turbulent, as inflation increases, organizations must become more proactive rather than reactive to the environment. Thus it is the function of OD to assist the organization, as a system, to effectively serve the larger system of which it is a part and, in so doing, to become more adaptive and responsive.

This passage indicates that the focus for OD in the future is no longer only towards the internal workings of an organization but also towards its relationships with the environment. Moreover, if OD is to adopt the above role, its concern can no longer be restricted to interpersonal relations and organizational behaviour, but must extend to the identification of changes from an analysis of the organization as a whole. Current OD writings (which are recognizable as such) do not propose how such an analysis should be undertaken.

An issue-based analysis represents one way of identifying such changes and the behavioural techniques of the OD practitioner may be useful in extending the degree of change that can, in any one situation, be regarded as feasible.

The content of this section has been more concerned with the context in which an analysis of management control systems is undertaken than to present progress related to planning, monitoring, and controlling. However, some appreciation of what these processes are, and how they can be viewed, is a necessary prerequisite to beginning to question how an enterprise currently manages itself.

Management Control – Concepts

In Chapter 2 a general description was given of systems concepts and the notion of a human activity system was introduced. This is the basic element used to describe management situations and it provides a coherent language in which the modelling of such situations can be untertaken. The set of activities (or subsystems) which represent the system described by the root definition are the 'on-going' activities which must exist if the particular viewpoint and purpose of the system are to be realized. Thus, in general, a systems model must include both the operational activities and those activities concerned with the control of the operations. Thus how can such a model be structured in order to make the process of control explicit? To answer this question it is necessary to make reference to the general model of any human activity system, i.e. the 'formal systems model'. It has been assumed that our concern is with purposeful human activity and hence the root definition will contain some statement of that purpose (or objective, goal, etc.; whatever term the reader prefers). This will usually be apparent from the particular transformation process described. Also there will be some statement of the performance expectations set by some higher level

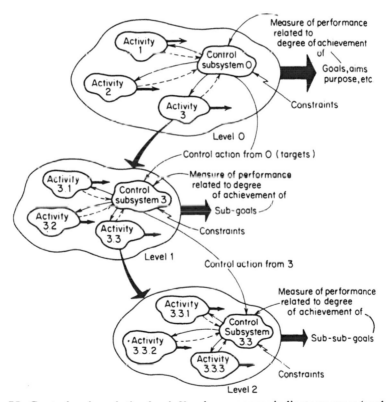

Figure 75. Control and resolution level. *Key:* heavy arrows indicate purpose (goals, aims, etc.); ⟶, control action; ‑‑→, performance monitoring information; ⟶ᴈ⟶, constraints

system (or wider system). Hence it is implied that within the system boundary there will be some decision-taking process that has authority to utilize the resources available to that system in order to achieve the purpose within the performance expectations specified, and within the constraints outside the control of that decision-taking process. Since this model is completely general it will be applicable to each level of resolution in the systems hierarchy. Having specified the level at which 'the system' is defined, the set of activities within the boundary become its 'subsystems'. At the next level, each subsystem can be expanded in exactly the same way, through the mechanism of a root definition, producing more detailed models of the system defined by the first root definition. Thus each systems model at each resolution level must contain control activities

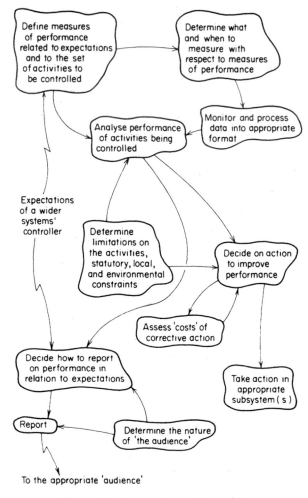

Figure 76. A general control model

relevant to the particular transformation process. The relationship between control and resolution level is illustrated in Figure 75. Here it is shown that the control action produced at one level becomes the target at the next level down. Here the word target is used to include both the specification of goal and the performance in terms of the degree of achievement of that goal. This is also the connection between one level and the next through which the span of control of the lower-level decision-taking process can be defined. It must be emphasized that these models are intellectual constructs for use in analysing a particular management situation; they are not a description of what exists or a specification for a design. It is frequently the case that when comparing the area of authority of a decision-taking process defined by the systems model with that actually occurring in practice, a considerable mis-match is observed. This feature is the basis of a methodology for reorganization discussed later. However, since a control subsystem is itself a human activity system, it is also capable of expansion and, whereas this should be carried out in relation to the particular activities being controlled, Figure 76 represents a general control model, which is applicable to the control subsystem at any level of resolution. This model specifies what activities must be 'on-going' in a management control system. It does not specify how each activity should be undertaken or whether each activity is the responsibility of a single decision-taker or allocated to a group. Once the set of operational activities has been identified that this control subsystem is responsible for, both that set and the control activities can be mapped on to an existing management situation in order to structure an analysis of its effectiveness. In essence, this means taking each activity in turn and questioning the actual situation in terms of 'how' each activity is done, by whom, and with what relationships.

A Methodology in Outline

Although the approach discussed here is believed to be applicable to the design of management control systems and hence is applicable to a 'green field' situation, it has never been used in its totality in such a situation. Parts of the methodology have been successfully applied to real management problems within the continuing 'action research' programme of the Systems Department at Lancaster. However, before a design methodology as such can be put forward, sufficient experience of its application to specific situations needs to be accumulated so that the necessary generalizations can be extracted with confidence. Hence what is suggested here is a tentative methodology based upon some experience of its application to the improvement of existing management control systems. The methodology is presented as a seven-stage process of analysis.

Stage 1 Derive an activity model of the total management system based upon a root definition of a system relevant to the expressed concern. This will need to recognize the characteristics of the transformation process (see Figure 72) and will contain detailed activity models of those systems required to (a) plan the transformation process, (b) operate the transformation process, and (c) control the transformation process.

This stage represents an application of the Checkland methodology. In this kind of application, however, the comparison of the model with the existing situation is used to make changes in the root definition and the corresponding conceptual model. This is repeated until a null comparison (between the model and the actual situation) is achieved may require several iterations. Thus a systems model is eventually derived which is a 'required activity' description of the area under review. This model is independent of any organizational structure; records what must go on within whatever organization structure is selected; it requires a root definition which expresses the 'manifest purpose' of the organization studied. This can be rapid set of iterations and is a very effective method for the analyst to learn about the particular situation. Thus what is required here is a primary task model, though as indicated earlier this may be preceded by an issue-based analysis to determine what to take as the primary task.

Stage 2 For this total system, define a portfolio of possible input disturbances and the particular activities perturbed by each input.

Stage 3 Define the decision process necessary to respond to each input and hence derive activity outputs. (These then become inputs to subsidiary activities until the output can be identified as an instruction requiring no further decision.)

Stage 4 Define information flows required to support the activities making up the total model of stage 1 through the decision processes derived in stage 3.

The purpose of stages 2, 3, and 4 is to ensure that, once the conceptual model has been generated in terms of activities, those information flows are identified which are essential to the functioning of the activities. These information flows are not defined in terms of the specific content of the information but in terms of the categories of information (defined later) required. The information input to an activity can be determined by first determining the decision processes contained within tht activity boundary and then by examining the information needs of each decision process.

The problem of 'how to define the decision processes' was tackled in one project concerned with the development of a management information system (Bowen and Wilson, 1971) by modifying an existing process control technique, i.e. impulse response testing. Thus, if the system is taken to be a black box and a disturbance is injected, by relating the output back to the input in an appropriate way (through correlation techniques in the case of process control), a description of the contents of the black box can be obtained. In the case of a management control system a portfolio of possible disturbances to the system can be obtained from past records. These can be inputs from higher levels such

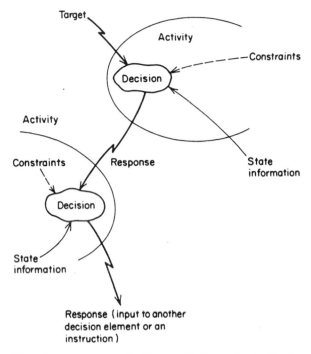

Target

Activity

Constraints

Decision

Activity

Constraints   Response

State
information

Decision

State
information

Response (input to another
decision element or an
instruction)

Figure 77. Decision elements: *target*, feedforward information indicating a desired state for some subsystem and performance expectations; *constraints*, factors outside the defined activity boundary (and hence outside the control of that decision-taker); *state information*, feedback information indicating the existing state of the system; *response*, output of the decision process, selected from a range of possible responses based upon some criteria

as a change in manufacturing policy or a change of product quality demanded, or inputs from the physical system such as reduction in raw material availability or unit breakdown. For each of these inputs, the level in the management hierarchy at which they enter can be established. If each input is then hypothetically introduced, the response that would have resulted can then be established. This response then represents the input into the next decision process and so on until the resultant response is an instruction requiring no further decision. Thus, given a model of a decision element, the information that was required to make the decision can be derived in each case. The model used in this project is illustrated in Figure 77. The particular production unit concerned was an intermediate set of plants B, C, and D using as raw material the output from plant A and feeding, as raw material, their outputs to plants E, F, I, and J. Considering the input disturbance to be a reduced flow of material from plant A, a decision network was constructed as in Figure 78, and the categories of information required to support the decision identified. If this kind of network is constructed for each input in the portfolio, it is found that some of the decisions become commong to several of the inputs. This is because in

174

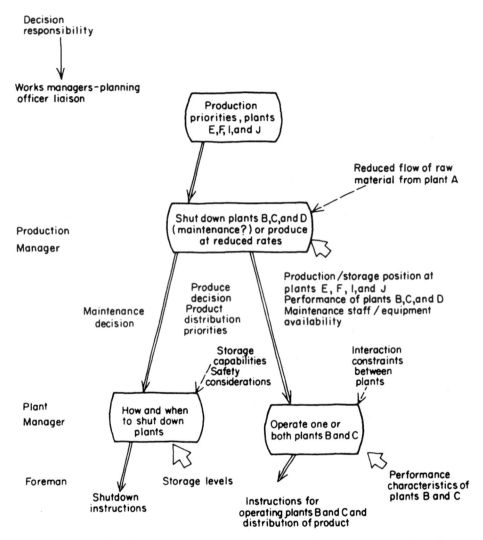

Figure 78. Part of decision network. *Key:* ⇒, targets/action; ←---, constraints; large open arrows indicate state information

one case the need for the decision arises because the target is changed: in another case it may be result of a change in state or a change in the constraints. The formal process of analysis described here can only be followed when analysing an existing situation. However, in a design mode, the broad categories of information required by an activity can be specified from an examination of the nature of that activity. This broad specification represents the start-up situation and the process described can then be used as the means of adapting and refining the subsequent information flows during operation.

On completion of stages 1–4 a model will have been derived which contains a statement of the minimum necessary activities to be undertaken by the system, together with the essential categories of information needed to support the decision-taking processes within it.

Stage 5 Define responsibilities for the set of activities.

During the construction of an activity model of a management control system several resolution levels will have been developed. Since the activities at any one level are seen as the set belonging to a system at the previous level, it follows from the definition of a system boundary that there is a decision-taking role already established. Hence, referring to Figure 79, it is implied that the authority of the decision-taker in system 'S' extends over activities, S_1, S_2, and S_3. Similarly

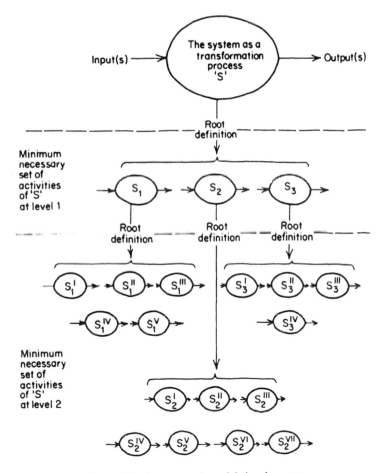

Figure 79. Conceptual model development

the area of authority of the decision-taker in system S_1 extends over activities $S_1^I, \ldots, S_1^V$ and so on. Thus, in a design mode, the decision-taking roles could be defined on the basis of the systems hierarchy. Role definition then becomes dependent upon the way in which the model is expanded. Further research is needed to established principles, or guidelines, for this kind of model development.

In the analysis of an existing situation the decision-taking roles established by the model (map 1) are used for comparison against the actual areas of decision-taking responsibility within the organization (map 2).

The comparison of the two maps is used to initiate debate about the appropriateness of any current responsibilities which are different from those derived from the systems model. The result of this debate is either to change the responsibility within the organization or to modify the systems model. Thus at any resolution level convergence is obtained between the systems boundary and the organizational boundary prior to development of the model at the next resolution level.

In a particular study it may be necessary to accept the existing set of management roles and structure as a constraint. In this case decision-taking boundaries are mapped on to the conceptual model to identify cross-boundary connectivity. Improvements to the functioning of management control are then restricted to procedural or information flow changes. A project in which this mapping technique was developed will be described in the section on reorganization.

Stage 6 Convert the 'activity-to-activity' information flows derived in stage 4 to 'role to role' information flows by using the groupings defined in stage 5 (or by accepting the constraint of the existing organization structure).

Stage 7 Define 'role-to-role' relationships (i.e. organization structure) by equating resolution level in the developed model to responsibility level in the organization.

In an existing situation, the process described in stage 5 generates role-to-role relationships, i.e. an organization structure. This is not dependent upon the way the systems model is initially developed, since the process of comparison against 'what exists' allows both the structure of the model and the existing organization structure to be modified. It is true that the resultant structure is 'acceptable' rather than 'optimum', but it is also argued that there are many possible structures and that 'the right answer' does not exist. Hence the concept of optimization in relation to a management control system is inappropriate. What should be aimed for is a structure in which there is no confusion over decision-taking responsibility and in which the process of evolution can take place in a more rational way (in that it can be related to an activity description of those operations which must be managed).

In a 'green field' situation there is nothing to compare the activity model against and therefore guidelines for the development of such models need to be established. As mentioned previously, significant research effort needs to be undertaken in this area before this stage can be proposed with confidence as part of a methodology for design.

In summary, stage 1 defines what activities must be on-going for the management system to function. Stages 2–4 define the minimum information needed to support these activities. Stage 5 defines who (in terms of role) is responsible for what set of activities (i.e. the job specification). Stage 6 defines the minimum information flow pattern, i.e. who is responsible for supplying who with what information (the purpose have been specified by stage 3). Stage 7 defines a rational organization structure.

A Methodology for Reorganization

The characteristics of a human activity system which are particularly relevant to this approach to organizational change are those which define boundaries and *responsibilities*. A system boundary represents that interface with the rest of the organization within which the decision-taking body has control over resources. Hence the system has a *decision-taker* who is *accountable* for the performance of the system. Also the system must have *objectives, measures of performance* which indicate how well the objectives are being achieved and *control mechanisms* which enable the decision-taker to allocate resources to improve performance.

To be completely general it would be more correct to say that a system has a decision-taking process, but for this application let us make the assumption that it will be useful to view that process as a single role-holder.

Again a seven-stage process of analysis is required. The stages will first of all be described and then illustrated by references to the project in which this approach was first used in which a mapping technique using transparent overlays was developed.

Stage 1 Develop a primary task model for the enterprise, or section of the enterprise, under review (first resolution level only).

Stage 2 Take each activity (subsystem) in the above model and, through a root definition, expand the model to the next resolution level. Draw the total set of activities at this level on a single sheet of paper (known as the base chart). Since, at this stage, we are only interested in the way the activities are grouped, the connectivity can be omitted.

Stage 3 Using a transparent overlay, draw the set of boundaries appropriate to the first resolution level model. Remove the overlay.

Stage 4 Using a second transparent overlay, annotate each second level activity with a symbol representing the existing responsibility in the current organization.

178

For both of the overlays it is helpful to use different colours to represent the different areas of responsibility.

Stage 5   Replace both overlays on the base chart and identify those activities whose current responsibility is different to that implied by the first resolution level subsystems' boundaries. These are termed 'island activities'.

Stage 6   Question the responsibilities for the 'island activities' and either move the activity to the appropriate subsystem or change the responsibility allocated to the activity.

An 'island activity' is seen as a potential anomaly in the organizational structure. Its existence is caused by there being a difference between the areas of authority defined by a systems grouping of activities (derived from the logical defence of a set of activities from a root definition) and the areas of responsibility that have emerged from the historical, political, and personnel changes within the enterprise itself. No assumption is made as to which is appropriate but, through a debate about the responsibility for these particular activities, a resultant set of boundaries can be derived in which all the activities, are the responsibility of a single decision-taker. Thus an acceptable definition of areas of authority can be obtained by convergence from two defined positions. Since

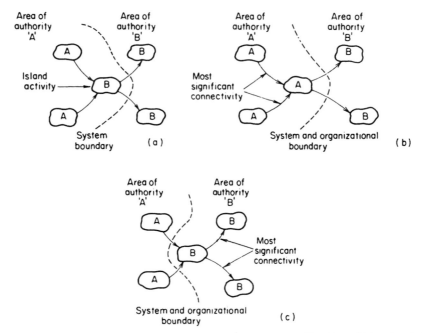

Figure 80. Resolution of island activity responsibility: (a) identification of island activity; (b) changed responsibility; (c) modified boundary

there is now no duplication of decision-taking responsibility within each boundary, these boundaries can be taken to be *both organizational and systems* boundaries.

Help in making the decision about acceptable responsibility for an 'island activity' can be obtained by developing the connectivity between these activities and those other activities with which they interact. This process is illustrated by Figure 80. A section of a notional model is shown in Figure 80(a), in which the mapping of existing responsibilities is illustrated by the annotation of each activity by the letters A or B. The position of the systems boundary (which defines the interface between the conceptual areas of authority relevant to A and B) highlights the central activity as an 'island activity'. The respective connectivities are also shown. If it can be argued that the connectivity between activities 'A' and the 'island activity' is more significant, in some sense, than that between the 'island activity' and activities 'B' then it may be beneficial to place the organizational boundary across the least significant connectivity, as in Figure 80(b). In this case the responsibility for the 'island activity' is changed and the boundary is taken to be co-incident with the systems boundary. If, on the other hand, the connectivity between the 'B' activities is the most significant, as in Figure 80(c), the responsibility for the 'island activity' is unchanged and the original systems boundary is modified.

This debate needs to be undertaken for each 'island activity' and in each case the critieria used for deciding 'most significant' need to be made explicit.

Stage 7 The process in stage 6 has defined a set of boundaries around the second level activities which can be taken to be the highest level organizational boundaries and, since they also represent the areas of authority of individual decision-takers, they can also be taken to be systems boundaries. Thus this stage consits of taking each subsystem at the level in stage 2 (though the boundaries will now be different as a result of the convergence process in stage 6) and redefining them in turn as 'the system' and repeating stages 1–6. Since the boundaries of these systems will have been modified, new root definitions will be required before they can be expanded.

The above seven stages represent a recursive form of analysis which can be applied in a 'top down' mode to any level of organization. Although the process of analysis is hierarchical, the resultant organizational structure need not be. A matrix type of structure can be the result. If, for example, the model contains activities of the form 'develop expertise in area A', the activity will be located within the area of authority of the manager with functional responsibility for area A. If there are also activities of the kind 'control the use of expertise in area A in the development of project X', this activity may be sited in the area of authority of manager B if that manager also has the project responsibility for project X, irrespective of his functional authority.

As an example of the application of this approach, the project in which the methodology was developed will be briefly described. The company concerned was in the telecommunication business. It was experiencing a major change in its market environment and the General Manager was asking the question, 'Are we appropriately structured to enable the company to respond to the changing situation?' He had assembled a team to examine this question consisting of himself, the Productivity Services Manager, and the manager of a section that he believed would be unaffected by any reorganization. They had issued a statement to all the managers within the company to the effect that a reorganization was taking place and that the rationale for the reorganization would be communicated to all concerned including the management unions. The statement also contained the assurance that there would be no redundancies as a result of the reorganization. The team designed an attitude survey and issued this to all the managers to form the basis for a set of interviews to establish current organizational problems. The initial approach to the reorganization consisted of reacting to the perceived problems. However, it was quickly realized that, although reorganization on this basis could be achieved, there was no rationale behind it. It was at this stage that we became involved and ISCOL were invited to participate with the aim of using systems ideas to provide the rationale. A full-time ISCOL consultant and I joined the team and were involved for a period of 8 months. During this time we undertook reorganization from general manager level down to foreman level. This necessitated the development of a number of primary task models containing around 800 activities for the company as a whole.

The first stage was to develop a primary task model appropriate to the company in total. The root definition taken was as follows (this has been generalized, however, to maintain confidentiality):

A system to obtain business, develop, manufacture, and distribute major and related products, to ensure technological leadership and good company image, and to carry out these activities at a long-term maximum profit consistent with company constraints.

This root definition was seen to imply five major activities (subsystems), each with a defined purpose but distinct in nature:

(a) *Develop markets* To develop an understanding of the future market needs and of particular major customers, to develop long-term commercial policies, and to ensure the projection of company image to enhance current selling effort.

(b) *Develop products* To develop, at an optimal cost, those products which will be commercially advantageous to the company, and to develop the associated manufacturing processes and plant.

(c) *Make products* To manufacture to customer requirements to specialized standard at an optimal cost.

(d) *Contact customers* To contact potential customers, to act as an interface between the company and actual or potential customers, in processing enquiries or orders, ensuring that requirements are understood and attended to, and ensuring customer satisfaction.

(e) *Control* To set company and subsystem targets, to monitor performance and take control action where necessary.

The titles of the subsystems were chosen to be unrecognizable as organizational groups within the existing situation in order that a clear distinction could be maintained between a subsystem (and the activity groups derived from it) and the departmental groupings that happened to exist in the company at that time. The activity model (at the level of expansion of the above subsystems) was drawn on a large base chart. Figure 81 is a representation of this model in terms of notional activities. The actual model contained about 80 activities, and it is impractical to try to reproduce it here. Figure 82 indicates the kind (and level) of activities included within the 'make products' subsystem.

Using a transparent overlay, boundaries appropriate to the original subsystems were drawn, using a different colour for each subsystem. This stage is illustrated by Figure 83.

What was now available was a definition of decision-taking responsibility for groups of activities based upon the conceptual (primary task) model. The boundaries of the 'develop products' and 'develop markets' systems are shown

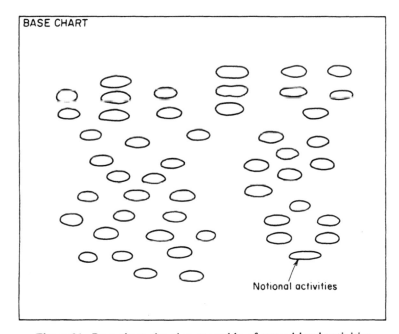

Figure 81. Base chart showing assembly of second level activities

Figure 82. 'Make products' system: second level activities (excluding connectivity)

to divide the top three activities. This arose because, in maintaining the condition of minimum, necessary activities when constructing the base chart, activities had been found which were identical and had therefore only been included once. These three activities had, in fact, both a technical and a commercial component and needed to exist in both systems. Complete separation could have been achieved by expanding the three activities to the next level of resolution, but that was not done until later.

The next stage was to remove overlay I and replace it by a second transparent overlay. For each of the activities on the base chart the following question was asked: 'Who in the present situation has the decision-taking authority for this activity?' The second overlay was then annotated with an appropriate departmental identification. The classification used in this example, together with the related subsystems, was as follows:

T   Technical Development Department (approximating 'develop products')

M   Manufacturing Department (approximating 'make products')

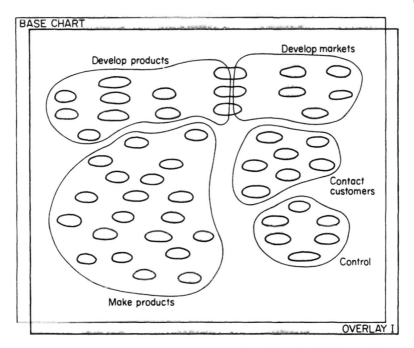

Figure 83. Base chart with overlay I, showing allocation of activities to subsystems

E   Production Engineering Department (no equivalent subsystem)
CC   Sales Department (approximating 'customer contact')
C   Senior Management Group (approximating 'control')

Once this annotation had been completed, the first overlay was replaced and the differences between the conceptual and the actual allocation of responsibilities became apparent, illustrated by Figure 84(a). The basic premise, in the systems allocation, was that there should be a single decision-taking authority for the activities contained within the system boundary. Where this was not the case, these activities were designated 'island activities'.

The identification of the island activities directed the debate to possible anomalies in the organization structure and at this stage decisions were made either to change the responsibilities or to move those activities to the sybsystems relevant to their allocated responsibilities. In some cases current responsibilities could not be identified (particularly in relation to some of the marketing-type activities; this was a reflection of the changing market situation) and in other cases the boundaries were seen to cross activities. This meant that, in this latter case, the activities had to be expanded to allocate them to one subsystem or the other.

In the former case, decisions were made as to the appropriate decision-taker (according to an analysis of the relevant connectivity) and the activities allocated

184

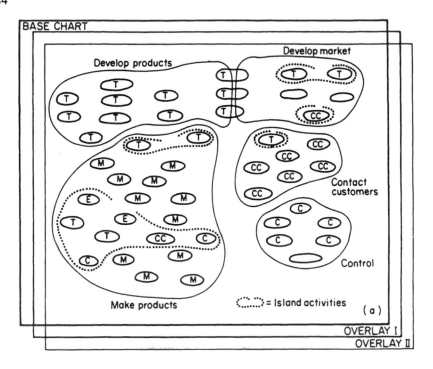

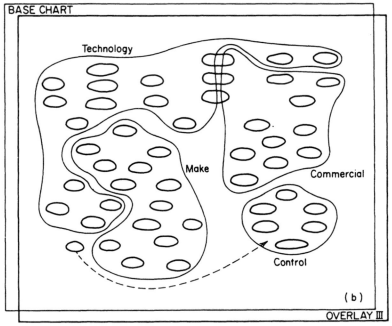

Figure 84. (a) Base chart with overlays I and II, showing identification of island activities. (b) Base chart showing assembly of second level activities and their allocation to the modified subsystems

to the respective subsystems. Figure 85 gives, as an example, a set of island activities within the 'make products' subsystem. These activities relate to the large 'island' illustrated in Figure 84(a), where the 'make products' subsystem contains several activities which were the actual responsbility of Technical Development, Production Engineering, Sales, and Senior Management. The solution adopted to this problem of diverse responsibilities was that shown in Figure 85, where some of the activities were moved to the Technical Development subsystem. The major decision activity on capital expenditure was retained within divisional control (Senior Management) and the responsibility for the remainder was firmly within the 'make product' subsystem.

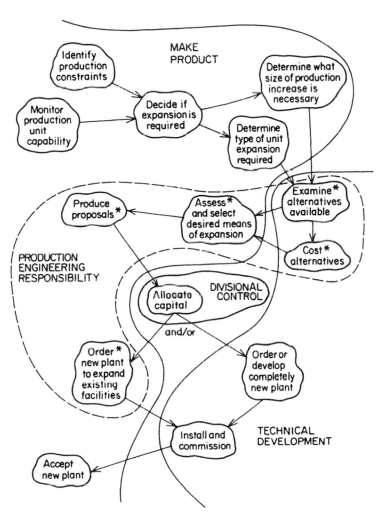

Figure 85. Island activities within the 'make product' subsystem
*Previously the responsibility of Production Engineering Dept.

The need for a separate Production Engineering responsibility disappeared and this reallocation fo responsibilities as indicated in Figure 85 was accepted as reasonable in this particular situation. The decisions concerning island activity responsibilities were made by the senior line managers concerned. Thus what emerged from the analysis was an organization structure that they had developed (and hence felt they owned) rather than one imposed either from outside or from above.

In the actual project there were many such islands and a considerable amount of time was devoted to examining the alternatives available. There was no right answer and the soluion in each case represented a compromise, but nevertheless, a compromise that the senior line managers believed to be the best and which represented a situation they felt confident to manage.

Once the 'island' questions had been resolved a set of subsystem boundaries existed which were consistent with acceptable organizational boundaries, within which there was no duality of responsibility. In carrying out this procedure the root definition of each of the subsystems had been changed by the addition or subtraction of activities and it was, therefore, necessary to redefine them and to ensure that they still represented a self-consistent set.

On completion of this stage of the analysis the subsystem boundaries were redrawn and overlayed on the base chart as in Figure 84(b). The names of the subsystems were changed to indicate that their root definitions had been modified by the reallocation of activities. This figure is displayed with Figure 84(a) in order to illustrate the degree of modification of the original subsystem boundaries. This is a true reflection of what occurred in practice.

These subsystems now represented new organization units, and the analysis proceeded by taking each subsystem, treating it as 'the system' and repeating the analysis at the next level of detail. For example, the 'make' subsystem was seen to consist of four major activities: (a) acquire raw materials, (b) plan production, (c) produce, and (d) ensure service utilities are available. In the particular project described, four such base charts were constructed (see Figure 84(b)), one for each new subsystem (or as it now was, new department) and the mapping process continued. The managers involved at all stages in the decision concerning island activities were those at appropriate levels in the organization. The analysis was carried out from general manager level to foreman level and since it involved the people concerned it became also the process of implementation.

The apparent reduction in area of responsibility of the 'make' decision-taker is as illustrated. This is what actually happened and the decisions about the responsibilities for island activities were made by the Chief Engineer and the Works Manager. The Works Manager argued that he would prefer to be held accountable for a smaller area over which he *really* had control than for a larger area over which he did not have complete control.

The connectivity between activities, which was used as the basis for debate about the allocation of responsibility for island activities, led, in this project, to three kinds of criteria:

(a) Activities with high interdependence were not separated:
 (i) Material interdependence.
 (ii) Information interdependence.
(b) Activities requiring common authority were not separated.
(c) Duplication of skills was minimized.

Examples to illustrate the appliction of these criteria can be taken from the 'make product' subsystem:

(a) (i) *Material interdependence* It was possible, in one part of the manufacturing process, to improve output at the expense of product quality. The effect of this was to introduce highly significant problems in the subsequent stage. Thus, since the quality requirements demanded particular material characteristics at transfer from one stage to the next, it was seen to be sensible to ensure common authority for the two stages even though the skills required in the two stages were very different.
 (ii) *Information interdependence* In the area of control of product quality, the information generated from product testing at various stages of the manufacturing process was vital to the control of process conditions. Thus the definition of information requirements, how it was collected and used, was taken to be the same responsibility as that for the manufacturing process.
(b) *Common authority* In the area of control of production it was unusual for the responsibility for plant maintenance to lie under the same head as that for production. However, it was argued that if there was a responsibility for achieving production targets (including a tight quality requirement) it would seem reasonable to give the same controller not only the control of 'up' time but also the control of 'down' time.
(c) *Duplication of skills* In this particular company a set of engineering skills had grown within the 'make product' subsystem, which effectively duplicated the work of similar groups within the 'develop technology' subsystem. The presence of this duplication also eroded the responsibility of both groups and led to an unclear definition of job responsibilities. The result of this particular duplication was a deterioration in personal relations due to the ineffective way in which tasks were carried out. Use of this particular criterion led to the reallocation of responsibilities illustrated in Figure 85.

The removal of island activities as potential organizational anomalies had led, at the first level of application of this methodology, to three organizational units: 'technology' (an engineering organization under the management of the Chief Engineer); 'commercial' (a sales and marketing organization under the management of the Divisional Sales Manager); and 'make' (the works organization under the management of the Works Manager). These three units could exist as departments, the job specifications of the respective managers being in

188

terms of the activities for which they were responsible. The control subsystem, on the other hand, was somewhat different in that it contained on-going activities which were the responsibility of managers at the same level as the Works Manager, Chief Engineer, and Divisional Sales Managers (first level line management) as well as periodic decision-taking activities that were the res-

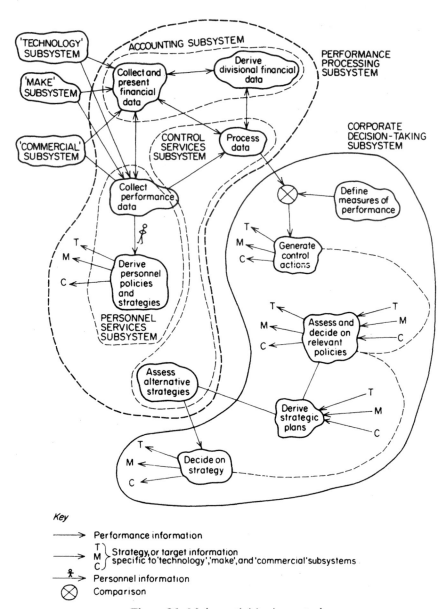

Figure 86. Major activities in control

ponsibility of a management board under the chairmanship of the General Manager. The control subsystem was expanded by taking the following root definition:

A system to define corporate policies and strategies; to set targets and define appropriate measures of performance and initiate control action as necessary; to ensure that adequate resources and services are made available to the company so that the various subsystems and the company as a whole can function effectively in pursuance of individual and corporate objectives.

This definition led to a model in which the activities were grouped as illustrated by Figure 86.

The activities were seen initially as of two kinds: activities to do with performance processing and activities to do with corporate decision-taking. Within the performance processing subsystem further groupings were made according to the following arguments. An accounting subsystem was formed on the basis that the collection, presentation, and derivation of financial data required a particular expertise which was independent of the nature of the business. The control services activities, on the other hand, required expertise which was business oriented and which was concerned with technical performance evaluation. Similarly, the evaluation of personnel performance and the derivation of strategies for personnel development and training required expertise of a particular kind which was different to the other two. Given these groupings the mapping process continued as described previously. The activities within the performance processing subsystem were related to the responsibilities of an accounting manager, a personnel manager, and a control services manager (previously called the Productivity Services Manager). The activities within the corporate decision-taking subsystem represented the area of responsibility of the Management Board.

The result of the application of the methodology at first and second line management levels is illustrated by Figure 87. This indicates the various management roles together with the structure relating them. The job specification of each role-holder was derived from the set of activities that each role-holder was responsible for: a straightforward task. The degree of change that was involved in the reorganization is illustrated by Figures 88, 89 and 90. In each case the organization structure prior to the project appears at the top of the figure and that proposed appears at the bottom. The boxes between the two structures represent organizational units that were either coming from, or going to, other parts of the organization. Those that are indicated as 'new' originated from activities in the original model for which an existing decision-taker could not be identified. The arrows indicate the degree of shift of responsibilities. These three examples represent the change in area of authority of the first line

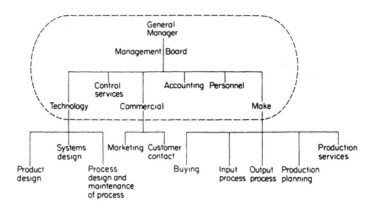

Figure 87. Company structure down to second line management level

managers, i.e. corresponding to the three subsystems in Figures 84(a) and 84(b). As would be expected from these figures, the greatest change occurred in the area of the Works Manager (i.e. the 'make' subsystem) as it was here that the greatest concentration of island activities occurred.

General messages which emerged from this project indicated that in changing the structure it is probable that operating procedures will also need to be changed and, before any implementation can proceed, these procedures must be identified and redesigned. The involvement of first level line management in deriving a structure down to second level line level is invaluable and similarly, if commitment and enthusiasm are to be generated, it is necessary to involve second level line management in the restructuring below them. Similar involvement for lower levels is necessary, dependent upon the complexity of the organization. The programme for implementation must be carefully designed so that the lower levels of management are involved at the right time, particularly if the restructuring implies a reduction in management positions. It may be necessary, to avoid problems of redundancy, retraining, or new appointments, to evolve a transient structure which will enable the company to move from its present structure to the new one in several stages rather than in one. The planning of the change process is as important as the design of the changes themselves, and it is worth considering the design of an 'implementation system' in which activities can be defined which are necessary to achieve the objectives, i.e. the implementation of the new structure. This is a transient system which only exists for the period of implementation but, derived in the same way as the organizational subsystem described previously, it provides a clear view of *what* needs to be done.

The transformation process appropriate to the implementation system relevant to this project was one which took, as input, information on the proposed new organization and the current structure, and produced, as output, the changes implemented. In this situation, the resources available to it were, in the first instance, those of ISCOL and the senior management team, and in

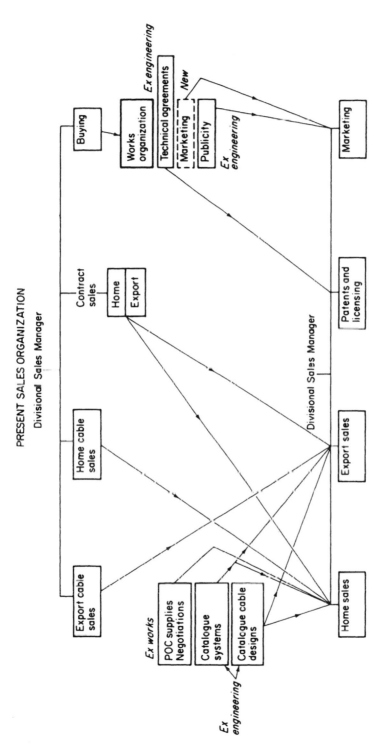

Figure 88. Sales reorganization

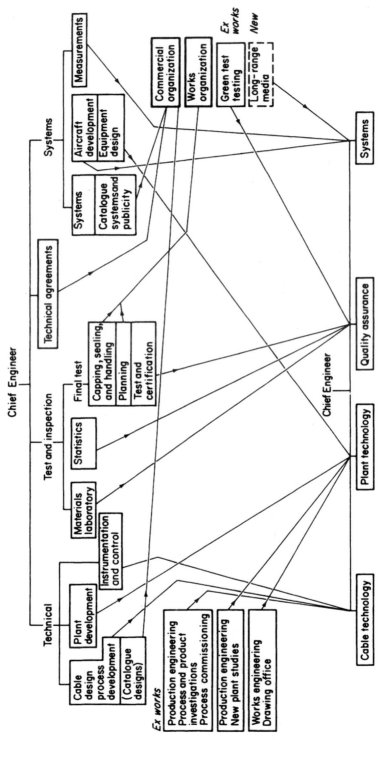

PRESENT ENGINEERING ORGANIZATION

Chief Engineer

Technical

Cable design process development

(Catalogue designs)

Plant development

Instrumentation and control

Test and inspection

Materials laboratory

Statistics

Final test

Capping, sealing, and handling

Planning

Test and certification

Technical agreements

Systems

Catalogue systems and publicity

Systems

Aircraft development

Equipment design

Measurements

Commercial organization

Works organization

Ex works

Green test testing

New

Long-range media

Ex works

Production engineering

Process and product investigations

Process commissioning

Production engineering

New plant studies

Works engineering

Drawing office

OUTLINE OF THE PROPOSED ENGINEERING ORGANIZATION

Chief Engineer

Cable technology

Plant technology

Quality assurance

Systems

Figure 89. Engineering reorganization

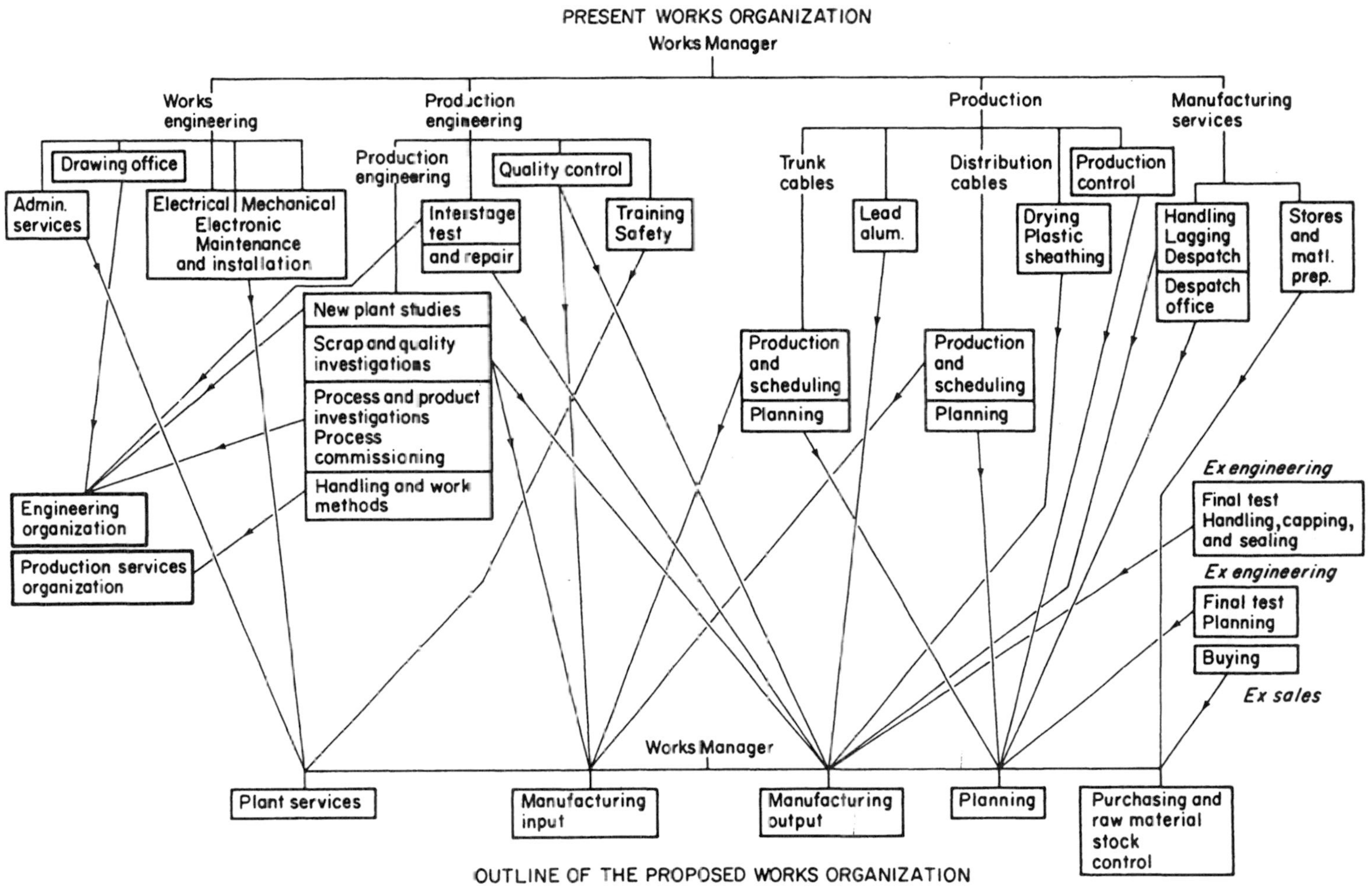

Figure 90. Works reorganization

the longer term the total management group. The constraints that it had to operate within were as follows: (a) a declared policy of no redundancy (implying a need to develop intermediate structures to cater for the existing staff complement); (b) a declared intention to consult with staff unions before reaching decisions about change; and (c), because this organization was part of a larger group, defined salary grades which any new structure had to recognize and accept.

Given the method of analysis used, it was necessary to develop an implementation system which operated in parallel with the analysis and was fully integrated with it, particularly in terms of the timing of some of the implementation activities. For example, since the management team (at any on level) was involved in the decision-making process associated with the analysis, this was also the mechanism through which commitment to the resultant changes was obtained; the gaining of this commitment being a necessary part of the implementation process. The conceptual model, of the system required, contained the following activities (see Figure 37 in Chapter 3):

(1) Identify the changes from the current organization and the problems of procedure and personnel that may result.
(2) Plan the sequence of changes (i.e. intermediate structures) that will minimize these problems.
(3) Take steps to investigate and overcome any constraints to implementation that might come from company and union policies.
(4) Gain commitment from second level line management.
(5) Develop lower levels of structure.
(6) Gain commitment at all lower levels of management affected.
(7) Identify and design any new procedures that might be required.
(8) Plan total sequence of change implementation.
(9) Allocate physical requirements (office accommodation, etc.) appropriate to the desired sequence.
(10) Initiate the change programme.
(11) Monitor progress and response and take control action as necessary.

This listing of activities is not intended to imply a once-through process or indeed a sequence. Activities (5) and (6) must follow activity (4), and activity (11) must be continuous and the control action necessary at any stage (of both analysis and implementation) may involve significant iteration.

A number of instances occurred in which it was necessary to develop transient structures to cater for existing personnel, some of these only existing for as long as it took to provide additional training. Some instances occurred in which it was necessary to change the restructuring based upon the response (i.e. as a result of doing activity (11)). One of the most significant of these arose during the reallocation of the island activities within the 'make' subsystem and is an example of power being exercised at lower levels than were being considered at the time.

Figure 91 attempts to illustrate the relationships that existed prior to the reorganization. The design office responsible of the Chief Engineer produced the designs for both the products and the process through which they were to be manufactured. Because of a very tight linking between product quality and the process, the Chief Engineer had responsibility for quality assurance and it was the case that, if the process was properly designed, the quality would be as required. The production engineering design office had grown within the Works Manager's area because of the way in which maintenance was done. If, for example, it appeared that a different design of part of the manufacturing machinery might produce fewer breakdowns or make maintenance easier, the production engineering design office produced the required designs. This they did through informal discussions with the maintenance fitters in order to determine the best way of designing the components from a maintenance point of view. The result of this *ad hoc* design activity was incremental change in the manufacturing process with resultant change in product quality. Thus when the Chief Engineer investigated complaints about reduced quality he discovered that the manufacturing process was no longer that which had been designed.

The reorganization suggested was to combine the two design offices (though not changing the site of the one within the works) and to make the manager of the production engineering design office responsible for this combined office. All managers at the levels affected were in agreement with the change. The only real effect of the change was to make this design office manager responsible to the Chief Engineer instead of to the Works Manager and to give him more responsibility.

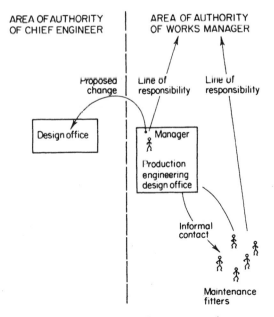

Figure 91. Design office restructuring

196

The maintenance fitters, who were responsible to the Works Manager through a different line structure, heard about the change and immediately stated that, if the change took place, they would work to rule and cease to operate the informal contact. The ensuing deliberations resulted in a modification to the proposed change. The two design offices were left as they had been and a small organizational unit was established consisting of representatives of the two offices but under the authority of the Chief Engineer, and this had the responsibility of assessing any proposed modifications to the machinery. If the proposed modification could have an effect on production quality, the design would be undertaken by the design office in the Chief Engineer's area. If it did not have an effect on product quality, then the design would be undertaken by the production engineering design office. This was, organizationally, a rather inelegant response to the problem, but it was the only solution that turned out to be acceptable to all concerned.

We were surprised by this particular response of the fitters as we had not imagined that a rearrangement of managerial responsibilities that had no effect on their day-to-day working would be of concern to them. However, they had apparently been losing status and this opportunity was seized as a way of regaining status. Thus, in retrospect, the episode was seen as a testing of power by the fitters rather than a specific comment on the design office reorganization. Using hindsight it is possible to rationalize the response, but I doubt if we could have anticipated it. The episode does illustrate that a significant part of rich picture buildng at the start of, or during, a project should be an examination of the power structures in the situation of concern.

The methodology represented by the seven stages discussed earlier can be summarized as in Figure 92. This identifies the assembly of activities that led to restructuring at the first resolution level. Once convergence of systems and

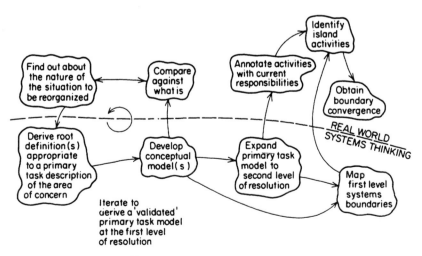

Figure 92. A methodology for reorganization

organizational boundaries has been achieved (through the debate about island activities), the process of analysis is repeated for each of the first level organizational units (they are now both systems and departments).

The methodology for reorganization presented here worked in a situation in which engineering managers were seeking a rational way of undertaking restructuring. It may be appropriate elsewhere and has, in fact, been used in other projects. However, there are, no doubt, situations in which it would be completely inappropriate. Thus, like methodology in general, it needs to be tailored to fit the nature of the enterprise in which it is to be used.

Chapter 6
Analysis of Business Information

Any discussion of information systems analysis must be preceded by a discussion of information and information systems. This whole area is confused by numerous interpretations and misconceptions. What is information? Is data neutral? Is information science the body of knowledge about information? Is information theory relevant? etc. These and many other questions spring to mind. However, it is not my intention to answer the above questions here, but at least, to make my own assumptions and interpretations clear so that what follows is hopefully self-consistent. Let us start with the first question; what is information?

It is my preference to adopt simple definitions where possible and in this case a definition which helps to explain the approach to analysis that is used later is to take information to be data plus the meaning ascribed to it. Thus as an example: the number of hours worked by an employee on a specific task is a piece of data. It becomes one kind of information when ussed by a salaries clerk in computing the amount of pay due to the employee. It becomes a totally different kind of information when used by a production scheduler who is concerned with allocating resources to tasks.

Given this definition any approach to analysis must start by determining what uses data is put to and hence by identifying the meanings relevant to an organization.

This will be explained in more detail later but for now let us return to the definition. It effectively makes the assumption that data is neutral. If this were not the case and we took a piece of data to be a fact plus its interpretation we would end up with a highly complex and unusable definition for information, i.e.

$$\text{information} = (\text{fact} + \text{interpretation}) + \text{meaning}$$

It would be impossible to sort out the difference between the interpretation and the meaning. Thus it is more userful to take data to be factual. Of course any piece of data needs to be bounded. So that if we are taking the number of hours worked by an employee on a specific task to be data then we need to define when the task starts and when it is completed and whether or not the time is employee dependent.

On the basis of the above definition, what are commonly referred to as information systems are really processed-data systems. They only become information systems when someone makes use of the output. Thus an information system must include the user as illustrated by the following diagram.

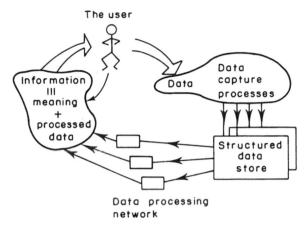

TOTAL INFORMATION SYSTEM

Here it is known that a user receives information and by virtue of performing some kind of activity (on the basis of this support information) he produces a data output. The cycle continues by capturing this and other data which is then stored as basic (or raw) data. The data-processing systems access the store and having operated on the data in some way present it as processed data for further use. The meanings are attributed to the processed data by the user on the basis of the activity that the information is to be used to support and it then becomes information. The cycle thus continues.

A number of approaches to information systems analysis make the assumption that data is a corporate resource. The first stage in the analysis is to identify the entities appropriate to the particular company (customers, products, personnel etc.) and then to define the data associated with each entity. This is captured and then stored. It is stored in some kind of database and the further assumption is made that if all the data relevant to all the entities is stored then all the information needed to perform the company operations will be available and can be processed.

This approach is effectively proceeding round the above cycle in a clockwise direction. The approach that I will describe later takes the opposite view and makes the assumption that it is information that is the corporate resource. Thus the analysis proceeds in an anticlockwise direction around the above cycle.

This yields the logical flow of the analysis to be:

$$\text{meanings} \longrightarrow \text{information} \longrightarrow \text{data}$$

Taking information to be a corporate, or company, resource may seem to be a reasonable assumption but this is not reflected in actual practice.

Few managers would disagree with the statement that information is a *company* resource, yet few organizations plan the development of, manage, and dispose of information systems, in the way that they organize other resource. The role of 'Company Information Managers' (interpreted in the above terms) is still a rarity. Most organizations have financial managers, personnel managers etc., and a large number of organizations have data-processing managers (under a variety of titles). This latter role tends to demonstrate an over-concern for the technology rather than a concern for the interpretation and use of the 'processed data' produced by the technology.

'Company information' is difficult to define. It is usually left to individual managers to specify their information requirements as if they owned the information and were the sole users of it. The trend towards end-user computing, through the proliferation of personal computers, is tending to reinforce this attitude. It may not seem unreasonable, since the managers have the responsibility for the management of an enterprise, that they should be in the best position to know what information is required for their support. However, asking individual managers the question 'What information do you need?' may elicit three kinds of answer:

(a) *I want this set because I have always had it* – it is difficult to dispose of something that may be useful; the human animal is traditionally a hoarder.

(b) *I want this set because I need it to do my job* – this is useful to know but it is seldom expressed solely in these terms.

(c) *I want this set because I have not received it before, it tells me more than I need to know but it may be useful* – information is power and opportunity to acquire more may be seized.

The analysts asking this question do not know which kind of answers they are getting; they may be getting a mixture of all three. In addition, the manager is answering the question from his, or her, particular role within the current organization structure. Hence the answer does not recognize role-to-role interactions or anomalies, which, in any case, will change as reorganization occurs (a not infrequent activity in most companies).

If we take the view that the emphasis should be on the information needed to support the *activities* undertaken within the organization rather than the management requirements (as perceived by the individual managers) then an approach can be derived which:

(a) Specifies the crucial information needs in a way that is not constrained by the particular organization structure at any time.

(b) Attempts to recognize and accommodate the multiple perceptions of the aims and mission of the enterprise relevant to the particular group of managers.

(c) Defines the management needs through a mapping of role responsibilities on to the above information needs and hence recognizes the role-to-role interactions.

(d) Produces a resultant information network that is coherent and is robust to changes in organization structure.

At the symposium of the International Federation of Information Processing in 1979, a Norwegian worker in this field, Leif B. Methlie suggested that there were two perspectives that could be adopted to the problem of information systems analysis. These he named as:

(a) A datalogical perspective.
(b) An infological perspective.

In the paper, given at the above symposium, he defined them as follows:

> The datalogical perspective regards the existing data flows as satisfactory represent-ations of the information needs in the organization. The aim of the change task is to find more *efficient* ways of processing the existing data. A common solution is to computerise manual procedures and data files. The benefits of this approach are primarily of the cost-saving type. This perspective is the traditional computer application view and is still common in current systems work.... The infological perspective of information systems design looks at the organization as an information processing system. Thus communication and control aspects are in focus. Information is the knowledge, communicated between individuals and groups, needed to perform tasks. The focus of this perspective is to find an *effective* information system for the whole or part of the organization to which the information is to give service

The difference between these two perspectives is essentially in terms of what is being taken as given. Although significant technological problems may exist in an approach which takes the datalogical perspective, the specification of the information that is required already exists. Given an infological perspective, all that is taken to exist are the specific tasks that the organization needs to perform. Information requirements are derived in order that the resulting information systems can provide the support necessary for these tasks.

The adoption of an infological approach suggests a view of the 'process' of information system development contained in the following diagram (a particular interpretation of this process also appeared in Figure 2 of Chapter 1):

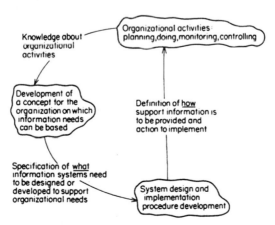

This is a conceptual distinction and, in reality, a single individual could do all three activities; i.e. a manager responsible for a particular area of a company could derive and design his own support information systems. In large organizations is is more usual for the organizational activities to be undertaken by what are termed 'the users' and the design to be the responsibility of a data processing, or management services, department. Frequently the information provided by the information systems that have been designed do not match the requirements of the users. Apart from the obvious reasons, that either the users do not correctly specify their requirements or the data-processing personnel provide only what they think is needed, a more significant reason is that the central activity in the above diagram, i.e. the explicit development of a concept for the organization, is omitted completely. This concept needs to be independent of any organizational structure if the resulting information systems are to be relatively robust.

Information systems that are derived for the needs of particular managers fulfilling particular roles, will require immediate modification if those roles change as a result of reorganization.

The major problems, of course, are to determine the necessary activities for the organization and, once these are obtained, how to derive the support information.

In this chapter I would like to present a methodology which aims to tackle these problems based upon the concept of a human activity system and, in particular, upon primary task modelling. I will describe two projects from which the ideas were derived. The first is concerned with improvements to an existing manual system (similar to a datalogical perspective except that the analysis started with activities and not existing information). The second project is concerned with information planning, with particular reference to developments in a large and complex information network. This latter project gave rise to a device, known as the Maltese cross (Wilson, 1980a), which has turned out to be a useful tool within the overall methodology.

Project 1: FABLE Industrial Group

The company in which this project was carried out wished to maintain confidentiality and hence the name used is entirely fictitious. It was part of a large group of companies concerned with food products which we will call Food and Bulk Liquid Enterprises (FABLE). Our particular client was the Works Manager of a manufacturing site which produced liquid products supplied in bulk to other industries such as the brewing industry and other food processing plants. This site was situated in the north of England with an associated head office in the London area. The manufacturing site was essentially concerned with satisfying customers' orders for products, though there was a sales department which undertook all of the short-term activities associated with order processing such as pricing, invoicing, and contract negotiations. This department was functionally responsible to the marketing department at the head office. Within the head office there was also a central accounts department and a data processing department which maintained all of the operational data concerned with customer records and stock levels. This data was stored and processed on a mainframe computer connected to the manufacturing site via a GPO line. However, as all the payroll calculations were also performed on this machine, the operational data received a lower priority in the event of computer malfunction (a not unknown occurrence).

The scope of the project was agreed at the outset and was concerned with 'deriving recommendation for improvement, where appropriate, in any aspect of the system of dealing with orders, including any restructuring which would lead to more economic use of computer and manpower resources'. The system referred to here was 'the system of paperwork and comunications which arose in order to ensure that orders could be met from material in stock, new production or acquisition through purchases, and that invoices had been submitted'. The only constraint that was placed on the study was that the work was not to be concerned with the manner of scheduling of production but would be concerned with the information enabling scheduling to take place. Increases in manpower as a result of recommended changes would not have been acceptable and hence that was also taken to be a constraint. The interdepartmental data flows that existed are illustrated by Figure 93.

In order to identify what these data flows were, we decided to follow a particular customer order from receipt to delivery and to document what actually

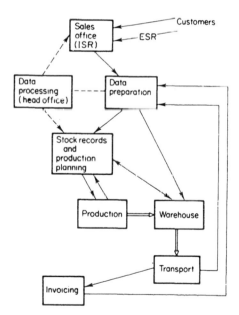

Figure 93. Data flows. *Key:* ISR, internal sales representative; ESR, external sales representative; ----→, data transmitted via GPO line; ——→, internal data flows; ⟹, product flow

204

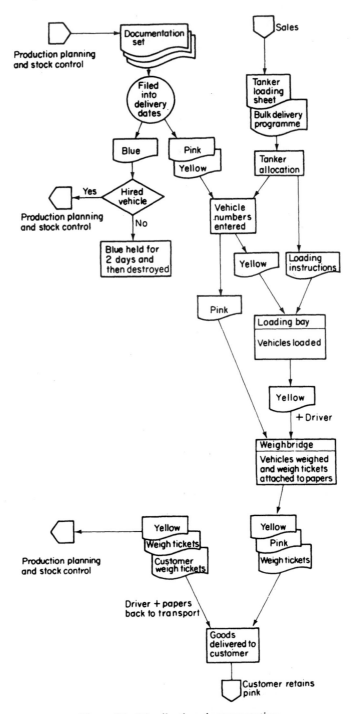

Figure 94. Distribution documentation

happened at each stage. The result of this process was an extremely complicated diagram, a small section of which is reproduced in Figure 94.

This figure illustrates the point made earlier that what is observed in practice is the particular 'how'. As such it gives no guide as to 'what' is happening at each stage and hence no indication as to why it is happening. Although this was useful later at the phase of comparison it was clearly not the place to start. The phases of analysis that we followed were:

(1) Derive a model of a system of communication based upon a primary task description of the activities initiated by a customer input (enquiry or order).
(2) Use this model to compare against the present methods of communication.
(3) From this comparison identify those features of the communication systems which significantly affect performance.
(4) Identify those changes to the present methods which could appreciably improve performance.

Derivation of primary task model

The primary task model was derived by, first of all, making the assumption that a 'customer response' system would be composed of the five subsystems illustrated by Figure 95. For each of these systems a root definition was derived

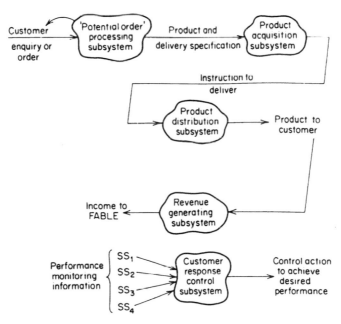

Figure 95. 'Customer response' subsystems

to enable further expansion to be made. As an example, consider the 'potential order' processing system. Here it was assumed that the input could be either an enquiry or an order and that the output was either the specification of the order to be met or, in the case of an enquiry, the terms under which a subsequent order would be met. The root definition taken was:

> A FABLE owned system which converts customer enquiries for a range of bulk liquid and other (associated) products into accepted orders which are to the continuing benefit of the company, also to produce instructions to the product acquisition system, and relevant information to the customer, resulting from the conversion, and to undertake these activities in such a way that is consistent with the performance expectations of the company.

The model resulting from this definition is given in Figure 96. Models of this kind were developed for each of the five subsystems (now redefined as 'systems' for further expansion) and the total model was 'validated' as a primary task description of the activities initiated by a customer input, leading to products delivered and income received. The system of communication was then derived by identifying the information required to support each activity in the model and the information generated by doing each activity, i.e. the system of communication was represented by the 'activity-to-activity' information flows.

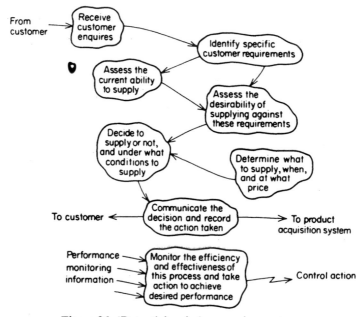

Figure 96. 'Potential order' processing system

Comparison and identification

Having validated the model, in the sense that each activity was legitimate, the activity-to-activity information flows represented the minimum necessary flows that were crucial to the operation of the customer response system. It was therefore these flow that were compared against the communication processes that had been identified earlier. The comparison was done formally through the construction of a table for each activity. The information required as a support for the activity was listed, the present 'how' of providing it was identified and comments were included on the performance and appropriateness of the existing 'how'. The form of comparison is given in Figure 97 and examples are given of the actual comparison, for activities within the 'potential order' processing system, in Tables 6.1, 6.2, and 6.3.

The results of this comparison (as represented in the third and fourth columns) was an understanding of the relationship of the current 'hows' to the activities they were supporting and an identification of the problems associated with the current 'hows'. The major areas of concern were summarized as follows. *Problems were observed in relation to the preparation and transmission of information particularly in terms of the large amounts of manual effort involved.*

As an example, delivery orders were prepared in the sales office and carried to the transport department, which was situated some distance away. Frequently the weather in the north of England is unconducive to exposed travel between the two departments. Thus delivery orders were allowed to accumulate and were transmitted in large rather than small batches. Apart from the time involved in transmitting them, this behaviour resulted in a 'peaky' input to

ACTIVITY A1

Information required	Present 'how'	Comment on performance	Comment on appropriateness
I_1	— — —	— — —	— — —
I_2	— — —	— — —	— — —
I_3	— — —	— — —	— — —
I_4	— — —	— — —	— — —

ACTIVITY A2

Information required	Present 'how'	Comment on performance	Comment on appropriateness
I_1	— — —	— — —	— — —
I_2	— — —	— — —	— — —
I_3	— — —	— — —	— — —

Figure 97. Form of comparison

Table 6.1. Receive customer enquiries (Activity A1)

Activity information requirements	Current information flows	Comments on present performance	Comments on present appropriateness
We want to know firstly who is enquiring: An existing customer A new customer The company The enquirer's authority	In FABLE the internal sales representative (ISR) for a particular product range deals with most customer enquiries. There is a separate routine if it is a new customer (a safety measure as they deal with food): the outside representative must call on them first. Enquiries are easily sorted into the various types by the ISR.	Enquiries are mostly by telephone and hence the enquiry is handled immediately in 90% of cases and hence response time is good. The quality of the response is determined solely by the ISR and so, since they are the company's voice, must be and are chosen with some care.	Since 90% of all customers are regulars or known, then the present system would seem appropriate. The problem is that since the telephone is used there is no immediate hard copy of the order from the customer, although one should come later by mail (but does not always).
Next we want to know the nature of the enquiry: Can you deliver? What, when, Please deliver! where, who pays? What's happened to my order? Complaint on: quantity, delivery, quality	The enquiries are around 90% from existing customers and around 90% by telephone, the rest by mail and telex. Customers are asked to confirm telephone orders by mail.		
The communication medium could be any of: Letter Telephone Telex Personal caller Remote computer terminal	The external sales representatives visit the customers for various reasons. During this visit they may take enquiries of any nature. These are reported to the ISRs, by whom appropriate action is taken.	This is difficult to assess as they very seldom obtain new customers from this activity. They keep customers happy, advise on storage and products, etc.	

Table 6.2. Assess the current ability to supply (Activity A4)

Activity information requirements	Current information flows	Comments on present performance	Comments on present appropriateness
Information needed to do this activity is to know Activity A3 plus	This information is obtained by the ISR by telephoning (internally) to	Information is acquired quickly by telephone. Reliability is in question, however, due to	The ISR relies on the person at the other end of the telephone to interpret the situation and so the quality of the final decision is very much dependent on the quality of the people in the chain
Know stock position and production forecast with respect to products	Production planning and stock control to find out the stock position and production forecasts		
Know the projected delivery situation		Human erros	No one person is responsible for making the decision. Bits of it are made by various people
Know the basis on which prices are determined	Transport to see if they could deliver when the customer required it and if not, when	The validity of the source due to the updating of information procedures	
	Prices are determined by reference to	Actual performance is difficult to assess as it is not monitored	
Use all of the three above to determine what to supply, when and at what price	The customer's file to see what he was charged last time		
	The price manual to check for any price changes (although he usually knows in his head of any changes)		

Table 6.3. Communicate decision and record action taken (Activity A6)

Activity information requirements	Current information flows	Comments on present performance	Comments on present appropriateness
The decision taken in Activity A6 must be communicated to the customer by telephone, mail, telex, remote terminal, or personally. The communication must tell the customer the exact conditions of sale or the reasons for no sale.	The decision to supply or not is communicated at present mainly by telephone (90%), next by mail, and finally by telex. Delivery date, price, address, and product are agreed (or not) by telephone	Since the telephone is usually used, the decision to supply or not and the conditions are quickly transmitted to the customer. Again performance is dependent on ISR quality, which is highly subjective	The telephone is probably the most appropriate source of communicating their decisions due to its immediateness
Instructions to the product acquisition system must include Customer Product specification and quantity Delivery date Delivery address Invoice address	The decision to supply is communicated to production planning and stock control, invoicing, and transport by taking copies (seven) of the delivery order which contains customer, customer number, delivery address, invoice address, product specification, and delivery date. Some sections only get a half copy which does not include price	The time delay in producing and transmitting this information can be quite long (subjectively up to half a day for information to get from sales office to production planning and stock control and data processing, which are the first links in the chain and then a further half to one day to get to transport and invoicing; transport can take a further half to one day to pass on the warehouse copy to the warehouse). Performance poor	It does seem very doubtful that the information transfer via seven copies of paper through a chain network that can introduce delays of up to 2 days is an efficient means of information transfer. The updating of a central order file that all other activities can use seems more immediate and more efficient in terms of manpower
All these must be recorded in some way	All the above information is recorded in the customer file or wallet		

the transport department that inevitably produced difficulties in transport scheduling.

In addition:

(a) Crucial information about the stock situation, ability to deliver, customer credit, product specification changes, etc., was obtained by the internal sales representatives through informal channels and its reliability was questionable (see Tables 6.1 and 6.2).

(b) Production of invoices was time consuming and generally inadequate.

(c) All updating information tended to be untimely.

Updating was done by the data processing department at the head office and, due to delays on a number of occasions, the production planners and stock controllers used to maintain and update their own records. Thus the process was duplicated and the time to do this updating was time wasted. Also the two sets of records were seldom in agreement.

It was apparent, as a result of the complete analysis, that small changes to the existing methods would not overcome the major problems and so, to produce significant benefits, a fundamental redesign was undertaken.

The above assessment was based entirely upon a comparison of activity-to-activity information flows and hence was independent of any organization structure. For a complete assessment an organization mapping was undertaken using the overlay technique discussed in Chapter 5. The reason for doing this was to highlight any activities which, although contained within the systems boundaries, did not reside within the boundary of the appropriate department (i.e. island activities). It was accepted initially, within the defined scope of the project, that some change in the current departmental responsibilities might ease the communication problems. This was not the case with most of the activities. However, the major exception was related to those data processing activities that were located at the head office. A recommendation was made, and was accepted, that the authority for these activities should reside within the boundary of the manufacturing site. With this proviso the conclusion reached was that the existing departmental boundaries were satisfactory since changing them would have brought minimal improvement in operating efficiency.

By mapping the (agreed) organizational boundaries on to the primary task model it was a simple matter to convert the activity-to-activity information flows into role-to-role information flows. This was done by identifying the cross-boundary flows. Thus, following this stage, it was known who required what information and also who generated it (and hence had the responsibility for its updating). Altogether there were 46 information categories which were grouped together to form nine sets of like data. The nine sets were: stock, delivery, price, production capability and forecasts, sales forecasts, raw material situation, company policies, export requirements, and customer constraints.

Having defined the requirements of the communication system, the next stage was to explore design alternatives. It was decided to design an initial system

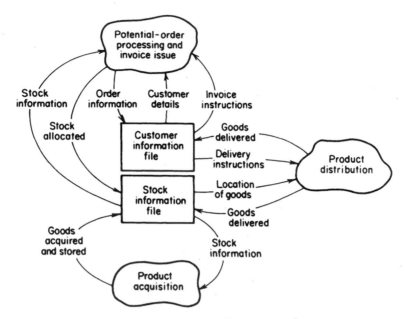

Figure 98. Relationships of data files to operating systems

based upon two files: (a) customer file and (b) stock file. A design was chosen which would facilitate further development once operating experience had been gained. The major interactions between the four operating systems and the two files is illustrated by Figure 98. This diagram shows the basic information to be assessed by these systems together with the necessary information inputs from the systems to update the files.

The next stage was to establish the content and extent of the two data files using the data groups already defined together with performance requirements. These were as follows:

(a) *Stock file*: (i) list of all stock by product, product grade, subgrade, and specification; (ii) identification to show if product is allocated or unallocated; (iii) stock location; (iv) allocated stock against order number; (v) physical and planned stock; (vi) item lead time necessary to produce respective product; (vii) special requirements, i.e. minimum stock levels, etc. *In addition*: (ix) updating of stock to be possible by ISRs quoting order number; (x) stock information to be instantly available to ISRs to enable them to respond to customer enquiries by telephone; (xi) above information needs to be accessed by production planners, (xii) above information needs to be accessed by stock controllers for updating purposes; (xiii) this file should provide transport department with the location of goods to be delivered, (xiv) production planners should be automatically notified when stocks reach predetermined minimum levels;

(xv) the files must be capable of upating the mainframe computer at head office.

(b) *Customer file*: (i) customer's name, address, account number, VIP codes, etc.; (ii) delivery address(s); (iii) invoice address; (iv) customer history, i.e. a listing of previous transactions containing product, grade, etc., price, order number, delivery date agreed and actual, payment indication; (v) special delivery instructions; (vi) customer intelligence. *In addition*: (vii) above information needs to be instantly available to ISRs to enable them to respond to enquiries by telephone; (viii) ISRs to be able to update file when new order is placed; (ix) delivery instructions to be sent to transport department; (x) updating of file when goods are delivered is necessary; (xi) information provided by file for invoicing; (xii) file to indicate credit status of customers; (xiii) the file must be capable of updating the mainframe computer at head office.

The definition of these requirements led to an examination of the alternative forms of implementation. No commitment to either manual or computer-based operation had been made by this stage. An assessment of the advantages and disadvantages of the two methods led to the adoption of a computer-based solution using a dedicated minicomputer. At the time most of the information specified for the customer file was already maintained manually in customer record cards; however, practically all the stock file would have been a new addition. If this was to have been a manual file it would have required additional staffing and the necessary improvements in the timeliness of data acquisition would have been difficult to achieve. The ISRs believed that to be able to respond to a potential customer while he was at the other end of the telephone was a significant advantage. Although this had been achieved in the past, it had been at the expense of using unreliable information. Thus, given this requirement, computer-based implementation was the obvious answer. Access to the two files was provided by visual display units (VDUs) and this provided an immediate and parallel access and updating facility. The storage requirement was of the order of 120K, including that necessary for the anticipated future expansion. The outline design is illustrated by Figure 99 and the operation of the system can be summarized as follows:

(a) To enable ISRs to respond to telephone enquiries they had access to both the current stock situation and the customer record file via the VDUs. On agreeing an order with a customer the ISR would then update the customer file with the details of the new order. He would also allocate stock by quoting the relevant order number and so update the stock file.

(b) The production planners accessed the stock information file, again via a VDU, in order to identify the current situation on stocks.

(c) The stock control department had responsibility for entering into the stock file data on the goods acquired via production and other sources together with their location. They also had the responsibility for the final

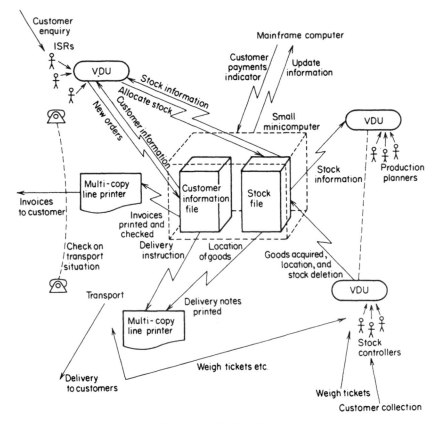

Figure 99. Outline design

deletion of goods from physical stock on receipt of despatch notes, weigh tickets, etc.

(d) All delivery instructions together with the location of the goods to be delivered were sent automatically to the transport department. The despatch notes were printed by a multi-copy line printer located in the transport office.

(e) All invoices were printed automatically on a multi-copy line printer in the sales department on notification of despatch/delivery.

(f) A telephone link was maintained as part of the system to enable the ISRs to check with the transport department on the up-to-date vehicle situation.

(g) The mainframe computer at head office was updated on a daily basis and this computer formed the main back-up for the minicomputer.

(h) The mainframe computer was also used as a performance monitor and had the additional duty of producing periodic digests of stock and sales information for accounting and business control purposes. It also

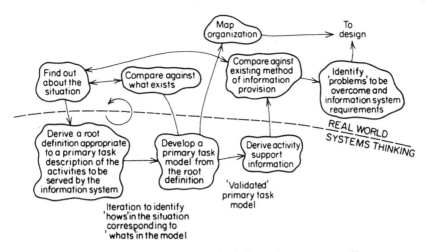

Figure 100. A methodology for information system audit

supplied a customer payments indicator to the customer file in order to update the customer's credit position.

This system of communication was implemented, first of all, in the basic form described here and was subsequently expanded. The method of analysis used provided a coherent set of arguments for the major change proposed and, as indicated, these arguments were accepted. The resultant computer system was relatively simple but, as a case example, it serves to illustrate the methodology used for the kind of situation in which a critical assessment of an existing system of communications is required. This methodology is illustrated by Figure 100.

Project 2: British Airways

Following the merger of BEA and BOAC into British Airways, there was a need to introduce considerable reorganization, particularly in the engineering maintenance activity. This was a major part of the organization, employing 14 000 people, with a responsibility for the maintenance of aircraft belonging to British Airways and other airlines. Information systems to support this highly complex operation had evolved independently in the two organizations and hence, following reorganization, there was a need to review information needs, to identify and define existing computer-based and manual data processing systems, and to define the total requirements for the new organization.

To tackle this problem British Airways had set up an organizational unit, the Management Systems Group (MSG), which consisted of representatives of various user and service departments and which met periodically to undertake the above reviews and to define and apply British Airways policy with respect to information systems development. This group had established a procedure for undertaking this development which relied, for its initiation, on a user-manager submitting a request for the design of a new information system. This submission contained details of the systems requirements and estimates of the expected benefits. If the proposal was acceptable (following iteration for more details if necessary) a subcommittee was established with the responsibility of progressing further development, through design to implementation. Since the procedure was initiated by a user-manager, the need expressed represented a limited view of information required, i.e. the need was defined by a single manager based on his position and responsibility within the management hierarchy. The problem that the MSG faced was, how to relate this stated requirement to all other requirements within this highly complex area of engineering maintenance? Various members of the group were requesting some sort of concept for engineering as a whole that would facilitate discussion about such relationships. It was about this time that ISCOL became involved and we were asked by the chairman of the MSG, who was the Chief Planning Engineer, to investigate the contribution to this problem that could be made by systems ideas. He established the investigation as a research project in which a number of students participated, with myself as the project manager. Throughout the project we had the participation of managers in a number of areas and worked closely with the systems co-ordinator, a senior manager from the management services department, appointed as the 'organizations' manager for the project.

The broad stages that we intended to follow were as follows:

(1) Develop a primary task description of a system relevant to the Engineering Department (BAED).
(2) Use this model (at whatever resolution levels are appropriate) to define information needs.
(3) Rank existing information processing procedures (IPPs) according to the resolution levels in the primary task model.
(4) Relate 'what is available' to 'what is needed' to identify interactions, or overlaps, between existing IPPs and those areas where further IPP development would be beneficial.

We did not know how we were going to do stage (4), but we felt that if we had some concept on which we could define information needs then, as part of the research, we would be able to develop a way of mapping what existed in terms of IPPs (in this case mainly computer packages). The expression 'information processing procedure' is used in order to remain completely general. The approach adopted here is equally applicable to manual as well as computer-based data processing. Stage (3) was included because the IPPs that existed at that time had been developed to meet the needs of the managers according to their responsibilities within the management hierarchy. There was no reason why the resolution levels implied by these IPPs should match the resolution levels in the primary task model. Obviously an IPP which produced, as output, the annual maintenance schedules (prescribed by Civil Aviation

Authority regulations) was at a higher organizational level (and hence lower resolution level) than an IPP which was concerned with material control. Thus the existing IPPs needed to be ranked in order of increasing detail of the activities they supported in order to know which were relevant to which resolution level in the model.

The first stage (stage (1)), however, was to develop a primary task model. Some discussion of this stage has already taken place in Chapters 1 and 2, in particular in relation to Figures 2 and 19. However, as the development of the model is an integral part of the approach, relevant parts of that discussion and related figures are repeated here for completeness.

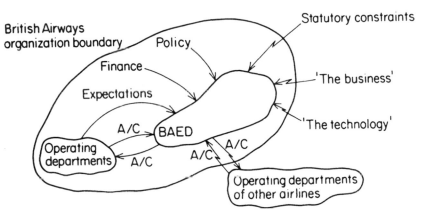

BAED and its environment. *Key:* ⟶, internal information flows; ⇢, information flows external to the system boundary; A/C, aircraft (part of Figure 2)

Figure 2 in Chapter 1 illustrated a simplified view taken of BAED within its environment and included some interactions which significantly affected the nature of the organization. Also identified are those parts of the environment with which these interactions occur. In reality there were many more interactions than are shown here but it was assumed that these were secondary to, or arising from, those illustrated. An examination of this situation gave rise to the following root definition of the organizational entity BAED seen as a human activity system:

> A British Airways owned system concerned with the continuously effective and efficient planned maintenance of aircraft belonging to BA and other contracted airlines, with a performance acceptable to operating departments but within statutory, local and BA-applied constraints.

This is a complex root definition, in the terms discussed in Chapter 2, and hence prior expansion into a number of distinct subsystems was made before development in terms of activities was undertaken.

218

The statement of ownership in this definition implies that activities must be included in the model which identify, and act on, the policy, constraints, and expectations of that owner. The words 'continuously effective and efficient' indicate a need to be well in touch with development in the methods and equipment associated with aircraft maintenance; also that the development and subsequent use and acquisition of these resources needs to be cost-effective. The existence of 'operating departments' and 'other contracted airlines' leads to the inclusion in the model of a set of 'marketing' activities together with the specific expectations of these 'customers.'

Thus, at a very broad level, it was argued that this definition led to the five major subsystems illustrated in Figure 19. Here the 'system concerned with obtaining business' interacts with the operating departments of British Airways and other airlines in order to establish expectations in relation to the maintenance task. This system operates within the business development strategies determined by the 'planning system' and it also identifies resource development needs appropriate to the nature and size of the business generated.

The 'system concerned with the development of technology and other resources' responds to this input and also to the messages it receives from its interactions with the general business area and technology of aircraft and other relevant maintenance.

The 'aircraft maintenance system' plans and executes the maintenance task and other associated engineering work in order that its resources are utilized in an effective and efficient way.

Finally, in order that all of these subsystems operate together in such a way that the total BAED system pursues its objectives in a way which meets the

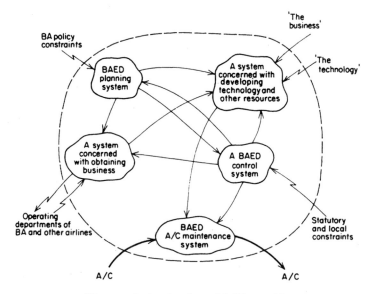

First resolution level model (Figure 19)

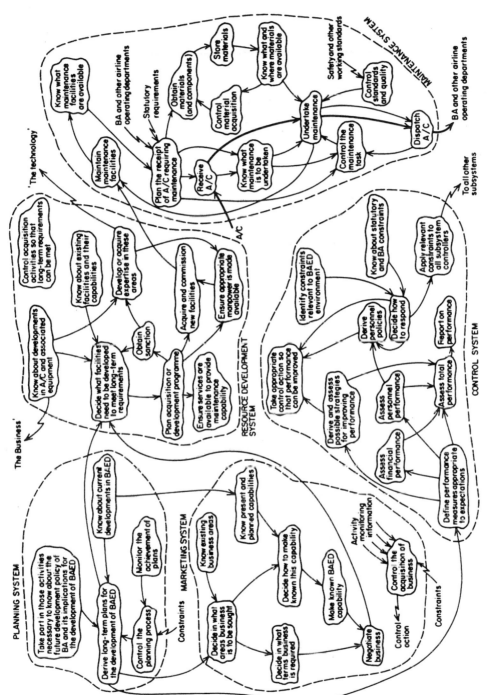

Figure 101. BAED model

expectations of British Airways, a 'control system' is needed to monitor the total set of activities and take control action where appropriate. This control system, however, must recognize the constraints under which the total system has to operate and ensure that these constraints are applied where necessary.

The model in Figure 19 is at too broad a level to be of use in the derivation of information flows, but it serves the purpose of an intermediate stage in the hierarchical model development process. Thus each of the five subsystems were redefined as systems and again, through the mechanism of root definitions, models were produced at the next resolution level. These five models were then put together to produce a complete systems model of BAED at this higher level of detail illustrated by Figure 101.

Further expansion of this model took place, but for the purpose of an illustration of the approach adopted and the resultant ideas that emerged from the project, this level of detail will be used for further discussion and, in particular, reference will only be made to the 'maintenance system'.

Stage (2) is concerned with using this model to define support information requirements. Effectively each activity is being viewed as an information transformation process irrespective of its prime purpose. Thus, with reference to the following diagram, the information relevant to each activity can be defined in terms of the category to which it belongs and whether it is an input or an output: The term 'information categories', however, needs clarification. Information is taken to be data plus meaning (where meaning is derived from use, not in the sense that it can be obtained from a dictionary). Thus a specific piece of

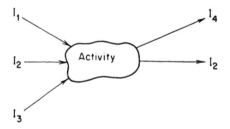

data, say the length of time taken to do a specific job, will become information of a particular kind when used by an accountant to do his payroll calculations and will become totally different information when used by a production planner to derive a production schedule. The piece of data is the same but the information is different because of the different meanings attached to it due to its use. Since, in this stage, we are deriving input information to support activities (i.e. what it is being used for), the meanings are implicit and hence we can talk about support information rather than support data.

Using this interpretation, an information category can be defined as 'the family name' for a particular set of data, i.e. the lowest resolution level description of the set. Thus, as an example, a category of support information required at the level of Figure 101 (the lowest resolution level taken for information derivation) was 'unscheduled defects'. This was taken to be the family name and hence the name of an information category. The members of the data set belong to the category can be defined by developing a data model for that set. Using a hierarchical model this particular data set is as illustrated by the partial model below:

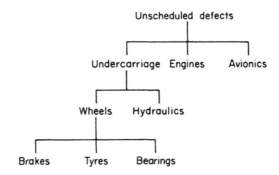

Returning now to stage (2) of the analysis, information categories were derived for all 13 activities in the maintenance system at this level. These are reproduced in the matrix of Figure 102. The elements of the matrix indicate whether the information category is an input to, or output from, the activity. In some cases the same information category is both an input and an output. This merely indicates that the data received as input has been operated on and may have had the format changed, but nevertheless still belongs to the same information category. Plotting the matrix in this form enables some rearrangement of information categories to be made to minimize the total number used. For example, 'stock locations' and 'stock movements' could have been combined into a single category 'stock information'. However, care and judgement need to be exercised otherwise the information categories can become too general. At the extreme all these categories could be combined to form 'engineering information'. This is certainly minimum but it would not be very useful. As a general rule, information which is about the same subject could be combined but that which in different in nature should not be. For example, it is obviously worth keeping 'Engineering Department plans' as a separate category to 'customer business'.

The approach adopted in stage (3) was to take each IPP to be an information transformation process in the same way that the activity was viewed in stage (2). Thus, referring to the diagram below, input and output information categories were identified (by relating the data, as far as possible, to the definition of the categories derived in stage (2)):

Information inputs/outputs	Maintenance system activities													Environment	
	Plan receipt of aircraft requiring maintenance	Receive aircraft	Know what maintenance is to be undertaken	Undertake maintenance	Dispatch aircraft	Maintain maintenance facilities	Know what maintenance facilities are available	Control the maintenance task	Control standards and quality	Obtain materials (and components)	Store materials	Know where and what materials are available	Control material acquisition	BA	Ext.
	1	2	3	4	5	6	7	8	9	10	11	12	13	BA	Ext.
Engineering Department plans and policies, constraints	I						I							O	
Customer business and needs	I													O	
Aircraft schedules, flying rates, plans	I													O	
A/C development programmes	I								I					O	
Available facilities and workload capability	I	I					O								
A/C system and component time/life requirements	I								O						
Aircraft maintenance schedules/aircraft	O								I						
A/C fleet maintenance plans	O	I	I/O	I				I	I		I		I	I	
Work area allocation		O	I												
Aircraft call-in		O										I		I	I
A/C and component inspection schedules			I						O						
Unscheduled A/C defects			O/I	I/O											
Scheduled A/C defects			O	I											
Flying statistical data			I	O											
Component repair and usage			I/O	O											
Material requirements for maintenance			O	O								I			
A/C and component history			O							I					
Material consumption				O				I			O	I			
Labour usage				O				I	I/O					I	
Jobs completed (work done)				O					I/O					I	
Inspection test results				O				I	I						
A/C despatch tests and results				O	I/O			I							
A/C delivery details					O									I	I
Facilities commissioned and capabilities						I								O	I
Facilities inventory						O	I/O								
Control maintenance action				I				O							
Proposals to change A/C and component maintenance								I/O						I	O
Reliability and performance of A/C components analyses								I	O			I			
Supplier of materials - information										I					O
Supplier performance									O	I					
Purchase orders and progress										O	I				
Delivery of materials - advice										I	I				O
Stock checks/locations											I/O	I/O			
Stock movement review											I	O			
Stock budgets and policy								O		I			O	I	
Performance report on maintenance system	I								I				I		
Personnel policies	I							I	I				I		O

Figure 102. Summary of information requirements of activities within the maintenance system. *Key:* I, input; O, output; A/C, aircraft, BA, British Airways

Also the activities that the outputs from the IPPs were supporting were classified as either primary or secondary. A primary activity was one at the resolution level of Figure 101 and a secondary activity was one at a more detailed resolution level of one of these activities. Hence the ranking that was done consisted of collecting together those IPPs whose output supported primary activities. Even if only one output of an IPP related to primary activities and the remainder supported secondary activities then that IPP was considered as relevant to this level. It was only those IPPs that provided outputs that were wholly secondary which were excluded from consideration at this level of resolution and retained until further expansion of the model had been made.

While doing this ranking it was also felt to be useful to identify the sources of the data input. The results of this stage were summarized in tabular form: an actual example is given in Table 6.4. However, there were a large number of such pages that made up the complete table and, at this stage in the project, we were faced with moving on to stage (4). It had been our intention to scan the final column and identify all those IPPs that provided support to the same single activity. For example, if MIPAC (major inputs planning and control, where an input in this context is an aircraft) had been the new computer package that a user-manager had proposed, then by entering it into this table alongside those other IPPs already in existence, we could identify its interactions through identifying the activities supported. One of these activities is 'plan the receipt of aircraft requiring maintenance'. Scanning the last column shows that this activity is also supported by the IPPs DISC and DOCK CONTROL (plus others from the remainder of the complete table). Thus in order to determine if this proposal made sense, in terms of efficient utilization of computer (or other) resources (in that it was not duplicating data processing that already existed), and that it represented coherent development in the IPP network in total, we would need to answer the following set of questions:

(a) Does the existence of more than one IPP providing an information input to an activity indicate a duplication of data processing?

(b) Could more efficient processing be obtained by utilizing data alread processed by one of these IPPs, rather than by processing raw data?

(c) Do the existing IPPs and their outputs fulfil the total information needs of each activity?

(d) Are the respective formats of the outputs of the IPPs supporting the same activity consistent, and is this format the most useful for the purpose of that activity?

(e) Is the data provided by the IPP required as a support to activities other than those for which it is to be, or was, designed?

These questions can, of course, be asked of an existing information processing network or asked when it is proposed to develop a new IPP for introduction into the network.

In this example the questions would have been asked of all the IPPs that were supporting the four activities (identified in the final column in Table 6.4) that MIPAC was intending to support. Given the complexity of the IPP network, this would have been a tedious task. An alternative was form of display was sought and the device known as the Maltese cross was devised.

Table 6.4. IPP data inputs/outputs – maintenance system

Information processing procedure	Data input		Data output	IPP supported activity*
	Source	Category		
Initial provisioning	Supplier Supplier Maintenance PUR	Provisioning data Technical data Operating data Operating data	Initial spares purchase Float size	(1) Obtain materials (2) Control material acquisition (3) Store materials
Defect information and serviceability control (DISC)	WAM CME CPPE DAS	Aircraft flying hours Aircraft unserviceability Work done Carried forward defects Selected system usage Planned maintenance inputs Coded listing of all defects	Listing of all defects daily Technical performance reports Operational performance reports Work performance reports	(1) Control standards and quality (1) Plan receipt of aircraft requiring maintenance (1) Know what maintenance is to be undertaken (1) Control the maintenance task (2) Obtain materials
Major inputs planning and control (MIPAC)	WAM CME CTCLS	Aircraft flying hours Aircraft modification status Component time/life status Scheduled maintenance requirements	Work package (work to be done) Material requirements Component time/life records Aircraft modification status	(1) Plan receipt of air craft requiring maintenance (1) Undertake maintenance (1) Obtain materials (1) Know what maintenance is to be done
Dock control	WIPAC WA	Work package Scheduled work done Unscheduled work done	Daily work progress report Unscheduled additions to work package Analysis of work and performance	(1) Undertake maintenance (1) Control the maintenance task (1) Know what maintenance is to be done (2) Plan receipt of aircraft requiring maintenance (2) Control standards and quality (2) Obtain materials

*(1) indicates primary; (2) indicates secondary.

The Maltese Cross – Structure and Assembly

In essence, the Maltese cross is a four-part matrix. The upper half contains the activities taken from the activity model (derived in stage (1) of the approach described in the previous subsection) together with an indication of the activity-to-activity information flows (stage (2)). The lower half contains a statement of the existing information processing procedures (IPPs). Although similar in appearance to the data matrix of Honeywell (Kanter, 1970) and the information cross of IBM (Orsey, 1982), it must be emphasized that the Maltese cross is different in both assembly and purpose.

Figure 103 illustrates the structure of the Maltese cross. The north axis is a listing of the set of activities making up the 'primary task' system relevant to the particular area or organization under review. The east and west axes are identical and contain the information categories deemed essential for the support of the activities at this level of resolution. The west axis (representing inputs) is the mirror image of the east axis (representing outputs). The south axis is a listing of the information processing procedures (automated and manual) and represents the existing state of the information processing network prior to the review. If the purpose of the review is to examine the potential for computer-

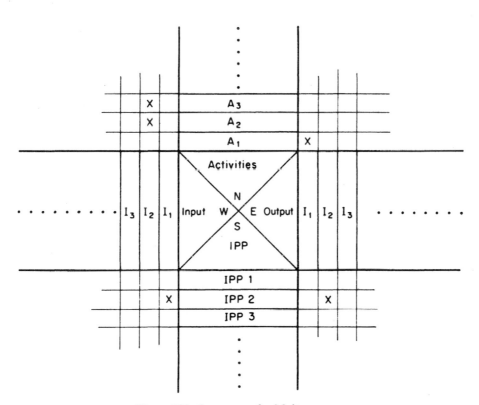

Figure 103. Structure of a Maltese cross

based processing of an existing manual network, the lower half of the Maltese cross will represent the complete manual systems, illustrating their scope and interactions. If the situation is entirely 'green field, the lower half will be blank.

Referring to Figure 103, the × in the south-western matrix indicates that data belonging to information category I_1 is used by IPP 2 to produce a processed output in the information category I_2, (south eastern matrix). The north-western matrix shows that this information category I_2 is requird as input to both activities A_2 and A_3. The × in the north-eastern matrix shows that the information category I_1 is produced by undertaking activity A_1, and hence this activity (or the manager responsible for activity A_1) has the capability of updating the category and thus providing timely data as input to IPPs. The significance of the two × s in the north-western matrix is that since they show that I_2 is an essential input to both A_2 A_3, the managers responsible for those activities must have access to this particular output of IPP 2. In practice this may not be the case, particularly if the development of IPP 2 had been initiated by only one of them. If the same manager is responsible for both A_2 and A_3, this is not likely to be a problem.

The Maltese cross is completed by filling in all the × s in the north-western and north-eastern matrices to give a complete picture of the activities and the activity-to-activity information flows deemed relevant to the area of concern in stage (2) of this process. In the south-western and south-eastern matrices a picture is obtained of all the IPPs used to process data and the data processed. Relating the bottom half of the Maltese cross to the top half will enable the set of questions to be asked of the whole information processing network directed by the existence of a potential lack of coherence indicated by a number of × s in the same columns.

Returning now to the project and our present concern only for the maintenance system, a Maltese cross was constructed and is partially reproduced as Figure 104. (It is partial because all of the information categories have not been included.) The horizontal axis represents the information categories; the right-hand half being identical to the left-hand half. The upper vertical axis lists the conceptual model activities contained within the maintenance system boundary and the lower vertical axis contains the listing of the relevant existing IPPs. The elements in the north-western matrix indicate those information categories which are inputs to the conceptual model activities and the elements in the north-eastern matrix indicate the information categories which are outputs. Thus the upper half of the Maltese cross represents the conceptual model and its information flows. The bottom half represents the IPPs. The south-western and south-eastern matrices represent the inputs and outputs of the IPPs at this particular resolution level. Thus a mapping of the IPPs on to the conceptual model can be achieved by relating the bottom half of the Maltese cross to the top half.

To illustrate the use of the Maltese cross, assume again that the IPP, MIPAC, is proposed by a manager as an addition to the existing network. It is entered into the maltese cross alongside the existing IPPs. The required input data

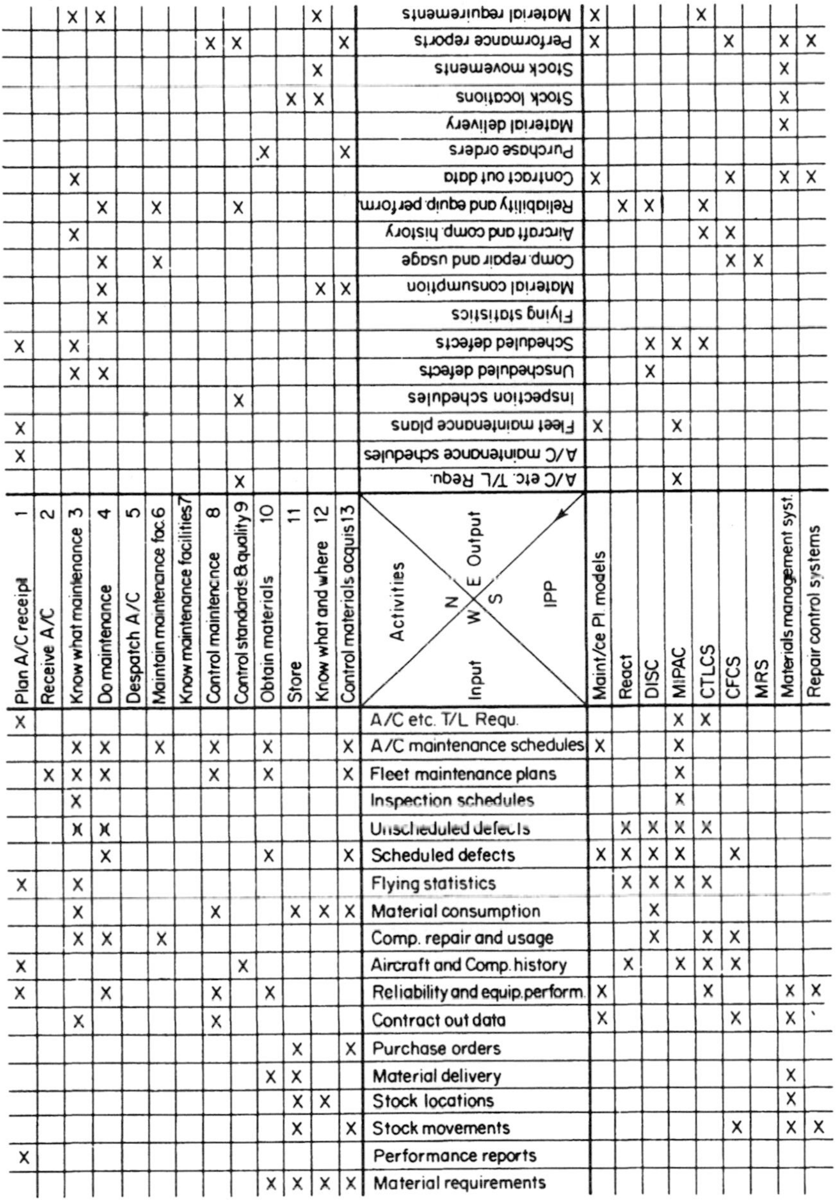

This is a Maltese cross diagram. The central box reads:

	Activities	
Input	N / W — E / S Output	
	IPP	

The diagram cross-references inputs (left columns and top-right columns) and outputs (right of center) against activities and systems (rows).

Top-left block — Activities (rows 1–13) against input columns. Input column headers (left to right):

Material requirements, Performance reports, Stock movements, Stock locations, Material delivery, Purchase orders, Contract out data, Reliability and equip. perform., Aircraft and Comp. history, Comp. repair and usage, Material consumption, Flying statistics, Scheduled defects, Unscheduled defects, Inspection schedules, Fleet maintenance plans, A/C maintenance schedules, A/C etc. T/L Requ.

Activity	Material requirements	Performance reports	Stock movements	Stock locations	Material delivery	Purchase orders	Contract out data	Reliability and equip. perform.	Aircraft and Comp. history	Comp. repair and usage	Material consumption	Flying statistics	Scheduled defects	Unscheduled defects	Inspection schedules	Fleet maintenance plans	A/C maintenance schedules	A/C etc. T/L Requ.
Plan A/C receipt 1	×						×	×				×						×
Receive A/C 2																×		
Know what maintenance 3						×			×	×	×		×	×	×	×		
Do maintenance 4						×			×			×	×		×	×		
Despatch A/C 5																		
Maintain maintenance fac. 6								×								×		
Know maintenance facilities 7																		
Control maintenance 8					×	×			×					×	×			
Control standards & quality 9							×											
Obtain materials 10	×			×		×			×				×		×	×		
Store 11	×		×	×	×	×				×								
Know what and where 12	×		×	×						×								
Control materials acquis. 13	×		×			×				×		×		×	×			

Bottom-left block — Systems (rows) against the same input columns:

System	Material requirements	Performance reports	Stock movements	Stock locations	Material delivery	Purchase orders	Contract out data	Reliability and equip. perform.	Aircraft and Comp. history	Comp. repair and usage	Material consumption	Flying statistics	Scheduled defects	Unscheduled defects	Inspection schedules	Fleet maintenance plans	A/C maintenance schedules	A/C etc. T/L Requ.
Maint/ce Pl. models							×	×					×					×
React									×			×	×	×				
DISC										×	×	×	×	×	×			
MIPAC									×			×	×	×	×	×	×	×
CTLCS							×	×	×			×		×				×
CFCS		×				×			×	×			×					
MRS																		
Materials management syst.			×	×	×				×	×								
Repair control systems							×			×								

Top-right block — Activities (rows 1–13) against output columns. Output column headers (left to right):

A/C etc. T/L Requ., A/C maintenance schedules, Fleet maintenance plans, Inspection schedules, Unscheduled defects, Scheduled defects, Flying statistics, Material consumption, Comp. repair and usage, Aircraft and comp. history, Reliability and equip. perform., Contract out data, Purchase orders, Material delivery, Stock locations, Stock movements, Performance reports, Material requirements

Activity	A/C etc. T/L Requ.	A/C maintenance schedules	Fleet maintenance plans	Inspection schedules	Unscheduled defects	Scheduled defects	Flying statistics	Material consumption	Comp. repair and usage	Aircraft and comp. history	Reliability and equip. perform.	Contract out data	Purchase orders	Material delivery	Stock locations	Stock movements	Performance reports	Material requirements
Plan A/C receipt 1	×	×			×													
Receive A/C 2																		
Know what maintenance 3			×	×				×		×								×
Do maintenance 4			×		×	×	×		×									×
Despatch A/C 5																		
Maintain maintenance fac. 6						×		×										
Know maintenance facilities 7																		
Control maintenance 8																	×	
Control standards & quality 9	×		×							×							×	
Obtain materials 10												×						
Store 11																	×	
Know what and where 12					×										×	×		×
Control materials acquis. 13					×							×		×			×	

Bottom-right block — Systems (rows) against the same output columns:

System	A/C etc. T/L Requ.	A/C maintenance schedules	Fleet maintenance plans	Inspection schedules	Unscheduled defects	Scheduled defects	Flying statistics	Material consumption	Comp. repair and usage	Aircraft and comp. history	Reliability and equip. perform.	Contract out data	Purchase orders	Material delivery	Stock locations	Stock movements	Performance reports	Material requirements
Maint/ce Pl. models			×							×							×	×
React													×					
DISC				×	×					×			×					
MIPAC	×		×			×												
CTLCS					×					×				×	×			×
CFCS					×	×				×			×				×	
MRS						×												
Materials management syst.										×					×	×	×	×
Repair control systems										×							×	

Figure 104. The Maltese cross for 'aircraft maintenance system'

(related to the appropriate information categories) are plotted in the south-western matrix and the proposed output data are plotted in the south-eastern matrix. Thus in the south-western matrix a row identifies the data set required at this level and a column indicates the interactions with existing IPPs. Scanning the columns in the south-western matrix will provide the requirements for the setting up and operation of various data bases. Thus, a large number of entries in a particular column could form the argument for the formation of a data base relevant to that particular data set, together with the definition of those IPPs which need to access it. An examination of the information categories in the north-eastern matrix defines those activities which have the responsibility for monitoring and updating the data content. If, for example, a data base had been formed for 'unscheduled defects' (fifth column from the central axis), it is immediately apparent that MIPAC would need to access it and, from the north-eastern matrix, it can be determined that activities 3 and 4 (or the particular part of the organization in which those activities lie) have the responsibility for providing the basic data. Hence questions would need to be answered in relation to that data base, such as

(a)   Is any modification to the data base, or the means of accessing it, required to accommodate MIPAC?
(b)   Is the data available in a format acceptable to MIPAC?

The most significant interactions with existing IPPs are identified through an examination of the south-eastern matrix. For example, the second output from MIPAC is seen to be in the data category 'Fleet maintenance plans'. This same category is also an output from the maintenance planning models, hence the set of questions concerned with potential duplication and consistency need to be answered before the new IPP (MIPAC in this example) is designed. Thus several entries in a column in this matrix represent potential duplication of data processing and need to be examined if an efficient development of information processing procedures is to be achieved.

The final question, in the set referred to previously, relates to coherence: 'Is the data provided by the IPP required as a support to activities other than those for which it is to be, or was, designed?' This can be examined by taking the output data set from the south-eastern matrix and by reading off (from the north-western matrix) those activities to which the same information categories form the inputs. Thus, using the previous example, 'fleet maintenance plans' represent output data from MIPAC; referring to this as an activity input category in the north-western matrix yields the set of supported activities – 2, 3, 4, 8, 10, and 13. Where an activity output is in the same category (north-eastern matrix), in this case activity 1, this means that the IPP is performing some, or all, of the activity itself. Thus the north-eastern matrix not only indicates the responsibility for maintaining and updating the data content in each category but, as in this example, it also indicates the responsibility for maintaining the IPP as well.

In addition to using the Maltese cross as a means of highlighting interactions and potential duplication of data processing, it can also highlight those areas in which information system development might be undertaken. Referring to the south-eastern matrix of Figure 104, it can be observed that none of the IPPs listed produce an output 'material consumption' (column 8). From the north-western matrix it is seen that this information category is a necessary input to activities 3, 8, 11, 12, and 13. Thus either the matrix is incomplete or this particular piece of information is transmitted informally. If the latter is the case, then it could be argued that essential information should be part of the formal information network. The argument is not being put forward that all information should be part of a formal network, but since the model represents minimum necessary activities, the information categories which appear on the Maltese cross are the minimum necessary to support the activities. Thus these are the information categories that should be provided by the formal information network. Informal communication would still be retained, of course, as this represents a significant feature of the social structure.

Within the project a number of Maltese crosses were produced. It is not necessary to produce one large cross that represents the highest level of resolution since this would become unmanageable. It is better to produce one for the lowest resolution level at which information categories are derived and then to produce subsidiary Maltese crosses for each of the subsystems at the next level. The relationships between them are maintained by the Maltese cross at the lower resolution level. Once the Maltese crosses have been produced, changes to the information processing network that are desired (for whatever reason), can be entered on to the appropriate crosses and the resulting interactions examined for coherence. We found it useful within the project to differentiate between information that was necessary to do the activity (operational information) and that which was required to control the activity (control information).

Separate Maltese crosses can be produced, and were produced, for each of these types of information. So far everything that has been produced, including the Maltese cross, is independent of any existing organization structure. Before the design of any changes to the IPP network can be undertaken, they need to be related to the needs of the managers themselves. The activity-to-activity information flows (represented by the top half of the Maltese cross) can be converted to role-to-role information flows by undertaking an organization's mapping (using the overlay technique) described previously. This I prefer to do on the primary task model itself and then to transfer the managerial responsibilities on to the Maltese cross either by ordering the activities in the north vertical axis according to responsibility or by annotating the activities by the symbols M_1, M_2, etc.

General Lessons

The Maltese cross has turned out to be a useful device, both for assessing an existing IPP network and for examining a number of groupings of potential

230

data processing activities prior to a definition of a set of computer-based data processing packages. However, within this project, another lesson emerged related to primary task modelling. During the organization mapping stage, it became apparent that a number of activities could be annotated with the initials of more than one manager. This could have represented a multiplicity of decision-taking resulting from a poor definition of responsibility, i.e. they could have been island activities. However, this was not the case. British Airways undertook certain maintenance activities at Heathrow (London) and at Ringway (Manchester). Also, the responsibility for maintaining British aircraft was different to that for maintaining American aircraft. Since the primary task model was in terms of *minimum* necessary activities, this real-world distinction was not apparent. It became necessary, therefore, to repeat these activities in the model if it was to relate to the particular 'how' at the stage of organization mapping. This process of expansion and mapping is illustrated by Figure 105. The particular 'how' is being taken as a 'hard' constraint, (softer constraints should not necessarily be taken as given). The existence of multiple activities of this kind in a primary task model usually means that other activities need to be imported into the model (concerned with their co-ordination), which cannot be defined on the basis of the original root definition, i.e. activity B in Figure 105.

It could, of course, be argued that the hard constraints could be built into the original root definition, but in a complex situation such a root definition would itself be highly complex, difficult to model and could only be derived from many iterations. It is suggested that a more useful process when faced with multiple activities of the same kind is to ask the question 'What coordinating activities are required to enable the group of identical (at the "what" level) activities to appear as a single activity in their relation to the rest of the model?'

Another feature of primary task modelling which emerged at the mapping stage is related to the existence of 'non-validated' activities in the model. One way of doing the comparison of primary task activities with the real situation

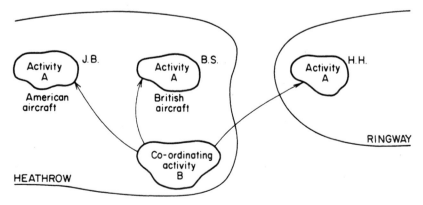

Figure 105. Organization mapping with multiple responsibilities. The initials J.B., B.S., and H.H. refer to managers with decision-taking responsibility for activity A

is through an identification of decision-takers. If all the activities in the model can be annotated with the initials of a real-world manager, then it is reasonable to take the model as 'validated'. It sometimes happens that a small number of activities cannot be so annotated. If their removal from the model would destroy its coherence, then it is useful to leave the activities in (perhaps with a dotted boundary) and to investigate the reasons for their omission. It could be that the decision-taking is so dispersed in the organization that the mechanism has not been identified, or maybe the activity does not happen. This latter reason is difficult to accept in a primary task model related to the current state of an organization, unless the activities concerned are associated with monitoring and control. These do not affect the existence of the output but merely affect the performance with which the output is achieved.

An Approach to Analysis

The purpose of the analysis phase is to answer the question, 'Who, in terms of role, needs what information for what purpose?' It is not, at this stage, concerned with how that information might be provided. The design phase decides whether the information is processed by computer or by manual methods, the source of the data, and such things as whether that data is contained within a central or distributed data base.

The approach described here, which has emerged from the projects discussed in this chapter, makes the assumption that it is sensible to derive the information needs on the basis of a model of the particular organization which is independent of the organization structure and then, and only then, to relate the information flows to the existing set of management roles.

In broad terms the approach consists of the following stages:

(1) Develop an activity description of the organization (or part of the organization) under review, i.e. a 'primary task' model. This process was discussed in detail in Chapter 3. Dependent upon the scale of the study, it may be necessary to derive a number of activity models at several levels of resolution in order to fully describe the information needs.

(2) Derive the categories of information required to support the activities in the models and the particular activities from which this information can be obtained.

(3) For a particular organization structure, define management roles in terms of the activities for which each existing role-holder has decision-taking responsibility. (If the organization structure is not a constraint, the overlay technique can be used to enable these roles to be defined on the basis of the description developed in stage (1) above).

(4) Use these role definitions to convert the 'activity-to-activity' information flows, in stage (2) above into 'role-to-role' information flows, i.e. define the particular information needs of a manager based upon this analysis of the activities for which he is responsible.

(5) Define the information systems needed to match the performance needs of the activities each system is supporting, so that a coherent network can be developed which makes efficient use of computing or manpower resources.

In summary, stage (1) defines what activities must be on-going for the system to function. Stage (2) defines the minimum information needed to support these activities. Stage (3) defines who (in terms of role) is responsible for what set of activities (i.e. the job specification). Stage (4) defines the minimum information flow pattern, i.e. who is responsible for supplying what information to whom (the purpose having been supplied by stage (2). Stage (5) defines the set of information processing procedures (automated and others) which represent efficient use of resources. This is a major stage which includes the design process. However, for the purpose of this description, the emphasis is placed on the process of analysis prior to design, since, unless this is done well, the final outcome will be less than satisfactory, however excellent the design.

It is seldom that a study of this kind is undertaken in a 'green field' situation; therefore some means will be required of displaying the information processing procedures already in existence, together with their interactions, so that fully informed decisions can be made about which procedures to keep, which to create or develop, and which to discontinue. It is for this particular decision process that the 'Maltese cross' display was derived.

Having described the Maltese cross as a device within this methodology, it is now reasonable to return to the whole process and illustrate where the device fits and the stages in its assembly. Figure 106 is a diagrammatic representation of the methodology as a whole (Wilson, 1987).

The first stage in this methodology is concerned with the derivation of a consensus primary task model as described earlier in Chapter 3. Since this type of model is used as the basis for the definition of information requirements they become highly particular to the organization for which they are derived. The set of activities within this model present a unique statement of what the organization is seeking to undertake based upon the analysis of the relevant set of Ws.

Each of these activities, in turn, is used to derive the input information needed to support the activity, together with the information output produced by doing the activity. The activities, the input, and the output information categories are then used to construct the top half of the Maltese cross. Two separate procedures are then undertaken. One is to treat the existing information processing procedures in the same way as the activities, i.e. to derive the information categories that the input data belong to and to relate the output data to the information categories. Once this is complete, the bottom half of the Maltese cross can be assembled. The second procedure is to map on to the primary task model the set of organizational roles that have decision-taking responsibility for the activities. This mapping can then be used to convert the 'activity-to-activity' information flows into 'role-to-role' information flows. The output of

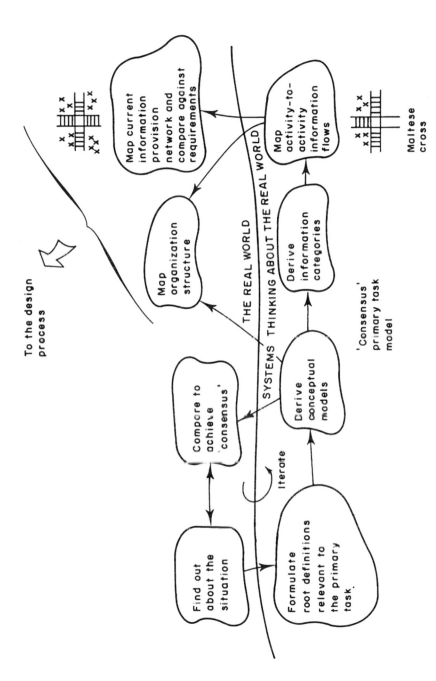

Figure 106. A methodology for information requirements analysis

the analysis phase is in the form of a definition of who (in terms of role) needs what information for what purpose (in terms of the activities for which they are responsible) and also who supplies it, together with a picture of how these needs are currently met by the existing set of information processing procedures. The design process is then initiated by the decisions which have to be made concerning how well the total needs are being met, what modifications to existing procedures might be made, those procedures that could be discontinued or replaced, and those procedures which need to be created. These decisions are made with respect to their effect on the total situation and hence will enable developments in the information processing network to remain coherent. As the development take place the Maltese cross is updated, and thus it always represents a picture of the current state.

Although the Maltese cross is shown here as an integral part of the methodology, whether it is used or not is a matter of judgement. It is no more than a tool which is available and the criterion for its inclusion must be its usefulness in the particular project. Work was also done within British Airways, concerned with control information in a particular area, where its use seemed inappropriate. The outcome of this part of the project was the design of an IPP which consisted of a prescribed set of meetings – daily, weekly, and monthly – in which the agenda items were precisely defined on the basis of the control information required and its frequency. Here the primary task model was derived and the organization mapping completed. The control information categories were defined on the basis of the constraints that affected the particular managers' freedom of action.

There were effectively no existing IPPs relevant to the information so defined and hence the construction of a Maltese cross was not seen to be a useful stage. In other projects, where the concern has been about the planning of information system development, the Maltese cross has been an essential element.

Maltese Cross

Since the development of this device within the British Airways project it has been used for a number of purposes and in order that this purpose may be appreciated it is worth going through the construction of a Maltese cross in more detail in order that it may be fully understood. A number of examples will be considered, each one representing an increase in complexity.

Supposing that a primary task model relevant to a manufacturing or a production process has been developed. It is likely that it will contain a group of activities concerned with the control of quality of the product. This could be as shown in the Figure 107.

The first task is to construct a table in which the information categories are listed corresponding to the inputs, the outputs and measures of performance related to each activity. The process of constructing this table is to complete the input row first, i.e. define all of the support information categories. The most obvious way of proceeding to fill in the remainder of this table is to proceed to a derivation of the outputs for each activity. This, however, could

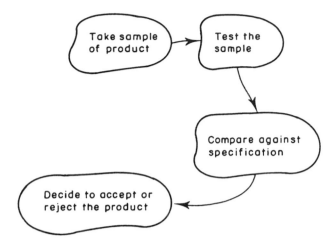

Figure 107. Sample–test system

be very inefficient as one may be spending considerable time thinking of plausible outputs from each activity that do not in fact get used. A more efficient way of proceeding, having derived all of the support information, is to focus attention on the information categories themselves and to ask for each one of them which of the activities would generate this information as output (it may also be the case of course that some of the information inputs are derived externally to the system and hence can not be located as an output of an activity). Thus in completing this process one is only identifying the information which actually gets used.

In the 'measure of performance' row the information categories are derived which satisfy the three Es (see 'Information for Control', later) though it is usually the case that the measure of effectiveness is not included for each activity but is a single measure required for the system as a whole. This is because each activity is a necessary part of the system on the basis of the root definition and hence it will be the case that if each activity is efficacious it will also be effective.

To return to the example of the sample test system the table of information requirements is given in Table 6.5.

In constructing the third row (i.e. the measures of performance) it may be the case that some of these measures may not be measurable in a practical sense. It is still important to include them since in practice one may be able to find indicators relevant to the measures which can be measured. For example, if the samples are taken by works inspectors and the tests are done in some quality control laboratory, 'sample routine discipline' (which itself is not measurable) might be reflected in complaints by the laboratories about the timeliness or standard of the samples that they receive. It is frequently the case that it is easier to collect information which indicates when an activity is being done badly than to collect information about when it is being done well. Thus complaints and complaint analysis are useful sources of control information.

Table 6.5 Information requirements

Information	Activity			
	Take sample of product	Test the sample	Compare against the specification	Decide to accept or reject the product
Inputs	• Product identification • Sampling routine (size, frequency, time) • Location of sample • Destination of sample	• Product sample I.D. • Test routine (parameters, process of testing etc.) • Destination and format of test results	• Product sample I.D. • Test results • Specifications • Standards	• Product sample I.D. • Comparison results • Criteria for decision • Destination for the decision
Output	• Product and sample identification (ID)	• Product sample ID • Test results	• Product sample ID • Comparison results	• Product sample ID • Decision to accept or reject
Measures of performance	• Sampling routine discipline • Cost of sampling	• Accuracy and completeness of test results • Timeliness • Cost of testing	• Accuracy of comparison • Timeliness • Cost of comparison	• Correctness of decision • Timeliness • Cost of making the decision

Tables of the kind described above provide a way of deriving and displaying the information needs of a set of activities. However, it is not a very convenient format when it comes to comparing this information requirement with the information (processed data) that is already provided. It is usually for this purpose that the Maltese cross is constructed. It initially looks complicated but it really just consists of four matrices put together. The upper two matrices are simply derived from a table similar to that illustrated above but split into two separate matrices. Matrix 1 describes the inputs to the activities and appears as the north-western matrix in the Maltese cross. This is illustrated below.

13	12	11	10	9	8	7	6	5	4	3	2	1	
	X		X					X	X				Decide to accept or reject
		X		X	X			X					Compare against specification
					X	X	X	X					Test the sample
								X	X	X	X	Take sample of product

Activities

Information categories

| 13 Decisions | 12 Comparison results | 11 Test results | 10 Decision criteria | 9 Standards | 8 Specifications | 7 Test results format | 6 Test routine | 5 Product/sample ID | 4 Destination information | 3 Location information | 2 Sampling routine | 1 Product ID |

Matrix – 1 inputs

Matrix 2 describes the outputs from the activities and appears as the north-eastern matrix in the Maltese cross.

Activities	1	2	3	4	5	6	7	8	9	10	11	12	13
Decide to accept or reject					X								X
Compare against specification					X							X	
Test the sample					X						X		
Take sample of product					X								

Information categories

| 1 Product ID | 2 Sampling routine | 3 Location information | 4 Destination information | 5 Product/sample ID | 6 Test routine | 7 Test results format | 8 Specifications | 9 Standards | 10 Decision criteria | 11 Test results | 12 Comparison results | 13 Decisions |

Matrix – 2 ouputs

If matrix 1 and matrix 2 are put together about the 'activities' axis the top half of the Maltese cross is formed and represents a display of the information that needs to flow into and out of the activities in the area of the organization of interest. Thus the top half of the Maltese cross displays 'the information requirements'.

The bottom half of the Maltese cross is similar in that it represents the input and output data (related to the information categories) but this time for the existing 'information systems'. Although it is common usage to refer to information systems what is actually being referred to are 'processed data systems'. these may be computing packages or they may be manual systems but they represent formal data processing.

As mentioned above it is assumed that the data (both raw and processed) can be identified with the information categories through their definitions. The bottom axis has a listing of the names of the formal information systems and hence elements can be identified in the south-western matrix corresponding to the data used by the information systems and elements can be identified in the south-eastern matrix which correspond to the processed data produced by them. The bottom half of the Maltese cross (and in particular the south-eastern matrix) represents the provision of processed data via the existing information network. Since the top half represents requirement, comparison of the bottom half with the top (i.e. provision versus requirements) provides the basis for changes to the the existing network.

Suppose that in the real world of the quality control system described above the following computer packages were used:

PRODSPEC: a listing of all products and their specifications.
PRODTEST: a listing of quality testing routines for all products.
SAMPROD: definition of sampling routines for all products.

then the completed Maltese cross would appear as given in Figure 108.

The south-western matrix shows that basic data about products, sampling and test routines and specifications is inputted to the three programs where it is stored and processed to provide the listings referred to and displayed in the south-eastern matrix. Thus the south-eastern matrix shows what is provided by the current formal 'information' systems and this can be compared with the information that is required. This is displayed in the north-western matrix.

The comparison shows that information categories in columns 3, 4, 5, 7, 9 10, 11 and 12 are not provided for. The analyst can question therefore whether or not these additional items of information should be designed into formal 'information' packages. It might be argued that location and destination information (columns 3 and 4) is 'one-off' and known about and would not change until the location of the works and/or quality control laboratories changed.

However it might be argued that it would be useful to modify PRODTEST to provide an output containing the results format (column 7) as well as the test routine.

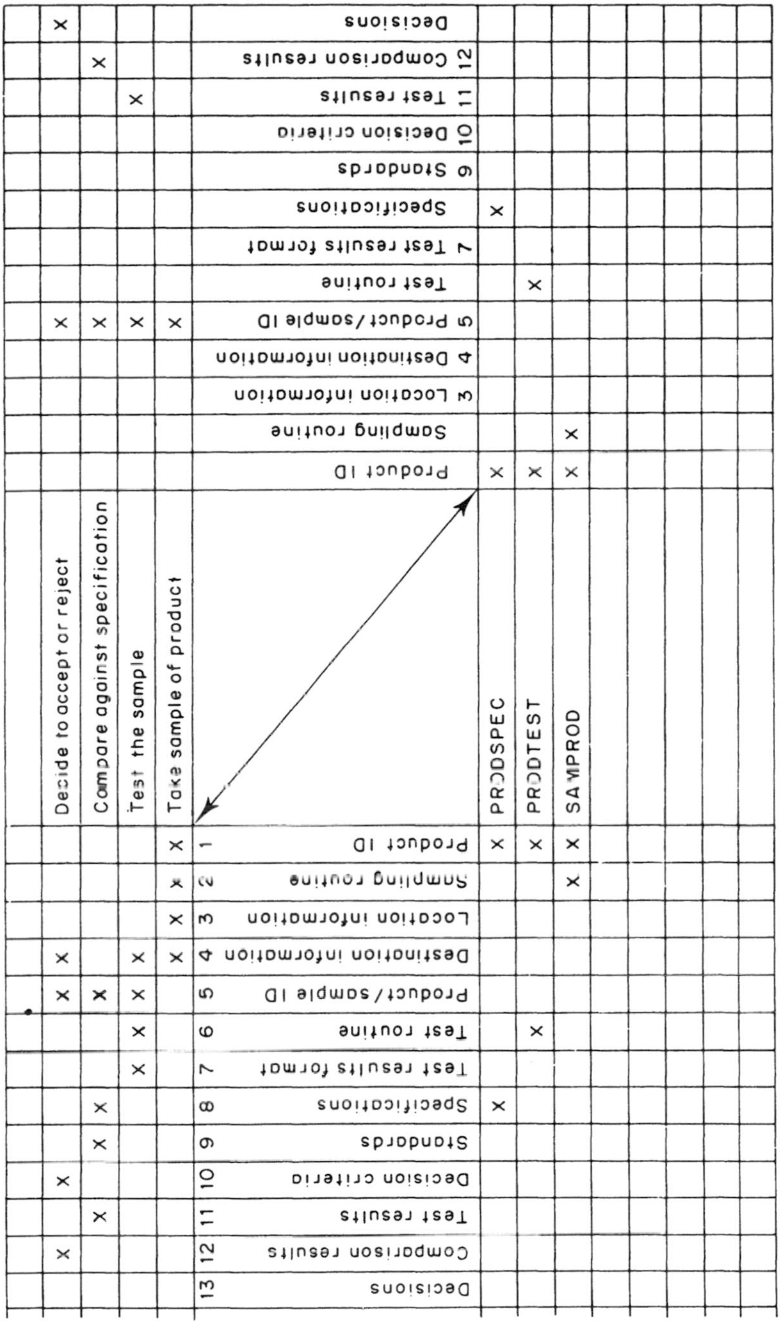

Top labels (reading): Decisions · Comparison results 12 · Test results 11 · Decision criteria 10 · Standards 9 · Specifications · Test results format 7 · Test routine · Product/sample ID 5 · Destination information 4 · Location information 3 · Sampling routine · Product ID

Left labels: Decide to accept or reject · Compare against specification · Test the sample · Take sample of product

Bottom-center list:
1 Product ID
2 Sampling routine
3 Location information
4 Destination information
5 Product/sample ID
6 Test routine
7 Test results format
8 Specifications
9 Standards
10 Decision criteria
11 Test results
12 Comparison results
13 Decisions

Center column labels: PRODSPEC · PRODTEST · SAMPROD

Figure 108. Maltese cross

It could also be argued that it might be useful to design a formal 'information' system to assemble historical records of comparison, test results and standards (columns 12, 11 and 9) along with product/sample ID (identification) (column 5).

Olympic Games

In this example consider a 'green field' situation in which, say, the International Olympic Committee (IOC) has requested an analysis of its operational information systems with a view to possible computerization. The first stage in an investigation of this kind would be to produce a consensus primary task model in the way that is described in Chapter 3. As it is the operational information that needs to be identified a useful starting point would be the derivation of a neutral primary task model.

In an attempt to make such a model both neutral and, at the same time, rich enough to represent the essence of the situation it is useful to start off by considering those features which need to be taken as given. So that for this to be the Olympc Games rather than any other athletic event a number of features can be introduced into the root definition through the E of CATWOE. These could be as follows: the events which form part of the games are prescribed and are international, the athletes who take part are all national representatives, the events take place every four years, and the location and timing are specified. It is also worth recognizing that gold, silver and bronze medals are awarded for each event. As the operations of the event are part of the total activity undertaken by the IOC it is reasonable to make them the owners of this system.

Given the above description a root definition of this operational system which seeks to be neutral is as follows (the CATWOE analysis and conceptual model are also given):

An IOC owned system to mount a set of prescribed athletic events in order to award gold, silver and bronze medals to nationally representative athletes on the basis of their performance in those events; the events to be international and held every four years at a time and place specified by the IOC.

C – nationally representative athletes
A – not specified
T – to mount prescribed set of athletic events
W – prescribed events are a suitable way of assessing performance and it is reasonable to award medals to recognize that performance
O – IOC (International Olympic Committee)
E – Events are prescribed and international, athletes are national representatives, the events take place every four years and the location and timing are specified.

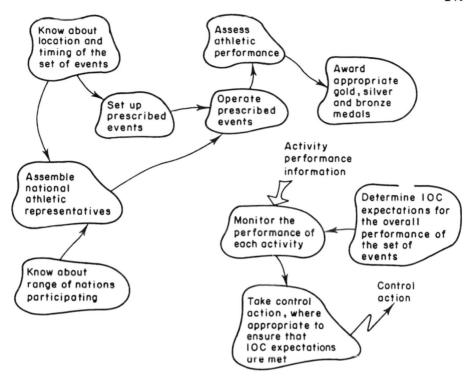

Using the operational activities in this model a table of inputs, outputs and measures of performance can be generated (illustrated below).

	Activities				
	Set up prescribed events	Operate prescribed events	Assess athletic performance	Award appropriate gold, silver and bronze medals	Assemble national athletic representatives
Inputs	• list of events • no. of entrants per event • location	• event details • officials' designation • participant location	• criteria for assessment • event results	• ceremony timing • recipient medal lists	• list of participating countries • participant list • residence details

(*cont.*)

	Activities				
	Set up prescribed events	Operate prescribed events	Assess athletic performance	Award appropriate gold, silver and bronze medals	Assemble national athletic representatives
Outputs	• event details, timing, no. of heats, location of event, athletes etc. • officials' designation judges, starter assistants etc.	• event results	• recipient medal lists	• summation awards/ country	• participant list/event/ county • participant location
Measures of per- formance	• event details unclear • lack of officials • cost of doing the activity	• evidence of mal- operation (delays, confused results etc.) • cost of doing the activity	• inadequate or missing lists • cost of doing the activity	• ceremony delays • incomplete awards • cost of doing the activity	• mal-operation of the assem- bly procedure • cost of operating the procedure

From this table a list of categories can be formed and for each information category the data model can be defined. The list is given below and an example of a data model is given for the first category.

List of information categories

- Event details
- Officials' designation
- Event results
- Criteria for assessment
- Medal lists
- Ceremony timing

- Awards summation
- Participating countries
- Participant list
- Residence details
- Participant location

Example data model

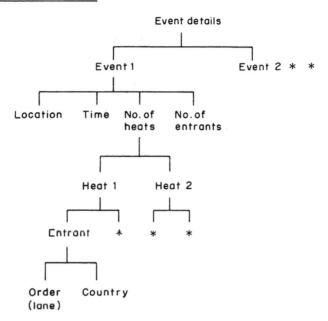

Using the information categories in the above list the Maltese cross can be structured (illustrated overleaf using operational activities only).

It was stated above that this was a 'green field' situation and hence the bottom half of the Maltese cross now presents a picture of each of the information categories needing to be produced, a data transformation process (DTP) for each one and the identification of the basic data which needs to be accessed by each DTP (south-western matrix). The design process which follows needs to examine the algorithm contained in each DTP in order that the basic data input can be transformed into the processed-data output and can also examine how the DTPs could be combined to form IPPs. For example DTP 6 and 7 could be combined as they use virtually the same basic data. Also DTP 3, 4, and 5 could be combined on the basis of the potential use of the output processed data. Other combinations could also be explored on the basis of various criteria. Such an exploration, however, could only be done in relation to the actual situation and with the actual people concerned. Three of the information categories are derived from outside the system boundary and hence their production is not of concern to this system: they are therefore specified as external inputs.

DTP 8	DTP 7	DTP 6	DTP 5	DTP 4	DTP 3	DTP 2	DTP 1		Set up events	Operate events	Assess performance	Award medals	Assemble participants
								Participant location		×			
×								Residence details					×
×	×	×			×			Participant list					×
	×	×						Participating countries					×
								Awards summation					
								Ceremony timing				×	
		×						Medal lists				×	
					×			Criteria			×		
				×				Event results			×		
						×		Officials' designation		×			
			×	×	×	×	×	Event details	×	×			
							×	Event details	×				
						×		Officials' designation	×				
					×			Event results		×			
External input								Criteria					
				×				Medal lists			×		
		×						Ceremony timing	×				
	×							Awards summation				×	
External input								Participating countries					
	×							Participant lists					×
External input								Residence details					
×								Participant location					×

Information Categories and Data

In all of the examples using the Maltese cross *information category* is a term that has been used to link both information and data. Initially this may appear confusing but it is only a mechanism used in the Maltese cross to enable a comparison to be made between requirements (for information) and provision of (processed data). As indicated earlier processed data becomes information when it is used for some purpose. It is not the form and content of the processed data that is changed, it is merely interpreted in a particular way. Thus it is reasonable to define the boundary of an information category by specifying its data content. This is done through the mechanism of a data model. If this argument is accepted as reasonable then the information categories axis in the Maltese cross represents a reasonable axis about which data provision and information requirements can be compared.

It is worth, at this point, introducing a brief discussion on the language of information categories. It is the case that the information categories are derived directly from the activities in a conceptual model and hence are themselves conceptual. Before comparison can be made with the real world processed-data provision translation needs to be undertaken in which the conceptual language (determined by the analyst) is converted into the real-world language (determined by the actual managers). Without this translation the analyst cannot be sure that he is comparing like with like. It is therefore important that when the information categories are inserted into the Maltese cross they are in the language of the real world. I find that this translation is best achieved at the stage of validation of the information categories.

Although it is possible to argue that if I were doing a particular activity (see below), I would need the information support from, say, I_1, I_2, and I_3.

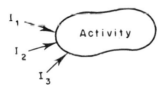

As it is probably the case that I have never done this activity I cannot be sure that these three information categories actually represent what is required. Thus the process of validation is one in which it is necessary to seek out the actual individual in the organization who does the activity and question him with regard to the information he requires. This leads to a number of possible situations:

(1) He apparently only makes use of I_1 and I_2. This may be due to a misinterpretation of the activity or a difference in the language in which he has described his information support. His description of I_1 and I_2 may encompass my definition of I_1, I_2, and I_3. Clarification of this will lead to

the appropriate interpretation and language being applied to the conceptual information categories.

(2) He may use information categories I_1 to I_4. This may also be due to a misinterpretation of the activity where a greater range may have been assumed which requires additional information support. It may also be due to the terminology difference or it may simply be that the analyst lacks sufficient experience of the activity to realize that this additional information is necessary. Again clarification of this will lead to validation of information support and the use of appropriate terminology.

The iteration described above seems to work well in practice. This is because the analyst will have given prior logical thought to the information requirements before questioning the actual individuals involved. Thus the analyst knows how to respond to the questions he is asking. This would not be the case if a simple question of the form: 'what information do you need?' was adopted.

Information for Control

I mentioned earlier that it was useful to make a distinction between the information required to *do* an activity and the information required to *control the doing* of an activity. This may be illustrated by considering the simple activity 'obtain stock'. The information required to do the activity would be in terms of the specification of what is to be stocked, the quantities required, and the time scale allowed for obtaining the stock items. The control information, however, would be dependent upon the definition that has been made of the measure of performance relative to this activity. Such a measure might include the availability of stock items and the value of stock held. Given this measure the information for control would be in terms of the number of nil-stock situations over a certain time period and the value of stock above (or below) a defined level.

Control information is, therefore, highly dependent upon the measures of performance selected for the range of activities within the area of responsibility of the controller. The activity 'define measures of performance' was seen as an essential activity within the general control model Figure 76. However, in order to define the total control information required, this model suggested that there were two aspects that needed consideration. These may be classified as 'inward looking' and 'outward looking'. The inward looking aspect is concerned with the information required by a controller to enable him to control those activities for which he is responsible (through the definition of measures of performance). The outward looking aspect is concerned with collecting and presenting that information required by a higher level controller in relation to the 'expectations' of that controller (or 'owner' in the context of a root definition of the system under consideration). By definition the 'expectations' are outside the area of authority of the decision-taking process of the system defined and as such represent 'given' or information generation constraints. Thus, to define total control information, such expectations need to be identified. This leaves the need, at any resolution level, to define measures of performance appropriate to the set of activities at that level. If our concern is with activities at an operational level, this is not too difficult. In a manufacturing process, for example, the measures of performance might be in terms of throughput or yield (measured as the number of products per unit time, tonnes per annum or conversion efficiency), product quality; (measured as customer complaints, recycle, and rejection costs), cost of resources (measured as man-hours used, materials and services consumed, etc.). The higher up the control hierarchy we move, the more difficult it becomes to specify measures of performance in such a way that the information can actually be collected. This is also affected by the nature of the enterprise itself. We discussed earlier problems associated with control in a school. At an operational level, the activities are concerned with teaching (excluding those concerned with material usage) and it is not clear what an acceptable measure of performance might be or what information might be collected to enable assessment of performance to be made. At the higher levels much of the information required is 'soft' and, hence, unquantifiable. This is one reason why much of the so-called control information is in financial terms: because it can be collected. However, although such information might be a useful indicator of the 'health' of the enterprise, it is not particularly useful as an indicator of the control action required.

Accepting these difficulties, I wish to argue, however, that help can be obtained in defining measures of performance on the basis of the set of activities that are being controlled.

In a general sense measures of performance can be related to what are known as the '3Es'. These are:

- Efficiency
- Efficacy
- Effectiveness

Relating these to the previous discussion; the first two, *efficiency* and *efficacy*, are what were described as inward looking and the third, *effectiveness*, is essentially outward looking. Other Es can be considered such as *ethical* (i.e. should the activity be done at all on moral grounds) but usually this question doesn't arise at the modelling stage since the moral issue would normally have been addressed in relation to the study itself. Some organizations also use *economy*, which is some relation between actual efficiency and theoretical efficiency. The one feature which *is* quite clear is that there is considerable confusion about the whole area of performance measurement. It is not my intention to enter into a detailed discussion of the various interpretations of performance measures as that would merely add to the confusion. Instead I will try to define a useable set of concepts of measures of performance to which we attach the labels given above as the 3Es.

It is frequently the case that it is easier to question whether something is done badly rather than to question if it is done well. For example a bad night's sleep could be measured in terms of the number of disturbances or the number of hours actually asleep compared against the norm. A good night's sleep, on the other hand, is more a question of fellings rather than of measurable indicators.

Similarly we can assemble concepts for the 3Es by asking the question (in relation to a purposeful activity within a human activity system), how could the transformation process (T) fail? (Forbes and Checkland, 1987).

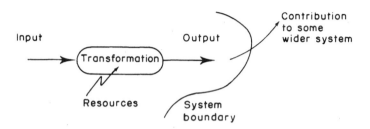

As the diagram illustrates the transformation consumes resources in order to convert a defined input into a defined output. The transformation represents the 'operational' activities within a system defined by the appropriate root definition and the 'owner' within that definition indicates the wider system to which the output is required to make a particular contribution.

The transformation could fail in three ways:

(1) It could fail to produce the output.
(2) It could produce the output but consume excessive resources.
(3) The output does not produce the required contribution to the wider system.

The first of these measures is *efficacy* and is essentially questioning the existence of the output. This is more than just its presence; it may also have to be a defined quality. The second measures *efficiency* and is concerned with the total 'cost' of doing the transformation. Cost will include the economics of the process but will also include time and effort. The third measure is the *effectiveness* and is questioning whether or not the transformation is the right thing to be doing to achieve the expectations of the owner. As an example consider the transformation: 'To dig the garden', i.e. to convert an undug garden into a dug garden. Using a lawnmower will not achieve the output, i.e. the means chosen will not work. Thus the transformation will not be efficacious. Using a knife and fork may well work but the time and effort required would be excessive; thus although efficacious the process would be inefficient. Using a spade could be both efficient and efficacious but if the 'owner' wanted the garden landscaping, the transformation process would be ineffective and the owner could choose a different process.

In the above example the 'what' was specified and different 'hows' were examined to question efficacy and efficiency. To question effectiveness, however, it is the 'what' itself that is under examination.

The 3Es represent a set of concepts, i.e. they are part of the systems thinking process. In order to illustrate them, as in the above example, it is necessary to map them on to some aspect of the real world. This is also the case when using them as part of the process of deriving the information requirements. Thus the question being addressed is: If this activity were to map on to part of the real world what would we have to measure in order to know if what was being done in the real world was efficacious, efficient and effective? If this question is being asked in relation to a consensus primary task model then effectiveness can be taken as given and the assumption made that if the real-world manifestations of the activities are efficient and efficacious then, by definition, they will be effective.

As a further example of the total process of information generation, on the basis of an activity model, let us take a less complex example to illustrate the distinction between measures of performance and control information. This is represented by Figure 109. Assume that a primary task model relevant to a hospital is that given at the top of the figure. The construction of the table is also illustrated within this figure. The row labelled, 'measures of performance' lists the components of the overall measure of performance relevant to each activity derived from considerations of efficacy and efficiency. Each one of these measures then defines the information represented by the set of arrows in the primary task model, labelled 'performance monitoring information' (or control

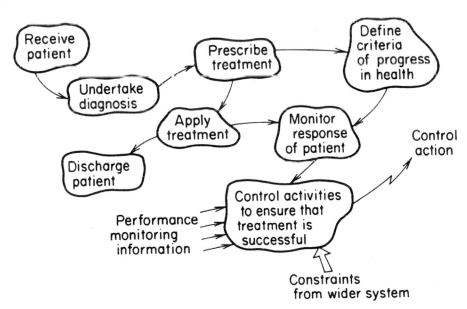

Figure 109. Information generation (see also table overleaf)

Figure 109 (*continued*)

Information	Receive patient	Undertake diagnosis	Prescribe treatment	Apply treatment	Define criteria of progress in health	Monitor response of patient	Discharge patient
			Activity				
Inputs	Patient identity; doctor's report	List of symptoms; specification of relevant tests; test reports	Priority listing of potential treatments	Specification of treatment; responsibility for application	Expected response; potential side-effects	Patient condition	Report of satisfactory condition; specification of any after-care required
Outputs	List of symptoms; bed allocation	List of potential diagnoses in order of selection; Priority listing of treatments	Selected treatment (spec.); re-sponsibility for application	Monitor patient condition	Stages of change in health; direction; time	Trends in health change; side-effects	Report to doctor; schedule of after-care
Measures of performance	Time from entry to installation; resources used; quality of listing	Resources used in testing; quality of diagnosis	Resources used in treatment; applicability of treatment	Discipline in treatment; resources used	Clear set of indicators	Discipline in identifying condition against indicators; resources used	Report accept-ability by doctor; satisfactory arrangement of after-care
Control information	Waiting time; costs; complaints about listings	Number of times diagnosis incorrect; costs	Trends in patient condition; costs	Comparison of what is applied with what was prescribed; costs	Existence of indicators that can be measured	Record of condition against indicators; costs	Doctor's complaints; patients' complaints

information). The controller uses this information to assess the degree of success that the system achieves in meeting its overall expectations (represented by the constraints). The expectations may be 'the successful removal of patients' symptoms of ill-health within a budgeted cost'. The assumption that the wider system would be interested in cost leads to the inclusion of an assessment of 'resources used' in all activities except for the definition of progress criteria (I have assumed that these would be negligible compared to the others). Assessment of the control information provided in this way enables the controller to direct control action to whichever activity is unsatisfactory. The decision on whether or not to take control action would be dependent upon the norms for performance set for each activity.

The information requirements specified by this kind of analysis can be related to the basic data available and the data processing activities specified. How these activities are then grouped will provide the specification for the information processing procedures required. If the situation is not 'green field', the Maltese cross provides a form of display for comparing requirements against what is available.

As an example of deriving an IPP, we could take the support information requirement for the activity 'undertake diagnosis'. This information is a symptom listing, specification of relevant tests, and test results. A set of specifications could be contained within a data file and, by entering the particular symptoms, the relevant set of tests could be identified and made available to the physician. Following the application of tests, the results could be entered into a program that computes the most likely set of diagnoses on the basis of a statistical evaluation. The actual data processing activities would be dependent upon the particular statistical routine used. Again this set of possible diagnoses could be made available to the physician, who would select that diagnosis that he believes to be the most likely (given his experience and other 'soft' information about the patient such as general appearance and history). Thus, in this example, computer-assisted diagnosis may be feasible; however, all of these data processing activities may be undertaken by the physician himself rather than through the particular 'how' of using a computer.

The data processing activities identified above could be combined with the output from the activity 'monitor response of patient' to provide the design specification of an IPP concerned with continuous diagnosis.

Although it is not the case in the model illustrated in Figure 109, some activity models can be expanded to the level at which data processing activities, themselves, become explicit. Take, for example, a primary task model relevant to a bookshop. One of the systems within this model may be concerned with the activity 'obtain books'. A model representing the expansion of this activity appears as Figure 110.

The activity 'decide supplier' requires, as support information, a listing of supplier names, addresses, type of books stocked, lead times on supply, etc. This need therefore implies a data processing activity associated with assembling and updating this list. The activities 'make order' and 'validate invoice', however,

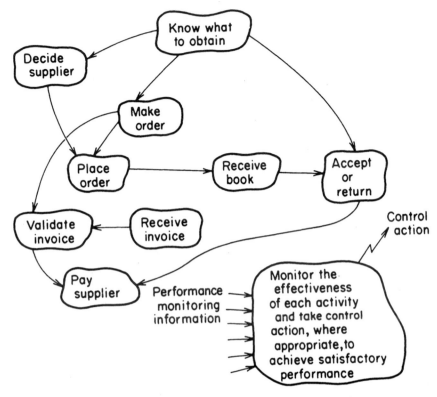

Figure 110. Model of 'obtain books'

are themselves data processing activities and, although they could be undertaken by a clerk, could also be automated.

The transition from an activity diagram of the kind shown in Figure 110 to the data processing activities required can be done directly as indicated. However, it is also possible to make the transition via a 'Gane and Sarson' type diagram (Gane and Sarson, 1979). Figure 111 represents such a data processing description relevant to the model in Figure 110. This illustrates the processing activities themselves, the sources and sinks, together with the relevant data files. What the diagram does not illustrate are the relationships that exist in terms of the use of the data within the business system. To do this a diagram of the form of Figure 110 is necessary and I prefer to develop that first as it is essential to know why data is being processed (i.e. with reference to its use) otherwise data may be generated, files defined, and files maintained which do not get used. Within Figure 111 the files, and data processing, associated with 'book order', 'books awaiting invoices', and 'invoices awaiting payment' are included as a result of the need to monitor the activities for control purposes. It has also been assumed that the activity 'pay supplier' is undertaken by the finance department so as not to complicate the diagram with the necessary financial data processing.

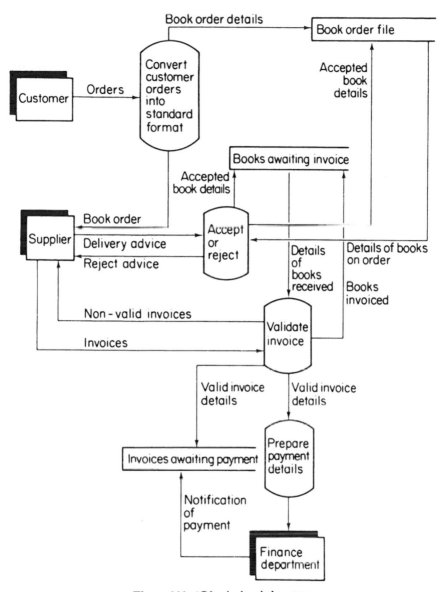

Figure 111. 'Obtain books' system

As a final example to illustrate the detailed application of this methodology, and also to return to an industrial context, I will refer to a recent project undertaken in Mexico and make use of part of the analysis that was done. The company was called Ponderosa Industrias SA (PISA) and was involved in the manufacture of wood products, chemicals, and paper and the cultivation of pine forests as the source of raw material. A number of companies made up PISA, and this in turn was part of a larger group of companies called the Chihuahua Group. This included non-industrial concerns such as banking and insurance. The project was centred on PISA and was concerned with the provision of information for the control of operations together with the supply of that information required for performance reporting to the larger group.

The first stage in the analysis was to develop a primary task model relevant to PISA, and to do this the following root definition was derived:

> A Chihuahua Group owned system for the conversion of a partially managed natural resource into defined end products and intermediates, for continued profitable sale, with a performance which meets the expectations of the owner but within group-, community-, and government-applied constraints.

Most of this definition is probably understandable, with the exception of 'partially managed'. This term was included because the short-term management of a forest was the responsibility of a local woodsman, who was not an employee of PISA, whereas the longer-term development of the forest, the provision and maintenance of machinery and roads was the responsibility of the forest group within PISA. The model resulting from this definition at the first level of resolution was in terms of a set of subsystems (since the root definition is complex, in the terms defined earlier) and is illustrated below:

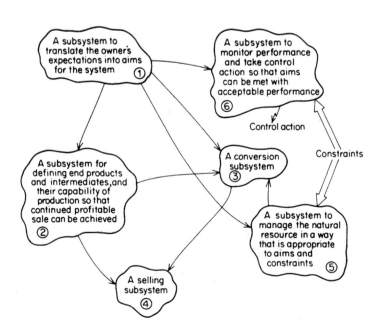

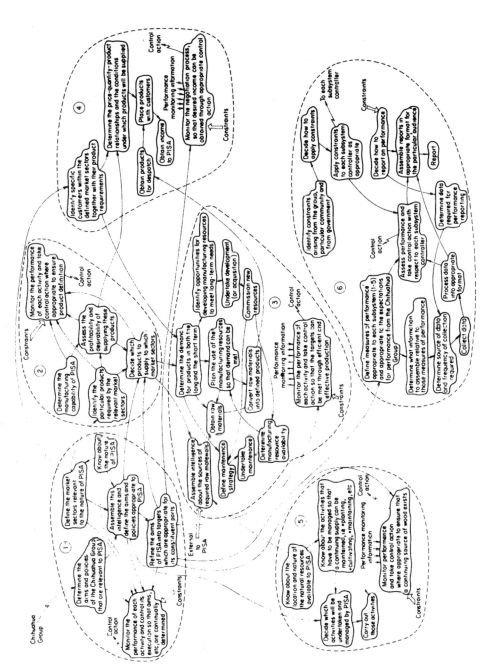

Figure 112. Primary task model relevant to PISA

256

A root definition for each of these subsystems was derived and the model expanded to the level illustrated by Figure 112. Further selected expansion was necessary to be able usefully to derive support information, but as an illustration of the process let us consider only the activity 'obtain raw materials' from subsystem 3. The root definition taken for this system was:

A PISA owned system to obtain and make available that material required by the conversion system so that production requirements can be met within inventory and other company-applied constraints.

The resultant model is given in Figure 113. For each activity within this model (with the exception of the activity 'know about material specification'), the input information categories were derived followed by the identification of those activities that generated these information categories as output.

An activity of the form, 'know about...' is itself an information generating activity and its input and output are identical. In this case the information category is 'material specification' and it appears as input to the other relevant activities. A 'know about' type of activity merely indicates that the actual source of the information is external to the system being considered.

Control information (i.e. that to be monitored) was derived from measures of performance that were determined for each activity.

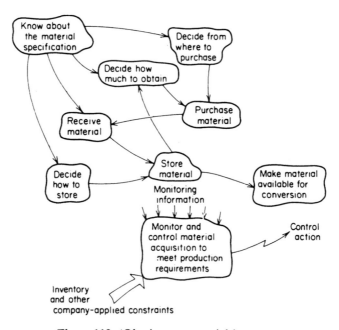

Figure 113. 'Obtain raw materials' system

The results of this stage in the analysis are contained in Table 6.6. It was at this stage also that the information categories themselves were defined. This was done through the derivation of a data model for each category. Two examples are given below for the information categories, 'supplier data' and 'demand':

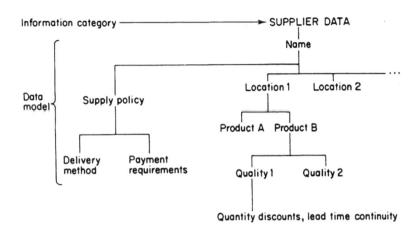

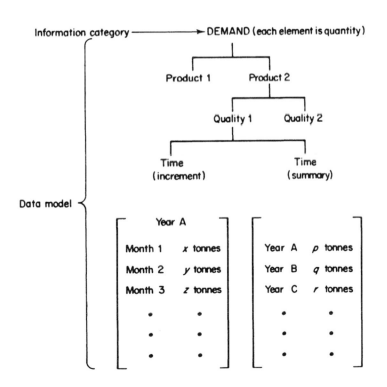

Table 6.6. Information categories

Information	Activities						
	Decide how much to obtain	Decide from where to purchase	Purchase material	Receive material	Decide how to store	Store material	Make material available
Inputs	Demand information Re-order policy Existing stock level Safety stock level Inventory constraints Product specification	Supplier listing Price per supplier Lead times Supplier performance history Price constraints	Supplier decision Quantity decision Price	Material specification Quantity ordered Price Timespecified Purchase order Advice note	Product location Storage availability Quantity to be stored Product specification	Product specification Location Quantity	Product specification Location Quantity
Outputs	Quantity decision Product specification	Supplier decision Price	Purchase order	Decision to accept or return to supplier Quality test results Time received	Product specification Location Quantity	Product specification Quantity stored Location	Quantity supplied Product supplied From where supplied Storage availability
Measures of performance	No. of times out stock Specific inventory cost Cost of activity	Quality of decision Activity cost Price obtained	Activity cost	Acceptability of material Timeliness Activity cost	Ease of location and acquisition Activity cost	Total inventory cost Stock turnover rate Storage cost	Access time Activity cost

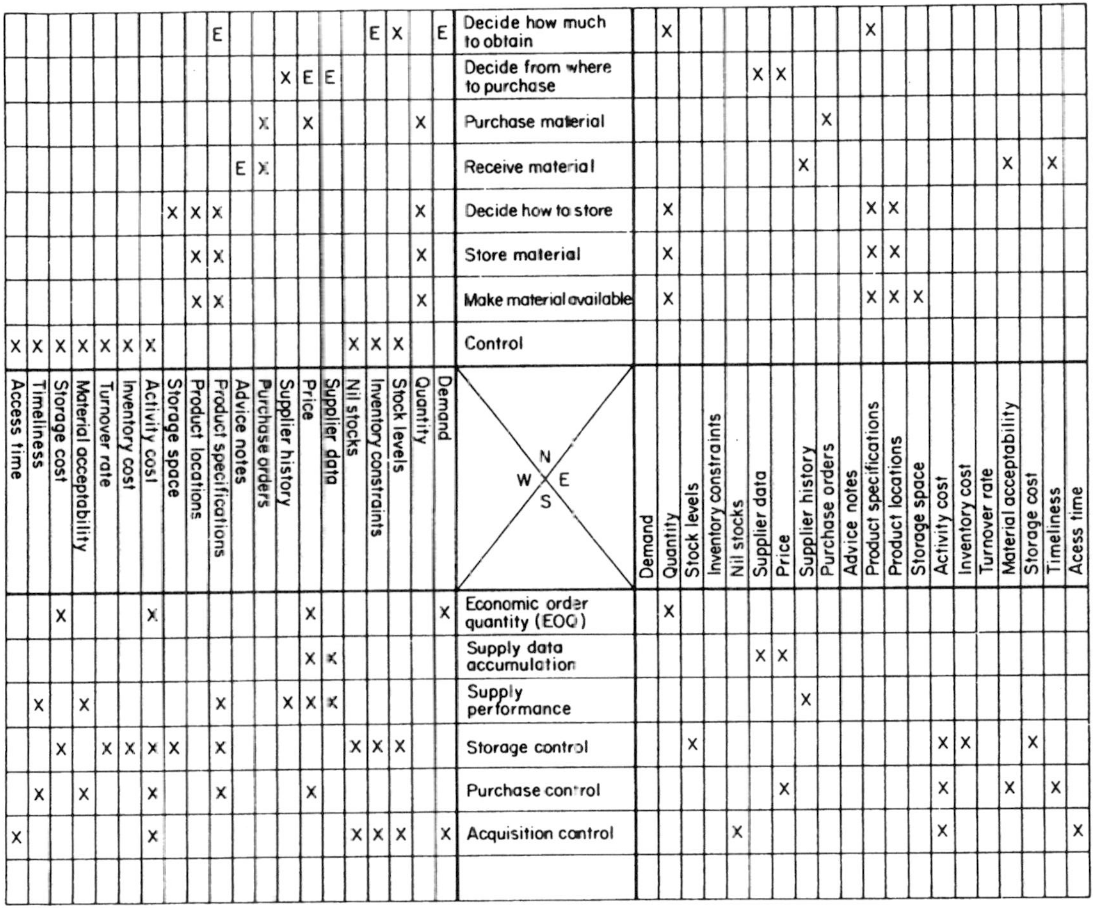

Figure 114. Maltese cross for 'obtain raw materials' system, E indicates a data source external to the 'obtain raw materials' system

As already stated, the information categories are effectively 'processed data' since the meaning is implicit in the use to which the data is put within the activity for which it is support information. Hence the data model is the device which translates information (derived on the basis of its use) to data (which is to be provided as the output of the data processing network).

Once this stage of the analysis is complete the activity-to-activity information flows can be displayed through the top half of a Maltese cross. Figure 114 is the Maltese cross developed for the system 'obtain raw materials'.

The next stage is to map on to the bottom half of the Maltese cross the existing information processing procedures (IPPs). This is done by identifying the information categories that the input and output data belong to (i.e. through the data models). Within the area of the organization represented by this model, the only IPP that formally existed was an operational package that calculated the economic order quantity (EOQ) of material to be purchased. The formula used was

$$EOQ = \sqrt{\frac{2RC_p}{iC}},$$

where R = demand, i = storage cost, C = item cost, and C_p = cost of ordering. This package is mapped on to the Maltese cross in Figure 114, as shown.

It is apparent from the top half of the Maltese cross that much more information than this is required and, since this derivation is based upon the minimum necessary needs, the information provision should be formal. An examination of the crosses in the north-western matrix will identify the support information categories that represents the processed data needed (as opposed to basic or unprocessed data). Here, processing refers to the manipulation, or the putting together, of several elements of data rather than simply the activity of updating. (Stock levels, for example, may involve the assembly of data about several products at several locations or it may be no more than the updating of single items of data without any accumulation.)

The specification of 'IPPs to be designed' requires several stages. Firstly, the above identification of support information categories needs to be carried out.

For each category required, the basic data available from which it can be assembled must be identified (this can be raw data or data which has already been processed). Once this is known the necessary data transformation process can be derived. This data transformation process may then be mapped on to the Maltese cross as illustrated on the next page.

The data transformation process so specified is not necessarily an IPP. It may be more efficient (in data processing terms) to group together several data transformation processes to form a single IPP. This is illustrated in Figure 114 by two examples. Firstly, to do the activity 'decide from where to purchase' it is necessary to have available a listing of supplier data (defined by the data model illustrated earlier). This listing will be assembled from individual data about individual suppliers and it may be necessary to access the listing by supplier, by

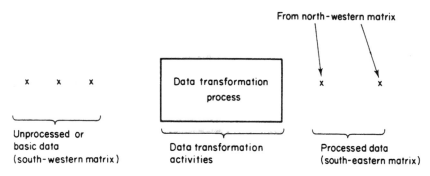

location, or by product. Thus a data transformation process is required to do this assembly in such a way that it can be accessed as appropriate. The decision about where to purchase may also be dependent upon the previous performance of the suppliers in terms of material acceptability, timeliness of delivery, etc. Thus an additional data transformation process is required which assembles this supplier history. An examination of the relevant columns (relevant to both these data transformation processes) in the south-western matrix indicates that some of the data elements are common and hence, to avoid duplication of data processing these two data transformation processes were combined into a single IPP to provide 'supplier intelligence'. A second example is included in the lower half of the Maltese cross in Figure 114 in which three data transformation processes to do with 'storage control', 'purchase control', and 'acquisition control' were combined with the existing EOQ package to form an IPP – 'inventory control'.

The final stage of the IPP specification, prior to design, is the construction of activity models of the data transformation processes themselves. For the EOQ package the model is obvious enough, i.e.

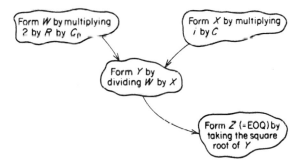

For the data transformation process 'supplier data accumulation', the model developed was as follows:

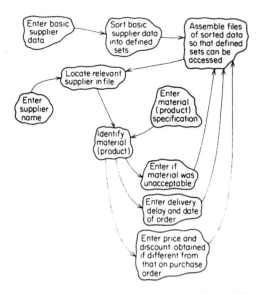

Figure 115. Activity model for IPP 'supplier intelligence'

Two further activities that it was necessary to consider prior to the design process were 'define what sets are required' and 'decide how to access'. These are not included in the model since they are not part of the data processing. The model for the data transformation process 'supplier performance' was added to this to give the total model for the IPP 'supplier intelligence' illustrated by Figure 115.

So far no decision has been made about *how* to do these activities. They may be undertaken manually or designed for computer processing. This decision is based upon the relative availability of computing and manpower resources, the costs associated with using either, and the preferences of the people concerned.

This assembly of examples (both from real projects and from exercises) has attempted to illustrate the various activities involved in undertaking an information oriented analysis. Thus they have illustrated the construction and use of a Maltese cross, the assembly of information tables and the definition of data transformation processes (DTPs) and information processing procedures (IPPs). It was stated earlier that the consensus primary task models, used as the basis for specifying information requirements, need not reflect the current situation. The following diagram attempts to summarize the various forms of information analysis required to support an organization in transition.

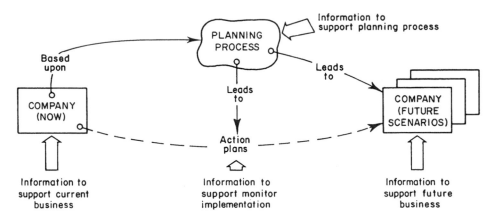

Before discussing the relationship of this methodology to others available. I shall provide a more detailed summary of the stages involved than was given earlier.

Information Systems Methodology (a Summary)

Stage 1

Undertake an issue-based analysis in order to determine what to take to be a primary task description relevant to the situation (current or future).

Stage 2

Develop a consensus primary task model to whatever level of resolution is appropriate. If 'validation' in relation to a current is required, two methods have been found to be useful (other than a general questioning based upon the existence, or not, of the activities). These are as follows:

(a) Determine what the output of an activity would be and look for the existence of that output in the real-world situation. If the output exists, then the activity producing it must exist (e.g. a production plan cannot exist without there also being an activity 'plan production').

(b) Through interviewing, develop individual primary task models relevant to the areas of responsibility of the individual managers. A mapping of these individual models on to the overall primary task model will enable validation to be achieved. This method also helps in the final stage of organization mapping.

'Validation' in these terms means that the activities in the model are legitimate. It can only be undertaken, of course, in relation to an analysis of current information requirements.

Stage 3

Treat each activity in the primary task model as an information transformation process and identify (a) the information required as an input to each activity so that the activity can take place; (b) the information produced as an output by doing the activity (but only that that is used as an input by some other activity); and (c), for the control activities, the monitoring information needed based upon a definition of the measures of performance for each activity. At this stage define each information category in terms of its data content (i.e. by deriving a data model).

Stage 4

Use the results of stage 3 to construct the top half of a Maltese cross.

Stage 5

Take each existing information processing procedure (IPP) and identify the information categories that the input and output data belong to (through the data models). Map the IPPs on to the bottom half of the Maltese cross.

Stage 6

By examining the support information categories required by the activities in the primary task model (north-western matrix), identify any omissions or potential duplication in the existing data processing network by translating requirements (north-western matrix) into provision (south-eastern matrix).

In a 'green field' situation, or where omissions are to be rectified, establish 'data transformation processes' necessary to fulfil provision requirements. Investigate alternative ways of combining these 'data transformation processes' into IPPs to avoid duplications and further omissions in the resultant data processing network.

Stage 7

For each IPP (new or modified), develop the activity model for the transformation of basic, or unprocessed, data into the required processed data. These activities represent a definition of *what* it is necessary to do to the basic data to provide the support information categories required from the IPP network. Decisions are then required to determine *how* to do the activities, i.e. manually, by computer, or by a combination of both. It is also necessary, at the same time, to determine how to capture and store the basic data.

Stage 8

Convert the activity information requirements into 'role' information requirements by doing an organization mapping on to the primary task model. (*Note*:
This may have been done already if the method (b) of doing the validation has
been adopted.)

In conclusion

Stages 1–7 determine what IPPs are required, within an efficient data processing
network, to provide the essential information requirements of the organization
in total (or in part). Stage 8 determines who (in terms of role) needs the particular
sets of processed data produced by the IPPs.

User Experience

The methodology for information requirements analysis forms an important
core element in the MSc course in Information Management at the University
of Lancaster. Soft Systems Methodology (SSM) is taught as a support module
within this course and numerous exercises are undertaken by the students
as a means of developing basic practitioner ability in its application (see
Appendix III).

WHAT WE HAVE LEARNT

DEVELOPING CONFIDENCE IN USING SSM AND
UNDERSTANDING OF IT
-LEARNING HOW TO CREATE A CONCEPTUAL
MODEL

THE EXERCISE HAS HELPED US IN THE INTELLECTUAL PROCESS OF DISTINGUISHING SYSTEMS THINKING FROM THE REAL WORLD – IT WAS DIFFICULT AT FIRST TO PREVENT OURSELVES THINKING ABOUT THE REAL WORLD WHEN SHAPING THE CONCEPTUAL MODEL

- **PROBLEM-OWNERS MUST SUPPORT SSM TEAM** – INTERVIEWS AND CO-OPERATION NEEDED EVEN ON THE MALTESE CROSS.
- **FURTHER UNCERTAINTIES ARISING** – e.g. HOW WILL OUR RECOMMENDATIONS BE PHYSICALLY IMPLEMENTED?

INCREASED CONFIDENCE WITH INPUT/OUTPUT TABLE NEXT TIME THE APPROACH TO HOW WE TACKLE IT WILL DIFFER AND CONSEQUENTLY ALSO THE TIME SPENT ON IT

INCREASED APPRECIATION OF THE USE OF THE MALTESE CROSS

IT WAS NOT UNTIL THE END OF USING THE MALTESE CROSS THAT WE
SAW ITS USES-IN MAKING US AWARE OF DUPLICATION OF DATA,
INFORMATION GAPS ETC.

AWARENESS THAT DIPLOMATIC AND
INTERPERSONAL SKILLS NEEDED
- WHEN CONSULTING THE PROBLEM-OWNERS
- IN DISCUSSION WITHIN THE TEAM

The library exercise in Appendix III is used as a means of bringing together the ideas of complex situational analysis and the development of information requirements. This is undertaken in groups, and members of staff within the department play the roles of some of the library staff.

Thus it attempts to mirror an actual consulting assignment with the student groups adopting the roles of consultant, undertaking interviews, and trying out potential recommendations on the role-playing library staff as users. The experience occupies several days and is concluded by the students making presentations to myself and other staff. The presentations cover 'what was done' in the exercise but, more importantly, they cover 'what was learnt' from what was done.

I am indebted to Richard McCarthy, a student on the 1989/90 course, for the light-hearted summary of what his group learnt; both about SSM and the difficulties of working in groups on this kind of analysis.

Relation to Other Approaches

Finally, in this chapter, I would like to place this particular methodology in the context of others that are available. At first sight the scene looks rather confusing since, for example, IBM's business system planning (BSP) approach, Nolan and Norton, Gane and Sarson, Jackson, etc., all tend to be referred to as if they are the same. The confusion arises because some of the approaches are actual alternatives to doing the same process of analysis while others may be regarded as complementary because they are actually concerned with a different part of the overall process from analysis to design. It is not my intention to describe each approach in detail; the references given will enable the reader to pursue those that he is not already familiar with, but I will attempt to classify

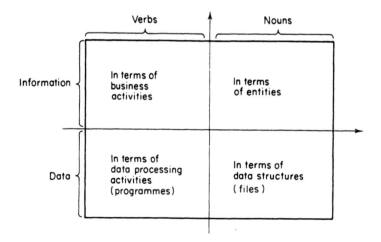

Figure 116. Classification scheme

them in a way which illustrates my view of their intended use. With one exception, the more familiar approaches can be placed within the scheme shown in Figure 116 (after Taylor, 1982). The vertical axis separates those approaches that use a language in terms of either verbs or nouns and the horizontal axis separates those that are concerned with data or information (using the interpretation that I have already defined).

The approach that I have described maps initially into the top left-hand box in that it uses a primary task description of business activities as a means of defining information requirements. However, it also exists in the bottom boxes since the analysis extends to the identification of data processing activities, and data models as illustrated in the examples and projects described. Other approaches which may be regarded as alternatives are the business systems planning (BSP) approach of IBM (IBM, 1975; Orsey, 1982), ISAC (Lundeberg et al., 1979), BICS (Kerner, 1982), SPM (Statland, 1982), and SSADM (Downs et al., 1988).

The systems planning methodology (SPM) was evolved primarily as an approach to define the data required to support management control decisions. The approach adopted is essentially problem oriented in that it is initiated by identifying 'a given business problem' and then by relating this problem to the affected organizational units through a standard reference model. The information needs and the resultant data processing activities are identified through a series of interviews with the personnel within the affected organizational units based upon a standard questioning format (Odepts). The terms in this mnemonic are described as follows:

O (output)	Information that the user perceives as necessary to perform a single information handling activity
DE (data elements)	Data elements or groups of data elements that are referenced during the production of the output data
P (processing)	The set of verbs required to convert the data elements into the desired output
TS (transaction sources)	The events that initiate the information handling process together with the source (i.e. the initiating organizational unit or an external source)

In this approach considerable attention is given to the design of this standard (Odepts) form, as this provides the basis for assembling the data elements, or groups, into data segment groupings to eliminate redundancy among data segments and also to sequence the processing steps into compatible data processing functions. This provides what the author calls a 'conceptual design systems flow chart' which is then converted into an information systems implementation plan including estimates of development effort, skills required, etc.

Little attention is given to the initial stage of defining business problems and to the difficulty of assessing the significance of the multiple perceptions of the personnel within the various organizational units.

ISAC (information systems work and analysis of change) is also a problem-orientated approach to information systems analysis. It was developed by a research group at the Royal Institute of Technology and University of Stockholm and was first published in 1978. The first stage in this approach is to undertake what is called 'change analysis'. The purpose of this stage is to identify changes, or improvements, that are needed to overcome problems experienced in undertaking the activities of an organization. These changes may be of an informational kind or related to other areas such as product development, process or organizational development, etc. Thus there is, initially, no commitment to any particular kind of development measure and the need to pursue an analysis in terms of information systems change emerges from a wider consideration. Change analysis is undertaken by identifying problems and interest groups (similar to the organizational units referred to above) and by grouping the problems and relevant interest groups into 'problem tables'. These tables are used to define the area of interest related to sets of interrelated problems. Once the areas are defined, current activities are described in terms of 'A-graphs' and 'property tables'. The A-graphs describe the relationships between sets of people, material, and messages, and the flows of these items between the sets. The property tables describe in more detail the current characteristics of the elements in the A-graphs. Thus, if an element in an A-graph is 'material provision', an entry in the property table relevant to that element could be 'current service level'. An evaluation of the current situation is carried out by determining the goals related to the activities described by the A-graphs, by assembling these into a table, and by then comparing what is wanted (the table of goals) with what exists (the problem table and the description of current activities). Mismatches identified by this comparison represent the potential needs for change. These are given priority according to the values of the relevant interest groups. Thus, if what is identified is a change to the existing information provision, the procedure is repeated through an hierarchical decomposition until the level of detail is reached at which information processing appears as an activity.

The underlying philosophy of this approach is that information systems are only developed when a need has been identified and when they can be shown to facilitate or improve some activities. It is intended that the users, themselves, undertake the change analysis and control the development of the resultant information systems to support their activities. In this mode of working the 'systems analyst' adopts the role of 'catalyst' to provide technical support when needed.

Both the BSP (business systems planning) and the BICS (business information control study) approaches are the result of IBM work in the area of information systems planning. They aim to describe information requirements at the level of an enterprise and do not attempt to define the detailed data processing functions required.

BSP is well documented and was made available around 1970 and has been widely applied. BICS is a current development and its application is limited.

As well as the references given above, which describe each approach in detail, a useful additional source of clarification is contained in the article, 'business systems planning and business information control study: a comparison' (Zachman, 1982).

In essence, BSP makes the assumption that a reasonable way of describing an enterprise is through a stable set of business process (~ 60) and then seeks to define the information requirements on the basis of managing the life-cycle of resources used in carrying out these processes. The life-cycle is taken to consist of four stages: (1) requirements planning, measurement, and control; (2) acquisition or implementation; (3) stewardship; (4) retirement or disposition. The processes are identified by working through the four stages for the resources represented by money, personnel, material, and facilities.

The processes are then grouped to remove inconsistencies in the level of detail, and to combine common processes, to arrive at a description containing around 60 processes (though no guide is given as to how to do this other than logic). Once the business processes have been identified, they can be related to the organizational structure of the business to determine which parts of the organization are doing which business processes and hence which parts of the organization create or use the necessary data. This is done by constructing an organization–process matrix. This matrix identified who makes the decisions related to each of the processes and also who is involved.

The next stage is to identify the data created, used, and controlled by the business processes. In ths stage data is also taken to be a resource and associated with a similar life-cycle. Data classes are then produced by combining data related to business entities, where business entities are defined as those things that an enterprise is concerned about and hence keeps information about. Such entities are customers, products, personnel, and so on. They will include, but not be restricted to, the resources used by the enterprise. The way in which the data is then used is illustrated by constructing a 'process–data class' matrix. The elements within this matrix are annotated with either C or U to indicate creation or use of data. This matrix then becomes the tool by which the 'information architecture' of the enterprise can be formed.

Manipulation of the matrix to form information systems is a matter of judgement but the procedure is initiated by rearranging the data axis so that the 'create' elements fall on the diagonal. 'Systems' are then formed by boxing groupings which create related classes of data. The data flows (and hence the relationships between the information systems) are derived by linking the Us outside the boxes by arrows whch flow from 'create' to 'use'. The resultant information systems and their interconnections represent the 'information architecture' that is required to support the business and can be used as a means of developing the information requirements of the enterprise as a whole.

One of the major problems with BSP is in the initial derivation of the business processes. This is critical since the identification of support data is wholly dependent upon the final set used. No guide is given as to how they should be derived and hence they tend to be assembled from a general business model (which may or may not be relevant to the particular enterprise) or they are derived from a set of interviews with top management. In the latter case they are subject to the bias inherent in the particular roles occupied by these managers and also may not represent a coherent or consistent set of processes.

BICS takes a totally different approach to the analysis of information requirements, though it also uses the data classes required to manage the resources of the business. In addition, however, it derives data classes from an analysis of the orders that the business receives. This particular feature is derived from a technique known as 'business information analysis and integration technique' (BIAIT) (Burnstine, 1979; Carlson, 1979). This technique assumes that the total information handling characteristics of a business can be predetermined from an understanding of seven binary variables related to the response of a business to its method of order handling. Here an order is interpreted in terms of three kinds of entity: a *thing*, e.g. a physical entity such as a piece of equipment; a *space*, either temporary or permanent, such as a van or warehouse; a *skill*, such as physical strength or intellectual ability. The seven variables related to an order together with a possible set of responses are illustrated in the following table:

Yes	Variable	No
The supplier invoices the customer	BILL	The order is paid for with cash, or cash equivalent
The supplier keeps records of the order	FUTURE	The customer takes what is ordered
The supplier keeps records of customer transactions	PROFILE	The supplier keeps no customer history
Price is negotiable	NEGOTIATE	Price is fixed
Customer hires the product	RENT	The customer buys the product
Supplier keeps records of product destination	TRACK	Supplier loses interest once the product is supplied
Product made to specifications	MADE TO ORDER	Product supplied from stock

The assumption is made that, once the orders have been derived and the binary responses to the seven variables determined, a set of information handling procedures can be designed. For example, in response to the 'bill' variable, if the company invoices the customer, a set of credit control procedures must exist. Whereas if the company receives cash none of these procedures are necessary.

Thus the essence of the BICS approach is the development of a generic business model related to the set of 'yes' or 'no' answers to the seven questions related to the kinds of orders handled by the company. Standard data classes are extracted from this model and, together with the data classes required for the management of those resources which are independent of the kinds of orders the company receives, related to the particular organization in terms of who is accountable for the data. Again matrices are used as an aid to defining this relationship.

Since this approach is based upon the BIAIT technique, it is entirely empirical and, to a large extent, independent of the particular company to which it is being applied. It is claimed to be flexible but in effect may require the nature of the company to be altered to fit the model.

The structured Systems Analysis and Design Methodology (SSADM) is a widely used approach to information systems development. It was originally developed by Learmouth and Burchett Management Systems in conjunction with the Central Computing and Telecommunications Agency of the UK Civil Service (CCTA). This latter body has also developed a combined approach called COMPACT which takes a version of the Soft Systems approach described here as a 'front end' analysis tool (to specify requirements) leading to the use of SSADM as the design tool. A detailed description of SSADM can be found in (Downes *et al.*, 1988) but briefly it consists of a highly documented, data-driven approach with six well defined stages.

> *Stage 1.* Analysis of the Current System – problems with the existing manual or computer-based information system together with intentions or new requirements are discussed with users. This leads to the development of what is called an 'ideal model' of the present information system and options for enhancements. Strengths and weaknesses are also identified. (What is a strength and what is a weakness and whose ideal is defined is not clear. Based upon what has been described earlier these value judgements must surely be highly W-dependent.)
>
> *Stage 2.* Specification of the Required System – this stage basically produces a logical view of the ideal required information system. Many user views are considered at this stage but how they are consolidated into a single definition of 'ideal' is not clear.
>
> *Stage 3.* User Selection of Service Levels – technical options are formulated and various schemes for implementation are discussed with users. The options could range from considerations of batch processing, or on-line systems, distributed or centralized systems to detailed questions such as number and location of terminals.

These first three stages may be undertaken at a general level in the form of a feasibility study or at a more detailed level in a full study. Considerable involvement of users and others associated with the potential information system is encouraged from the point of view of gaining commitment and easing acceptance.

> *Stage 4.* Detailed Data Design – data structures and relationships are derived at this stage. Data requirements to satisfy screen layouts, input

and report documentation as required by users is determined and logical data structuring carried out to meet processing requirements. These two phases of the design are document driven and entity analysis driven respectively. The two phases are combined to produce a composite logical design (CLD) which leads to the database design and the creation of a data dictionary.

Stage 5. Detailed Procedure Design – the required functions of the information system (as agreed with the users in stage 3) are checked against the CLD to ensure that the logical design is capable of satisfying them. It may be the case that a prototype version will be produced at this stage and the functions rehearsed with analysts and users.

Stage 6. Physical Design Control – this stage essentially covers five requirements. A test plan is developed to check the design. Program specifications are created together with operating schedules. File or database definitions are created and the user manuals are written.

SSADM is a highly detailed design process which is well documented and widely used. Only a very brief overview can be given here but the interested reader is referred to the appropriate reference for a detailed account. It is mapped on to the matrix of Figure 117 for comparative purposes but as already stated some combination of this approach with the soft systems approach has been attempted via the approach called COMPACT.

The above approaches can all be seen as alternatives to the systems approach described here. The remainder are complementary.

The Gane and Sarson approach is concerned with the data processing activities that are necessary to convert the basic data into the processed data required. Their approach is essentially a method of displaying these activities and the data flows between them. They have devised their own modelling language and an example of this was given in Figure 111. Their approach does not concern itself about the derivation of information needs to support business activities, and hence it could be used in conjuction with the other approaches discussed.

The Jackson approach (Jackson, 1975; Bleazard, 1976) stemmed from an interest in program design and is concerned with the structure of data. Jackson makes the assumption that the structure of a user's application of data is reflected in the structure of the data itself, which, if properly represented, can lead to a simple and error-free program structure. He has developed a language which consists of three constructs:

(a) *Sequence*

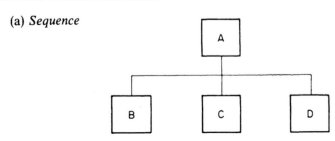

Here part A consist of part B followed by part C followed by part D. Thus if A were 'employee history' a possible breakdown could be

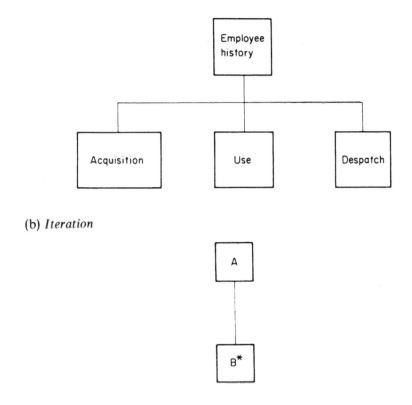

(b) *Iteration*

In this construct part A consists of part B, which is performed a number of times (Thus the asterisk represents multiplicity rather than iteration; however, 'iteration' is Jackson's term.)

(c) *Selection*

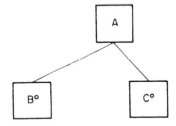

Here part A consists of part B or part C according to some selection criteria. The example of 'employee history' can be used to illustrate each construct.

Thus:

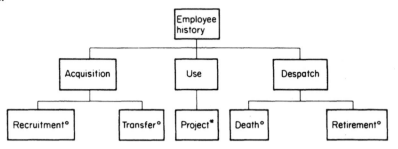

The language can be used as an hierarchical decomposition method, where each element represents an implied data processing activity within, in this example, an employee history program.

Both Jackson and Gane and Sarson are concerned with data rather than information and represent approaches which are situated at the interface between requirements specification and design.

Also at the level of data are those systems concerned with its organization and use, i.e. data dictionary systems (DDSs). These are not approaches to analysis in the sense used above since they make no pretence to derive data to satisfy business needs. They are aimed more at ensuring unambiguity and integrity of data used. Essentially a DDS is a tool for recording and processing information about the structure and usage of data. Although it is not necessarily the case, it

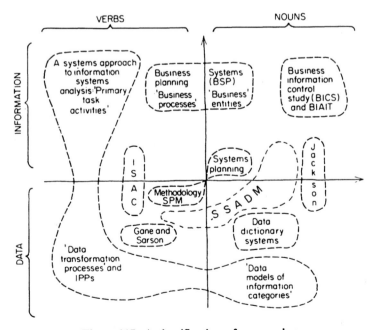

Figure 117. A classification of approaches

will usually be a computerized system. Much interest has been shown in DDSs over the last ten years, initiated mainly by the development of large cental data storage facilities. Further general reading about this area can be found in the report of the Data Dictionary Systems Working Party of the British Computer Society (1977) and Lefkovits (1977).

Returning now to the classification scheme proposed earlier, represented by Figure 116, it can be used to identify the relative interests of the major 'approaches' referred to above. Figure 117 is an attempt at such a classification. The one exception referred to earlier is the approach known as 'Nolan and Norton' (Nolan, 1979). The reason why this is an exception, and the reason for its omission from the map of Figure 117, is that it is one approach 'different in kind' to the remainder. Its concern is with the development of computing resources within an organization and is not at all concerned with the derivation of information needs of that organization. As an illustration of the particular emphasis of this approach, in the article referred to above, Nolan traces, graphically, a typical development path of computing in companies. This shows the level of expenditure on data processing resources as the various major growth processes within a company pass from initiation to maturity.

Thus, although Nolan and Norton are frequently spoken of as if they were an alternative to BSP and others, this is not the case.

Given the picture represented by Figure 117, an obvious question emerges: which approach should be used? Unfortunately there is no answer which would be recognized as the right answer by all concerned. Each approach has its strengths, weaknesses, and dangers, and an analyst should, at least, be aware of what they are before making a choice. The approach described in this book attempts to overcome a number of the dangers and weaknesses inherent in some of the other approaches and I believe that it is an approach that is generally applicable. Of course I am making that statement from a biased position, but, for me, it works and has produced results which the respective organizations believe to be useful. As I mentioned, earlier, in relation to methodology in general (see Figure 22), the assembly of systems concepts (i.e. methodology) needs to be appropriate to the situation and to the particular personality of the analyst, himself. Thus the analyst should choose that methodology which 'works' for him, and which, of course, produces results which the organization will agree are useful. Given this view, I cannot answer the above question. The analyst can only answer it for himself and only then, as the result of experience.

Chapter 7
Role Analysis

Introduction

Role analysis forms a significant part of most kinds of problem-solving related to management problems. The whole process of consensus primary task modelling is seeking to answer the question, 'What do we take the role of this organisation to be?' In Chapter 4 examples were given related to an exploration of the role of a headmaster and also to the derivation of a primary task model to represent the role of the Management Services Branch of the South West Region of the CEGB. These represented particular approaches which may be applicable in other situations. The concern in this chapter is with the implications arising from the creation of new bits of organizations as a result of the introduction of new technology.

In a number of organizations service departments have grown up which are essentially concerned with exploiting the developments in information technology (IT) and the particular boundary and responsibility of these departments is frequently of concern to the organization they are supporting. Models can be developed which aim to represent particular interpretations of the roles of these departments and can be used to explore the feasibility of adopting any one role in terms of an allowable area of responsibility and the necessary interactions with the rest of the organization which will be demanded by that role. It is usual for a role exploration such as this to develop primary task models whose boundary, and hence area of authority, is greater than that which is likely to be considered acceptable.

Potential role boundaries can then be mapped on to these models and the cross-boundary interactions examined in order to determine their significance, acceptability and feasibility.

Features which determine the role of an organization or the role of a department or even an individual role are many and varied. Organizations typically define objectives and mission statements which bear little relation to what actually happens. There is usually ample room for individual interpretation of what such objectives or statements mean. As hopefully illustrated in Chapter 3 the derivation of a primary task model which attempts to describe what the organization does, or intends to do, relies on a deep appreciation of this scope of

278

interpretation. Such a primary task model, once having progressed through the process of accommodating 'Ws' to a consensus primary task model represents a realistic definition of the organization's role. This description is more useful, and usable, than the pretentious mission statements or statements of objectives.

The introduction of new technology in organizations brings about two kinds of change. One is in relation to the changed methods of working of the individuals using the technology and the second is in terms of the creation of new departments and individual management roles concerned with the support and exploitation of this technology. Thus in a number of organizations information systems departments have been formed and questions are raised such as:

(1) What is the nature of such departments?
(2) What authority do they command?
(3) How are they to be structured?
(4) What information support do they require?
(5) What breadth of skills are required in order to produce an effective support?

Such questions as those above need to be addressed in a full role exploration though in the following examples the question of information support will not be illustrated as the process follows the methodology described in the previous chapter.

Role Exploration in BP

Within the Engineering and Technical Centre (ETC) of this company the use of personal computers (PCs) had proliferated to the extent that each engineer had his own. This was used to produce, analyse and store data which effectively became 'owned' by the individual engineer. It was the case, however, that the data was not the personal property of the originator; it belonged to the organization and was shared by others. The individual ownership of data produced particular structures and interpretations of it which led to inconsistencies and misuse when the data was shared. The company believed that a change in attitude was required which would result in a shift from individual to corporate ownership. Their response was to make a structural alteration and appoint a 'data administrator'. The intention was that the appointment of a central responsibility for data would bring about the desired change. The big question, of course, was 'what could the role and responsibility of this data administrator be?' within the current environment within the company. It is current philosophy in many organizations, as it was here, that 'each manager should have a PC on his/her desk' and if that is the case then this problem of individual data ownership is one that must be faced generally.

Within the situation described above, the role of data administrator could lie somewhere on a spectrum which extended from a mere store-keeper (with responsibility for the capture, storage, and availability of data) to the other

280

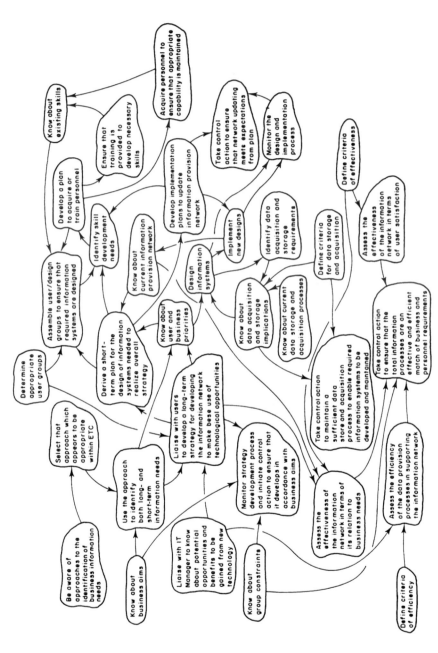

Figure 118. Activities appropriate to the role of data administrator at the end of the spectrum denoted by 'Information Management'

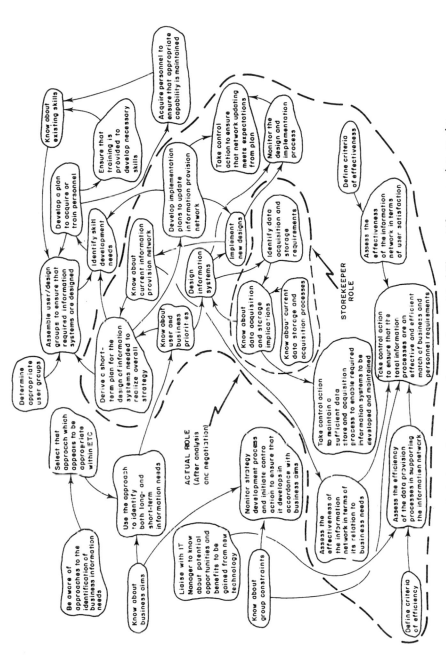

Figure 119. Role boundaries mapped on to the model of Figure 118 appropriate to the 'Storekeeper Role' and the actual role adopted for the data administrator

extreme of 'information manager' (with a responsibility for the planning, progressing, maintenance, and control of an information network).

We investigated this problem by developing a model of the task of information management (see Figure 118) and then used it to explore the implications of taking different responsibility boundaries for the data administrator (see Figure 119).

The model of Figure 118 consists of a set of activities connected together on the basis of their logical relationships and the whole model could be looked at as a job specification for the role of data administrator (if it were to reside at the end of the spectrum represented by the task of information management in total). Each boundary which is drawn on the model represents a reduction in the area of responsibility. The set of activities within the boundary could be seen as a reduced job specification and the interactions between these activities and those outside the boundary represent the communication processes and procedures that would need to exist to link the data administrator with those other managers who are undertaking these external activities (if, in fact, they are undertaken).

Each alternative was examined both in terms of its practical feasibility as an individual management task and also in terms of its feasibility within the existing culture of the Engineering and Technical Centre.

Having defined the activities within the role of data administrator it became important to turn attention to those activities outside the boundary. It is necessary to identify the existence or otherwise of those activities on which this role depends since, if they are not in existence, it will become difficult to undertake relevant activities within the role. The nature of the dependency needs to be examined. If the dependency is of the nature of information then the source of the information must be identified and ways must be discussed of ensuring that it can be provided. If the dependency is of the nature of a material flow then the absence of such material would represent a constraint on subsequent operations.

In this project the dependencies were of the nature of information. That which was immediately available was specified. The inability to obtain information from activities that did not yet exist would obviously reduce the effectiveness of the activities requiring such information. The identification of these activities represents the source of recommendations for the future adaptation of the information management role in total.

Role Exploration in Schools

The education scene in the UK is changing very rapidly, particularly at the level of primary and secondary education. At the academic level a national curriculum is being imposed and the local management of schools is being encouraged. The impact of both of these changes is to move the responsibility for education and its management away from the local education authorities to the school and its governing body. Those activities such as hiring and firing of teaching staff, the employment of services (such as catering and cleaning) and the detailed definition of school activities (within the constraints of the national curriculum) become the responsibility of the governing body and senior school staff. However, the precise definition of the responsibility of the governing body is not clear. The local authority still retain some responsibility (for example, the maintenance of buildings and the allocation of government finances), but precisely where their responsibility ends and where that of the governing body begins is problematical.

In my role of chairman of one such governing body I decided to use systems ideas to attempt to define the actual responsibilities of the governing body, vis-à-vis the other interested parties. Since I was in the multiple role of client and problem solver and, taking the governing body as a whole to be in the role of problem owner, it seemed important to me to ensure that we undertook those management activities that would produce maximum benefit to the school. If we could do this properly then we could be forward looking and proactive in our role rather than reacting to issues as they arose, which tended to be the *modus operandi* of the governing body. In order to define the management activities that we should be involved in it seemed useful to undertake a systems analysis relevant to the exploration of our role.

Figure 120 attempts to illustrate a view of the school as a particular kind of transformation process together with those organizational entities with which it

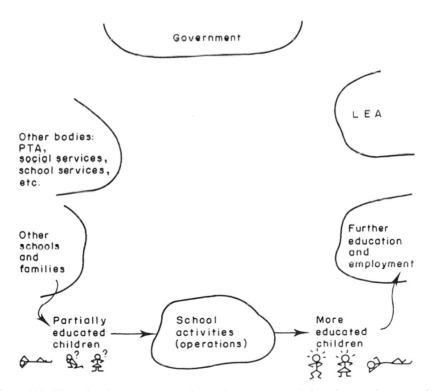

Figure 120. The school seen as a transformation process and the relevant elements of its environment

284

interacts. Children are seen to enter the school, some with enquiring minds, other less so. They are transformed into enlightened children though some may see less light than others. However, we will concentrate on the intended transformation. The other organizational entities are shown as elements of the school environment and in order to relate them to the school there needs to be a set of liaison and management activities as illustrated in Figure 121. These activities will be those required to do the following:

(a) To translate general government policy in relation to education and school management into that appropriate to the particular school.
(b) To process the relationship with the local education authority (LEA) in terms of information transfer, financial management, and those school services which will be available centrally.
(c) To process information related to other schools and families.
(d) To process information and the additional finances, contracts etc., with other bodies such as PTA and those services not centrally provided.
(e) To manage the provision of resources to the school, be responsible for the development and management of the school's strategy and to provide performance and other information as required.

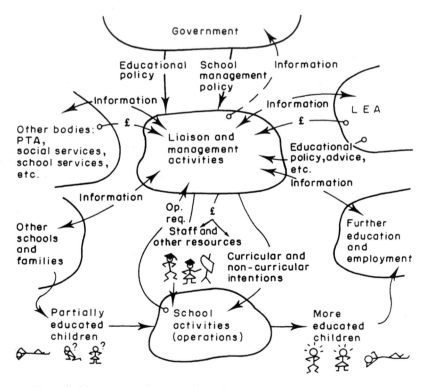

Figure 121. A linking process between the school and the environmental elements (op. req. = operational requirements)

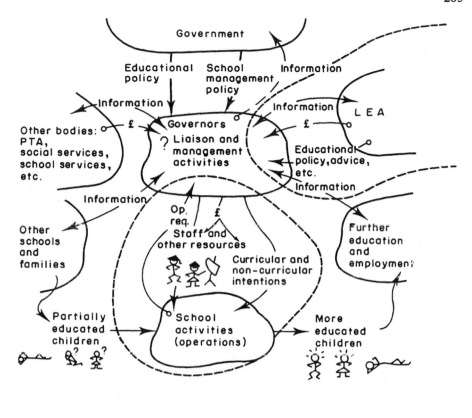

Figure 122. Mapping of authority boundaries

These liaison and management activities will *not* represent the area of responsibility of the governing body since the school and the LEA will have some involvement also. Figure 122 illustrates this situation by mapping the organizational boundaries of the LEA and the school. It must be the case that what is left can only be the responsibility of the governing body. The questions that we are seeking to answer therefore are, where do these boundaries lie? and, what is this remaining responsibility?

To investigate these questions a system relevant to these liaison management activities was derived. The root definition that was taken is as follows:

An LEA and community owned system for the translation of government educational policy into a formal secondary education process that meets the needs of the local community, while recognizing financial, LEA, and local constraints and the needs of the premises in which it takes place.

286

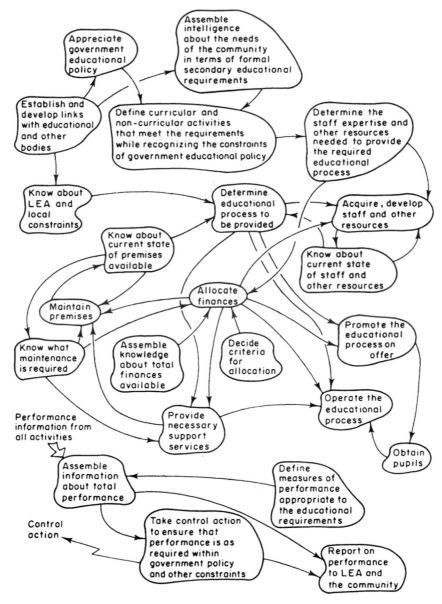

Figure 123. Conceptual model relevant to liaison management

The conceptual model is illustrated in Figure 123. This model served two purposes. Firstly, it enabled us to identify the expertise required to be able to undertake the activities. The mapping of Figure 124 shows a requirement for six different kinds of expertise. These are:

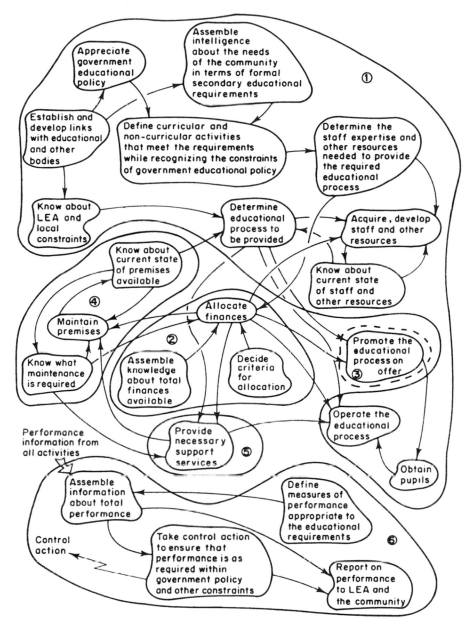

Figure 124. Expertise boundaries

(1) Educational and teaching skills.
(2) Financial management ability.
(3) Promotional expertise (i.e. advertising etc.).
(4) Buildings and ground maintenance.

288

(5) Contracting and job definition ability.
(6) Performance, assessment and reporting skills.

Traditionally members of governing bodies have been elected on the basis of their political and organizational affiliations and their general popularity with parents and staff. This can no longer be the case if the governing body is to be an

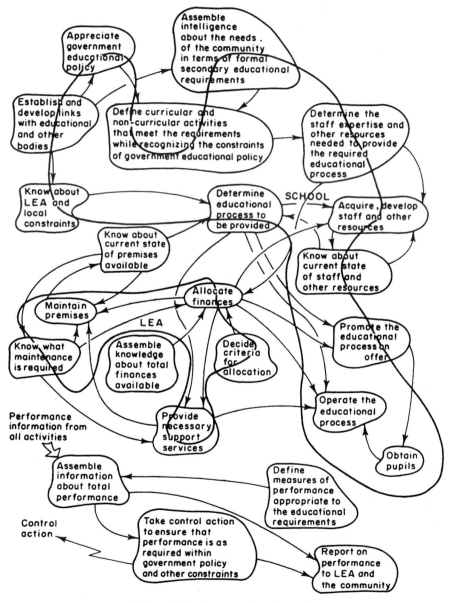

Figure 125. Organizational mapping

effective management unit requiring the above range of expertise. It will be necessary to ensure that this expertise is available through the elected members.

The second use of this model was to identify the role of the governing body, which was the original intention. The organizational mapping of Figure 125 illustrates the LEA and school responsibilities. What is clear from this mapping is that, whereas some activities are wholly contained within these boundaries a large number are shared between organizations. The argument is that what is contained within these boundaries is the responsibility of the relevant organization and what is outside these boundaries is the responsibility of the governing body. Apart from the activities associated with performance monitoring, control, and reporting all of the other activities are shared with either the LEA or the school. This is to be expected since, as illustrated by Figure 121, the role is essentially one of liaison.

Given the expertise requirements, as illustrated by Figure 124, a number of sub-organization units is suggested as a way of structuring the governing body. The governing body as a whole is responsible for the total task but these sub-units could usefully be seen as advisory bodies to the main body, each with their own particular expertise. Four such units (i.e. sub-committees) were defined on the basis of the mappings in Figures 124 and 125. Their responsibilities are as follows:

- *Education and promotional sub-committee:*
  To liaise with the school in order to
  (a) ensure a common understanding of government and LEA educational policy,
  (b) derive plans for the schools' curricular and non-curricular development,
  (c) define staff and other resource needs,
  (d) promote the school and its activities.
- *Buildings and contracts sub-committee:*
  To liaise with the LEA and other service providing bodies in order to
  (a) ensure that the premises meet the required standards from aesthetic, utilitarian, and health and safety viewpoints,
  (b) ensure that these support services (meals, grounds etc.) are provided in such a way that they are both appropriate and satisfactory.
- *Financial management sub-committee:*
  To liaise with the LEA and other fund generating bodies in order to
  (a) negotiate the allocation of responsibilities for financial control with respect to the school,
  (b) define criteria for the allocation of funds,
  (c) maintain an updated account of funds available and potential sources.
- *Performance and reporting sub-committee:*
  To liaise with the other sub-committees in order to
  (a) ensure that performance is consistent with overall aims,
  (b) identify problems or general areas of concern which prevent individual responsibilities from being achieved.

To liaise with whoever is appropriate in order to
(a) define measures of performance which represent the total educational process and its management,
(b) to report on performance (to the LEA, the school, the community, the governing body in total, potential employers etc.).

The committee organization has now been implemented though at the time of writing its effectiveness has yet to be assessed.

Centurion plc

Centurion plc (a pseudonym) is a large international company in the electronics and communication business with a high variety of products and services. A description of this probject is included in order to illustrate the use of the Information Requirements Methodology, in a *reverse mode* as a means of establishing a particular organizational role.

The company had established a policy of customer orientation which had given rise to a number of organizational and procedural changes in order that it could become more sensitive in its relations with the market. One of these changes was the introduction of 'a single point of contact' as a means of providing an effective interface between the large variety of customers that existed in the market-place and the highly complex organization internal to the company. Thus this single point of contact would effectively provide a 'shop window' to the outside world enabling effective communications to take place across the organization boundary. It had been decided to provide computer-based support to the operators of this 'shop window' so that by the use of multiple screen displays the effective routing of customer enquiries could be achieved rapidly. A suite of software was currently being designed and the implementation date had been fixed. In considering the implementation of this technology questions were being raised as to what the role and organization of the 'shop window' should be. Since the software was already being designed a concept for the 'shop window' must have been in the heads of whoever in the organization had produced its specification.

Our task in the project was to attempt to define this concept and hence identify what organizational change needed to be brought about in order that the software could be used effectively. Thus the route of our analysis, in some way, had to proceed from a 'taken-as-given' processed-data support to an identification of the set of activities that would have given rise to this requirement and hence to the specification of the role of this implied 'shop window'. We decided the most realistic way and hopefully the most useful way of undertaking the analysis was to postulate a number of systems relevant to customer service provision, identify their information requirements and see how these matched what was being provided by the suite of software. This suite of software is known as the 3S program (shop window support systems) and we needed to find out

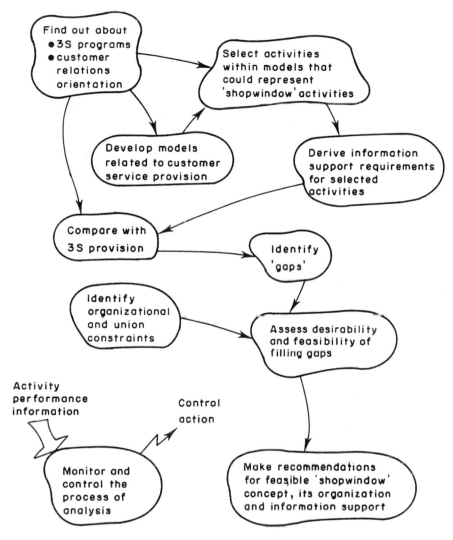

Figure 126. Problem-solving system

enough about its specification to be able to identify its total output. We also needed to know about the particular company's policy towards customer orientation. The approach adopted is presented in Figure 126.

As illustrated by this figure the two activities following the initial finding out were concerned with developing customer service provision systems and then with mapping potential role boundaries. An example of a customer service provision system is given in Figure 127. This was derived from the following root definition:

292

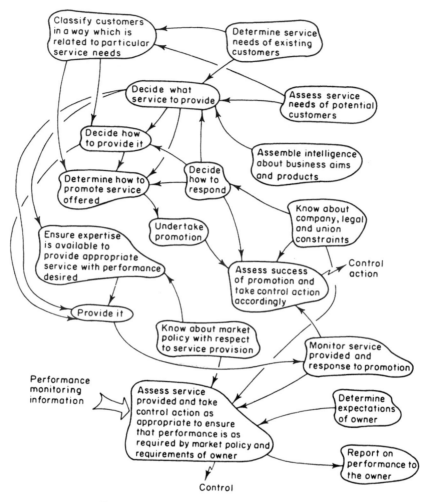

Figure 127. Customer service provision system

A Division owned system which ensures the provision of an appropriate service to its customers and potential customers which is related to its business aims and products with a performance dictated by its market policy but within company, legal and union constraints.

Figures 128 to 130 illustrate mappings of potential role boundaries representing differing levels of authority.

(a) Figure 128 – The shop window seen as manager and provider of the

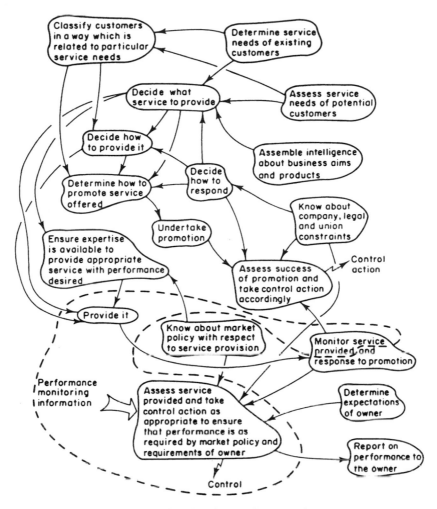

Figure 128. 'Provide service' mapping

service. What the service consists of and how it is provided are decided elsewhere.

(b) Figure 129 – The shop window seen as the provider only but with the additional responsibilities of resourcing and organization.

(c) Figure 130 – The shop window with both the sets of responsibilities defined by (a) and (b) above but with the additional authority for promotion and performance reporting.

The intention here was to compare a range of information support requirements from various customer service provision activities against the 3S provision. Through this comparison gaps could be identified. These gaps could be of two kinds:

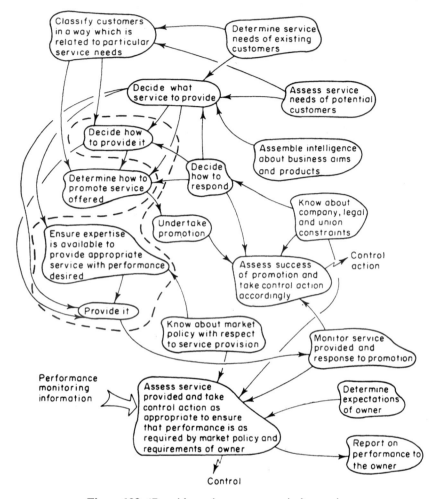

Figure 129. 'Provide and resource service' mapping

(1) Processed data produced as output from the 3S programs which was not used in any of our models. This would lead to further model development.
(2) Output from the 3S programs incomplete on the basis of a specification of information requirements from coherent models representing potential 'shop window' roles.

In the event the approach of Figure 126 did not turn out to be the precise approach adopted. A wide range of models were produced but the difficulty that we found was that no single model of ours matched (even closely) the underlying model that must have led to the 3S specification. Thus a conscious change to the approach was made following the identification of gaps.

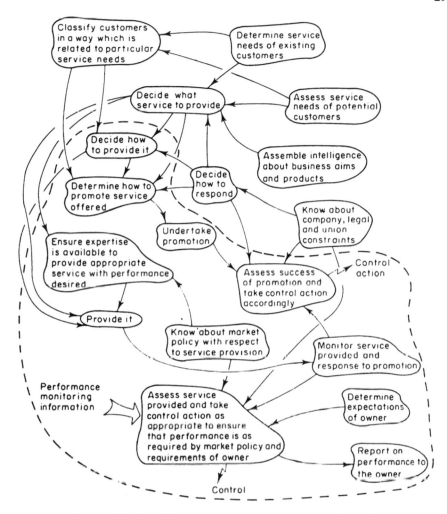

Figure 130. 'Provide, resource and promote service' mapping

Table 7.1 is an illustration of the way in which we undertook the comparison for one of the models derived, and what this demonstrates is illustrative of each of the models chosen. As the table shows few of the activities are actually supported by 3S provision. The next stage in the approach therefore was to select those activities from each of the models that was supported from 3S provision and to assemble these into a single model.

As is the case with the construction of a consensus primary task model this assembly may be no more than an aggregate. It is therefore necessary to go through the process of ensuring that the resulting model is coherent. This is done in exactly the same way as described earlier. In the case of this project the resultant model is given in Figure 131. The activities within the dashed

	Activities	Inputs	3S provision	Comments
1.1.	Know how to communicate effectively	Communication training		
1.2.	Know customer classification	Market segmentation; 'customer profiles'	Customer type; customer classification; revenue bracket, credit rating	Only sets variables; definition of content is up to area (according to consumer service policy, marketing policy, financial policy etc.)
1.3.	Know enquiry classification	Classification provided by PPS (procedure provision system)	User procedure manual	Can be allocated to 'business tasks' as defined by the area; 3S only provides classification for enquiries that fit the pre-defined 'sub-systems' and 'business tasks', other enquiries such as general complaints and general enquiries are not covered.
1.4.	Know criteria for decisions	Allocation rules provided by PPS	User procedure manual	As 3S is procedure-driven, the implicit criterion for decision is whether there is a 'transaction' to apply; if there isn't, the enquiry is referred
1.5.	Know specialist systems	Functional organization; 'job descriptions'		
1.6.	Receive customer enquiry	Customer communication −verbal −written		3S and telephone are not connected
1.7.	Communicate to identify special requirements	Additional customer enquiry information provided by customer	'Customer history'	
1.8.	Classify customer	'Customer profiles'; customer data	'Thumbnail sketch' only (as in 1.2.)	(As in 1.2.)
1.9.	Classify enquiry	'Enquiry classification'; enquiry details	(As in 1.3.)	(As in 1.3.)
1.10.	Decide whether to refer	Allocation rules; classified enquiry	(As in 1.4.)	(As in 1.4.)

1.11.	Decide on appropriate special system	Decision to refer; functional organization. Allocation classified customer enquiry	User procedure manual, complemented with area referral points
1.12.	Inform customer of referral	Address of specialist system; action envisaged	(As 1.11.)
1.13.	Brief system chosen about enquiry	Cost enquiry details; information given to customer	Up-dated records	All customer enquiry details which do not involve a 3S application are lost if not transmitted by extra-3S means
1.14.	Refer enquiry to system	Target system: customer enquiry details (brief)	(As in 1.13. in case of procedures-driven follow-up)	(As in 1.13.)
1.15.	Inform linking system	Classified customer enquiry address of special system action envisaged	Database information related to customer enquiry (as in 1.15.) and 'customer notes' to the extent they have been entered in different applications	(As in 1.13.)
1.16.	Know 'owner' expectation		
1.17.	Assess effectiveness	User feedback; action-system feedback; M. of P.* re-effectiveness	Feedback using message	Messaging is entirely sender-controlled; if no use is made of the facility, information was to be generated from scratch
1.18.	Monitor 'cost'	Resource-use information	
1.19.	Translate into M. of P.*	'Owner' expectations		No details on 3S provision available as yet
1.20.	Take control action	Assessment/monitoring information M. of P.*	
1.21.	Report to 'owner'	Gap-analysis: control-action decision	

298

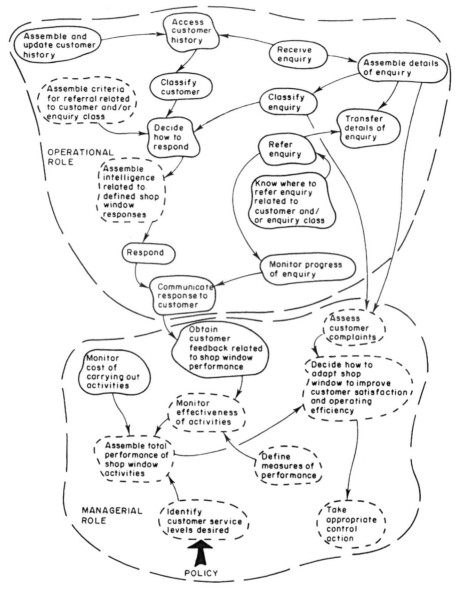

Figure 131. Derived role for the 'shop window'

boundaries represent those activities which had to be added in order to make the model coherent. Some of the activities in the individual models leading to this assembly (such as those in Table 7.1) were combined, in the aggregation, to form activities at a lower resolution level.

The activities in Figure 131 represent those activities that needed to be in place if the 3S programs were to be implemented. The activities are divided into sets appropriate to an operational role and a managerial role and it is apparent that in defining the 3S programs the original concept had consisted almost entirely of the operational activities. It is of course necessary that the 'shop window' be managed as well as operated.

This project has attempted to illustrate how system ideas can be used as a means of uncovering an implicit role or concept. It has effectively used the methodology in reverse since the starting point is a suite of computer programs, or in the terminology used previously, an IPP network. It was the case, however, that the Maltese cross was not seen to be a useful device in this project, since it was the set of activities that we were seeking rather than information provision.

Summary

The approach to role definition described here relies on two assumptions.

(a) A wider system, mission statement, job specification etc., relevant to (but wider in scope than) the role being explored can be defined and taken as given.

(b) An acceptable/feasible role can be identified by mapping various role boundaries on to the model developed from (a) above.

In the CEGB project described in Chapter 4, a wider system was developed from the South West Region's statement of objectives. It was then argued that somewhere within the resultant set of activities we could find some that could be taken to be relevant to the role of the Management Services Branch. In the BP project described earlier in this chapter a wider system of information management was developed. These are just two examples that represent the application of assumption (a).

In the CEGB project – the actual role definition (assumption (b) above) was obtained by mapping the concept of an adaptive control system on to the wider system model and by extracting those activities that represented the adaptive element.

In the BP project the process of mapping made use of transparent overlays on which various role boundaries were drawn. Each alternative was then explored in terms of its feasibility as a job specification and its acceptability to the role holder and within the existing culture of the organization.

A methodology appropriate to this form of role exploration is given in Figure 132 and as a five stage process as follows:

(1) Develop a primary task model relevant to the area of concern whose boundary is greater than that of the role under consideration.

(2) Map various role boundaries (using transparent overlays etc.).

(3) Assess expertise requirements for activities within each boundary.

(4) Explore the organizational implications of the activity-to-activity linkages across the boundaries.

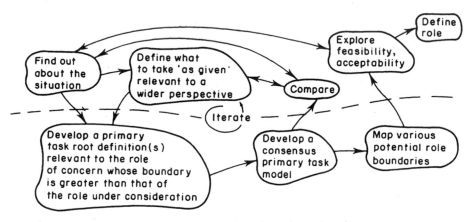

Figure 132. A methodology for role exploration

(5) Decide on those boundaries which are feasible and acceptable given resource and political considerations.

Stages 3, 4, and 5 represent an expansion of the 'explore' activity within Figure 132.

Conclusion

In the introduction (pp. 3–7) I discussed the process of inquiry and emphasized the distinction that needed to be made between activity that is in the real world and the intellectual activity that is concerned with the related thought processes.

I hope that the importance of this distinction has been brought home in the material of the seven chapters. It is precisely the maintenance of this distinction that has allowed me to learn from the variety of experiences that I have enjoyed since joining this department and to extract lessons from the action research programme that has been so central to our activity.

So to the final paragraphs of the book. I hope that this is only your beginning and that my reflections on my experiences will turn out to be helpful in your own particular journey.

It is on the subject of experience and on learning from experience that I would like to end. The process of learning from experience is, at the same time, one of the most rewarding, one of the most painful, and one of the most difficult of all intellectual processes. I don't think that anyone would disagree that it is one of the most rewarding. Few people would disagree that it can be very painful, but what makes it so difficult?

One component of this difficulty is related to time. Time is needed for the necessary reflection on experience. One cannot know, at the time that learning is occurring, what it is that is being learnt.

A second component is related to the language of description. If such learning is to be made explicit so that it can be communicated (even to oneself), then a language must be available in order to describe what has been learnt. This may turn out to be a major hurdle, since the appropriate language is dependent on what is to be described and 'what is to be described' is not known because the appropriate language is not available or understood. Some people may have difficulty in emerging from this closed system and hence may never know what they know (or don't know).

The third component is related to the opportunity for critical debate. Assuming that time is available for reflection on the experience that is accumulating, and assuming that a language is available for describing what has been learnt, without the opportunity for debating the outcome with a critical audience (i.e. seeking refutation) such learning may be superficial.

In the preceding sections I have attempted to describe the outcome of my own learning experience, which includes that knowledge gained from the learning experiences of others. The method of investigation has been 'action research' and I believe that this is the most appropriate method of investigation where the area of concern is activity in the real world. I hope that some of the ideas will prove to be useful, but they can only be the starting point. They can feed your own action research program, the essence of which is learning from experience. This was my reason for emphasizing the above three components of difficulty. Ensure that you allow time for reflection on your own experience. Attempt to debate the outcome of the process of reflection with a critical audience. Finally, ensure that the language of description is clear, appropriate, and unambiguous so that the debate can be fruitful.

Appendix I

Analytical Models and the Process of Model Building

In Chapter 1 a matrix was presented in which four classes of analytical model were displayed. This is reproduced below:

	Steady-state	Dynamic
Deterministic	Algebraic Equations	Differential equations
Non-deterministic	Statistical and probability relationships	Discrete-event simulation

As a means of illustrating each type of model, simple examples will be chosen and these are accompanied by selected references in which the reader can pursue to a greater depth those modelling techniques of particular interest.

Steady-state Deterministic

Any algebraic relationship can be considered to be a model of this type. Thus a model of a resistive element in an electrical network (assumed to obey Ohm's law) can be described in terms of the current I flowing through the resistance R giving rise to a voltage drop V, where $V = IR$.

A typical problem to which formulation in algebraic terms is particularly appropriate is that concerning the allocation of productive resources where many alternative choices exist, but where resources are limited. A technique to solve this type of problem has been well developed and is known as linear programming.

As an example, consider the problem facing the production manager of a company that produces two types of cloth, say type L and type M. Each type uses three colours of wool: red, green, and yellow. The material requirements for a metre length of cloth are given by:

302

Type (in grams)			Wool available
L	M	Colour	(in grams)
4	4	Red	1000
5	2	Green	1000
3	8	Yellow	1200

The final column in the above table represents the quantity limit on the raw material available for this particular production run. The profit that can be realized from the sale of cloth type L is 25 pence per metre and that for type M is 15 pence per metre. The problem is how much cloth of which type to produce in order to maximize the total profit.

The problem can be formulated as a linear programming problem as follows. Let X and Y be the numbers of metres of L and M produced respectively. Then

$$4X + 5Y \leqslant 1000,$$
$$5X + 2Y \leqslant 1000,$$
$$3X + 8Y \leqslant 1200,$$
$$X \text{ and } Y \not< 0.$$

The above inequalities define the feasible space for X and Y within the resource limitations and the final inequality specifies that neither X nor Y can be negative. The final equation which completes the model is that which defines the objective function to be maximized or minimized. In this case this is the total profit $P = 25X + 15Y$, which is to be maximized. The linear programming technique is one which searches the feasible space for X and Y and determines those values which yield the maximum value for P.

A good introduction to the technique of linear programming is contained in Gass (1964).

Steady-state Non-deterministic

This type of model is used when the mechanisms governing behaviour are not known but it can be assumed that certain variables are wholly or partially dependent upon others. It might seem reasonable, for example, that total electricity sales are a function of industrial production (they may also be a function of other things as well such as domestic usage, but we might argue that industrial usage is a dominant factor). To find out if this is the case, we could collect information on the electricity sales (as the dependent variable Y) and the index of industrial production (as the independent variable X) over a number of years:

Year	Index of industrial production, X	Electricity sales, Y (kW h)
1950	88.3	22.9×10^9
1951	91.3	25.4×10^9
1952	89.2	26.1×10^9
1953	94.3	28.0×10^9
1954	100.0	31.6×10^9
1955	105.1	34.6×10^9
1956	105.6	37.2×10^9
1957	107.5	39.3×10^9
1958	106.3	41.2×10^9
1959	112.6	44.7×10^9

A model which relates the two variables can be dervied by means of a technique known as regression analysis. There are a number of forms that this can take but the simplest is known as linear regression. This makes the assumption that there is a straight line relationship between the two variables given by

$$Y = a + bX;$$

values for a and b are derived which minimize the sum of the errors squared at the discrete data points. The values may be computed from

$$b = \frac{\left(\sum XY - \frac{\sum X \sum Y}{n} \right)}{\left(\sum X^2 - \frac{(\sum X)^2}{n} \right)} \quad a = \bar{Y} - b\bar{X},$$

where n is the number of data points and $\bar{Y}$ and $\bar{X}$ are the mean values of Y and X, respectively.

For this example,

	X	Y	XY
Sum	1000.2	331.0	33671.92
Sum of squares	100713.78		
Mean	100.02	33.1	

These values yields the relation

$$Y = 0.839X - 50.816.$$

An alternative way of displaying the data, which is also a useful method of determining if there is a relationship between the two variables and the form it

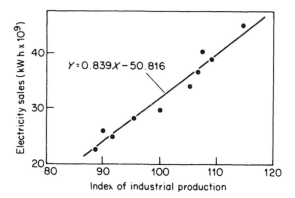

Figure A1. Scatter diagram

might take, is the scatter diagram. Here the two variables are plotted against one another and the resultant array of points is indicative of the strength of correlation. For this example, the scatter diagram takes the form of Figure A1.

Also plotted on this diagram is the resultant linear relationship, and it can be determined from the lie of the data points about the line whether or not the simple regression model is accurate enough for the purpose for which the model has been derived. If the scatter diagram had been of the form of Figure A2, it could have been argued that the two variables were unrelated.

As already mentioned, linear regression is the simplest form of regression model. Alternative forms include (a) non-linear (curvilinear) regression, e.g.

$$Y = a + bX + cX^2 + dX^3 + \ldots,$$

or, in terms of more than one independent variable, (b) multi-linear regression, e.g.

$$Y = a + bX_1 + cX_2 + dX_3 + \ldots,$$

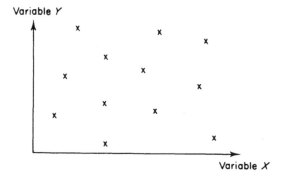

Figure A2. Scatter diagram

and (c) multiple regression, e.g.

$$Y = a + bX_1 + cX_2 + a_1 X_1 X_2 + b_2 X_1^2 + c_1 X_2^2.$$

For further reading on regression analysis and other statistical methods see Levin (1978). This is a book written for a management audience and therefore does not become too involved in detailed statistics.

Probabilistic modelling is also included in this section, though to be entirely accurate it should, perhaps, occupy the interface between deterministic and non-deterministic modelling.

A technique which has been developed using probability models is known as the Monte Carlo method. It is frequently referred to as Monte Carlo simulation, though this is a misnomer. There is a difference between probabilistic modelling and simulation. The Monte Carlo method is a way of evaluating a deterministic problem by means of setting up a probability problem on to which it can be mapped. The probability problem is then evaluated by repeated sampling trials and these trials are used to evaluate the deterministic process. Alternatively simulation, which uses random variables, is a technique used because of the analyst's ignorance of the underlying cause–effect relationships (Sisson, 1969; Tocher, 1969).

An example of the use of this method comes from my previous experience while with the United Kingdom Atomic Energy Authority. The problem is concerned with evaluating the dimensions and fuel characteristics of the core of a nuclear reactor so that the neutron chain reaction can be maintained over the lifetime of the core. The core can be approximately described as a right circular cylinder of about 10 metres diameter and 10 metres in height. Within this core there is a heterogeneous arrangement of moderator (graphite; or water) within a steel structure together with the fuel elements and control rods. The critical size of a reactor core is defined as the size for which the number of neutrons produced in the fission process just balances those lost by leakage and by capture in the non-fissionable components in the core. The critical size is not constant but depends on the isotopic composition of the uranium fuel, the proportion of moderator, the shape and arrangement of the materials, and the presence of various substances causing parasitic capture of neutrons. If an assembly is smaller than the critical size, i.e. subcritical, neutrons are lost at a greater rate than they are replenished by fission, and so a self-sustaining chain reaction will be impossible. It is essential, therefore that the size of the uranium–moderator lattice should be equal to, or larger than, the critical value, i.e., supercritical, if the fission chain reaction is to be maintained. In a reactor which operates at 'thermal' energies the process may be illustrated by Figure A3.

With reference to Figure A3, the fission fragments produced are frequently poisons in the sense that, at certain neutron energies, they are significant absorbers of neutrons and so remove them from any further possibility of causing fission. Xenon-135 is one such substance and since it is formed as a result of fission, its concentration in the core is a function of the way in which the fission process progresses over time and gives rise to the parasitic capture

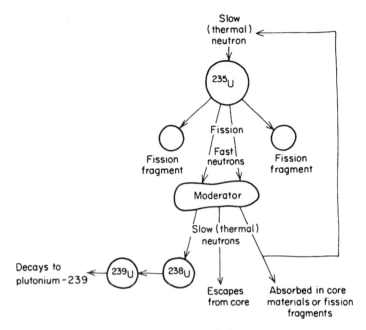

Figure A3. The uranium fission process

referred to earlier. The fast neutrons produced by fission emerge at an energy around 2 MeV and before they can cause further fission in ^{235}U, they have to be slowed down by collisions in the moderator to thermal energies (around 0.03 eV) and since in natural uranium ^{235}U is only present as one part in 140 compared to ^{238}U there is plenty of opportunity for the neutron to do other than cause fission.

I hope that this rather lengthy introduction to the example illustrates that the mechanisms of fission are understood and that an appropriate way of investigating the relationships, between the overall size of the core containing these mechanisms, the ratio of moderator to ^{235}U, and the ratio of ^{235}U to ^{238}U (this can be changed from that existing in natural uranium by a process known as enrichment), could be through a model which is in terms of the probabilities that certain events could occur.

The model is established by first of all (for a single incident neutron) ascribing probabilities to the production of fission products $(P_a, \ldots, P_z)$ and to the production of a number of further neutrons P. For each new neutron the following probabilities are sampled:

P_1 the probability that further fission will be caused
P_2 the probability of absorption in fission products
P_3 the probability of absorption in ^{238}U
P_4 the probability of absorption in core materials
P_5 the probability of loss from the reactor core

These probabilities are a function of the size and constituents of the core and so a large number of trials are carried out for each of a set of design conditions and the ratio of final neutron number is compared to unity for each set. That set is selected which gives the desired factor of supercriticality.

The Monte Carlo method, whch is useful as a design aid, relies on performing a large number of trials on a probability description of the mechanisms of the situation which are known. Another method which similarly relies upon repeated experimentation but where the mechanisms are not known has found considerable application in the chemical industry. Since the experimentation in this case is on the plant itself, the method is not useful as a design aid but is helpful in improving the performance of the process in operation. The method is known as evolutionary operation or EVOP (Box and Draper, 1969).

One of the problems associated with plant experimentation, where throughput, yield, or product quality are being assessed at different operating conditions, is that, for the duration of the experiment, off-specification product is likely to be produced. This is unlikely to be tolerated by the particular plant manager who has a programme of demand to meet. EVOP was developed in order to overcome this problem while still using the plant itself as the generator of real data. The basic philosophy underlying EVOP is that it should be possible to generate data on how to improve the product at the same time as producing it. Thus, if any forced perturbations in the system are small, they will not affect production nor produce an off-specification product *but yet* provide the data from which improvements can be made. Because these perturbations produce effects which are of the same magnitude as the random variations which always exist within the process (process noise), they must be repeated a sufficient number of times so that the effect of the noise is averaged out.

The experiments provide the data on which a particular model of the process is constructed. The model consists of a mapping of operating of conditions in the performance space of the process. Suppose that the measure of performance of the process is in terms of the cost of producing the product, i.e. the total raw material costs factored by the yield plus the cost of utilities, then at the current set of conditions the cost can be calculated. This provides one point in the map. Generally, the space is multi-dimensional but for illustration assume that only two variables affect performance. Figure A4 illustrates the formation of the map. The central cross, P_0, represents the starting conditions so that with variables X and Y at their current values, the cross represents the cost of production, i.e. one point in the performance space. Symmetrical changes in the variables are chosen so that ΔX and ΔY are small enough not to affect product quality.

An experiment is initiated by introducing the perturbation $Y_0 \rightarrow Y_0 + \Delta Y$, $X_0 \rightarrow X_0 - \Delta X$, say. The process is left for a period to allow any initial transient effects to settle out and the new measure of performance is calculated. This single perturbation will be contaminated by the process noise and so the procedure needs to be repreated a number of times and the average value of the measure of performance taken, yielding a second point on the map, P_C. It

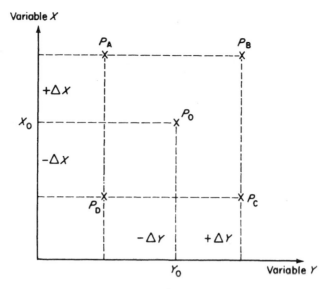

Figure A4. Performance map

is assumed that if the process noise is truly random then it will cancel itself out in the averaging process. The first stage of the mapping is completed by introducing the remaining sets of perturbations in order to identify the values of P_A, P_B, and P_D.

Since the measure of performance chosen in this example is product cost, the aim of the investigation will be to minimize this value within constraints (such as product quality). The relative values of P_0, P_A, P_B, P_C, and P_D are compared and, if, say, P_B is less than the remainder, the process conditions are changed by making the new values of X and Y, $X_0 + \Delta X$ and $Y_0 + \Delta Y$. This point P_B now represents the starting conditions for another four experiments centred on these values for X and Y. The result of a whole series of such experiments is that incremental improvement in the process performance is achieved without producing off-specification product. The procedure is rather like that of a blind person climbing or descending a hill by making tentative steps in four directions centred on their present position. The performance assessment is in terms of height gained or lost. The shape of the performance map obtained will then represent the shape of the relevant part of the hill.

In the case illustrated the map becomes the performance model of the plant under investigation. An actual case example which uses this method of modelling is given in Appendix II.

Dynamic Deterministic

The most common modelling language used in this particular category is that of differential equations. A convenient tool for solving such equations is the general

310

purpose analogue computer, though similar methods of solution are now available for digital computers. The underlying principle of solution will be briefly described in relation to this kind of model, though the reader is referred to other texts for a more detailed introduction (e.g. Welbourne, 1965).

Consider the following simple first-order linear differential equation:

$$A\frac{dx}{dt} + Bx = C,$$

where x is the variable of interest, A, B, and C are constants, and t is time. The method of solution is obtained by assuming that the highest derivative is known and rewriting the equation to show how it could be derived i.e.

$$\frac{dx}{dt} = \frac{1}{A}[C - Bx].$$

In an analogue or digital computer, functions are available which perform the mathematical operations of integrating, summing, and scaling (these are usually accompanied by sign changes in an analogue computer but these are ignored for ease of illustration). Given the derivative dx/dt as input to an integrator, one can obtain as output the variable x. If this is scaled by $- B/A$ and summed with C/A, the right-hand side of the above equation can be derived. This is equal to the derivative dx/dt and hence can become the input to the integrator. The circuit to solve this simple equation is therefore

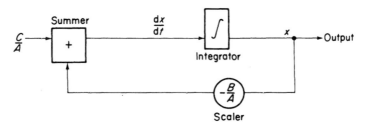

Once the circuit is assembled, an input of C/A applied to the summer will yield the output, which is the time-dependent solution of the original differential equation. Thus, if the situation to be modelled can be expressed in differential equation form, an analytic model (in Ackoff's terms) can be assembled. By applying the above principle it can be converted into an analogue form of model.

As an example, consider the problem of designing a suspension system for a vehicle. The design difficulty is to decide the damping coefficients for the shock absorbers and the spring stiffnesses required to give a particular damped performance on various road surfaces. To simplify the analysis let us make the assumption that the vehicle has a symmetrical weight distribution and that a study of the dynamics of one wheel and its suspension will be representative of the total vehicle behaviour. An analytic model can be derived by considering a force balance on the various components but the initial production of an

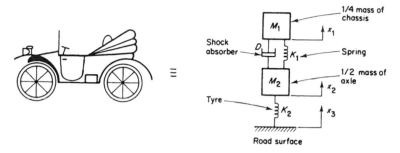

Figure A5. Mass–spring analogy

analogue model of a particular kind will help. Consider the wheel and its suspension to be represented by a mass–spring analogy as in Figure A5. It is assumed that the tyre can be represented by a simple spring with stiffness K_2, where $K_2 = f_2(P)$, P is the tyre pressure, and hence K_2 is some function, f_2, of this pressure. The main design parameters to be determined are K_1, the main spring stiffness, D, the shock absorber damping coefficient, and P, the tyre pressure. The displacements x_1, x_2, and x_3 can be considered as deviations from rest conditions so that, in the steady state (i.e. at $t = 0$),

$$x_1 = x_2 = x_3 = 0.$$

The model is derived by considering a force balance for each mass in turn:

$$[\text{mass} \times \text{acceleration}] + [\text{damping} \times \text{velocity}]$$
$$+ [\text{stiffness} \times \text{displacement}] = 0.$$

For mass 1,

$$M_1 \frac{d^2 x_1}{dt^2} + D\left[\frac{dx_1}{dt} - \frac{dx_2}{dt}\right] + K_1[x_1 - x_2] = 0. \tag{A1}$$

For mass 2,

$$M_2 \frac{d^2 x_2}{dt^2} + D\left[\frac{dx_2}{dt} - \frac{dx_1}{dt}\right] + K_1[x_2 - x_1] + K_2[x_2 - x_3] = 0. \tag{A2}$$

The final equation required to complete the model is that for the independent variable x_3, i.e.

$$x_3 = f_1(t). \tag{A3}$$

This equation is known as the forcing function since the whole behaviour is dependent upon the form of x_3 and it is this function which disturbs the system from rest. It could take the form of a step change to stimulate the vehicle

312

mounting the pavement or it could be an impulse to simulate the vehicle going over a pothole in the road or any other form the designer wishes to consider.

The set of equations (A1), (A2), and (A3) represents the analytic form of model for a continuous time solution. Conversion into an analogue form requires rewriting equations (A1) and (A2) in terms of the highest derivative. The rewritten equations are

$$\frac{d^2x_1}{dt^2} = \frac{K_1}{M_1}[x_2 - x_1] + \frac{D}{M_1}\left[\frac{dx_2}{dt} - \frac{dx_1}{dt}\right], \tag{A4}$$

$$\frac{d^2x_2}{dt^2} = \frac{K_2}{M_2}[x_3 - x_2] + \frac{K_1}{M_2}[x_1 - x_2] + \frac{D}{M_2}\left[\frac{dx_1}{dt} - \frac{dx_2}{dt}\right], \tag{A5}$$

and the corresponding analogue model is given in Figure A6.

Typical solutions for x_1 (the vehicle body) following a step change in road surface are of the following form:

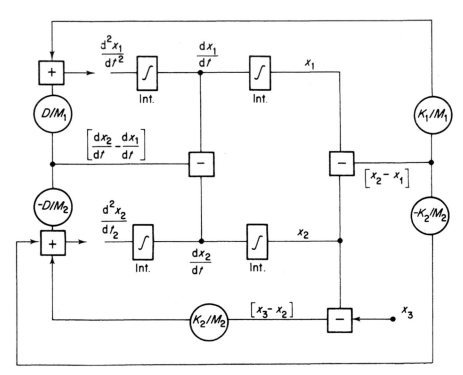

Figure A6. Analytic and analogue models for vehicle suspension analysis (see also equations (A4) and (A5))

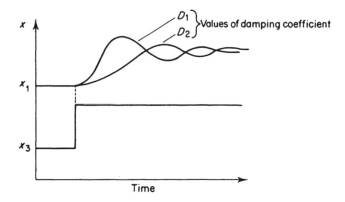

A number of books have been produced which consider, in detail, the principles of analytic modelling related to continuous time simulation. One which is a good introduction to the use of mass–spring analogies is by Shearer *et al.* (1967) and one which concentrates on the use of physical principles to develop models related to various processes is by Campbell (1958).

Dynamic Non-deterministic

The previous section has briefly described the use of differential equations for the simulation of situations in which the time-dependence can be represented continuously. The most obvious way to deal with discrete-event simulation is to replace the differential equations by difference equations. In this case, the data available is in the form of a time series or a sampled form of continuous data. In addition, sufficient knowledge is usually unavailable about the actual mechanisms which govern behaviour. Thus, a model is constructed which takes an input series and reproduces the appropriate output series based purely upon a statistical correlation between the two. For example, if a model is required to represent a production process, data may be available in the form of past weekly demand figures and past weekly production. These two time series can be used to construct a set of difference equations in which the parameters are adjusted to minimize the sum of the square of errors computed over, say, a couple of years' worth of data. A number of techniques are available to perform the correlation (Anderson, 1976), but one of the most successful is that due to Box and Jenkins (1970). As an example consider the problem of controlling fibre concentration (basis weight) in paper production. Figure A7 illustrates the basic production process. A water–fibre mixture (thick stock) is fed via a stuff valve into a flow box. The mixture (about 90% water) is fed on to a fine mesh wire and, as the mixture travels down the wire, water is extracted. At the end of the wire the paper (still 40–50% water) is fed into steam-heated dryers prior to reeling. The main control problem is that the measurement of basis weight can only be made after the dryers, whereas the control action is initiated at the stuff

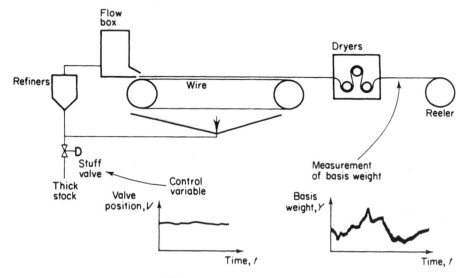

Figure A7. Paper machine

valve. Because of the large time delay that exists between the measurement and the point of control, feedforward control must be used based upon a prediction of the likely result of some control action. Hence a model must be used that represents the dynamics of the whole process.

The essence of the Box–Jenkins approach is that two models are produced. The assumption is made that the output time series is a combination of effects produced by the process, and hence correlated with the input, together with a noise component which is uncorrelated with this input. The following diagram illustrates the way in which the noise and process models are related:

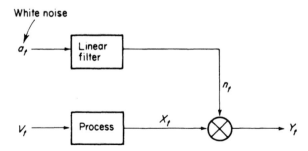

In the case of the paper machine, the variable Y_t is the measure of basis weight and V_t is the valve position (both at time t). Sophisticated correlation programs have been developed which estimate the parameters in both the linear filter (which takes white noise as its input) and the process model so that the error in the actual output Y_t compared to the predicted value are minimized. The structure of the models is also varied as the analysis proceeds in order to make use of the simplest formulation. In this particular case the form of the two models was as follows: the prediction of process output at time t for a control input at time $t - 1$ was given by

$$(1 - \delta_1 B + \delta_2 B^2)X_t = g(1 - \delta_1 + \delta_2)V_{t-1},$$

where δ_1, δ_2, and g are constants and B is the backward shift operator. The noise model was of the form

$$(1 - B)^r Q(B)n_t = P(B)a_t,$$

where r is a constant; P and Q are polynomials in B.

Thus, for actual operating records, which are time series in both V and Y, the constants can be estimated and the polynomials derived.

Another form of discrete event simulation, which is frequently used, makes use of random number generation as a means of deciding whether an event takes place or not. A simple example which illustrates this approach is related to an investigation which was concerned with the reliability of process units used in cement production. There are three major units which take the raw material from a quarry and produce cement. These are the crusher, the raw mill, and the kiln. The problem was to investigate the effects of various maintenance strategies on the availability (and hence production) of the units as a whole. The time to fail for any unit can be taken to be a statistical variable and the time to repair can also be represented by a statistical distribution. In simplified terms the simulation can be represented by Figure A8. It is assumed that, unless a unit has failed, full output is being achieved. The random number generator is initially used to determine the time at which a unit fails and also

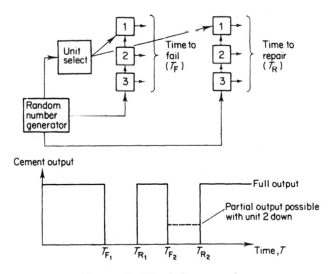

Figure A8. Simulation procedure

selects which of the three units it is. For each unit there will be an associated distribution of time to repair and the random number generator is again used to determine a repair time from this distribution. During the repair time it is assumed that no output is achieved. However, for one of the units (because of in-process storage) reduced output is possible during the repair time. The plot at the base of Figure A8 shows a typical response. Here full output is achieved until unit 1 fails after a time T_{F_1}. The time to repair is $T_{R_1} - T_{F_1}$ and during this period no output is produced. Following the repair, full output is again achieved until unit 2 fails at time T_{F_2}. During the repair time $T_{R_2} - T_{F_2}$, partial output is possible. Various maintenance strategies can be assessed by changing the time to fail, though they will have little effect on the time to repair. The time to repair distributions can be adjusted according to the level of maintenance resources available.

For further information on discrete event simulation the reader is referred to Tocher (1964) for an introduction to the subject, Anderson (1976) for time-series analysis, and Pritsker and Kiviat (1969) for a description of a particular simulation language.

The Process of Model Building

The real world is extremely complex and models of situations in the real world cannot be expected to reproduce that degree of complexity. *A model is always a simplification.* Because of this, any modelling activity must include an explicit statement of the assumptions that must have been made about the real world in order that a model can be derived at all. These assumptions may be about the choice of boundary of the situation being modelled, or that certain variables may be ignored in relation to others, according to the nature of the investigation. Our concern may be about the dynamic behaviour of a part of the real world in which mechanisms are known to exist, which affect dynamic behaviour, but which change over very different time scales. The problem of critically in a nuclear reactor, referred to earlier, is one such example. Once the reactor core is installed as the primary heat source in a power station, the control of power output is achieved through the control of criticality in the core. This is achieved by moving absorber (control rods) into, or out of, the core to produce subcriticality (if it is desired to reduce power) or supercriticality (if it is desired to increase power). Power does not change instantaneously and the degree of criticality change is a complex, time-dependent mixture of changes in the various probabilities for absorption and fission. The probability that fission will occur is a function of the temperature of the uranium fuel and this changes relatively quickly (with a time constant of the order of seconds), whereas the probability that absorption will take place in one of the fission products (xenon-135, for example) is a function of its concentration, and this changes slowly (with a time constant of around 24 hours). Thus, if the model which is being developed is to be used to investigate the design parameters for the control rod system, the short-term effects are important. Hence the concentration of xenon-135 can be assumed to be constant, while the equations, which determine fuel and moderator temperature changes, need to be expressed as time-dependent differential equations. On the other hand, if the model is being developed to investigate the long-term power behaviour, it is reasonable to assume that the fuel and moderator temperatures change instantaneously (and are, hence, described by algebraic equations), while the dynamics of fission product generation and decay are described by time-dependent differential equations.

The conditions under which certain effects can be assumed constant, or under which certain effects can be assumed to be instantaneous, are, of course, a matter of judgement on the part of the analyst, and this judgement needs to be justified. Thus a necessary part of the model development is the setting up of tests of the model to demonstrate that the assumptions are, in fact, reasonable.

To return to the general process of model building, Figure A9 attempts to clarify the major considerations and their relationships. Figure A9 is itself a model of the development process. This is shown to be an iterative process in which the testing of the model leads to some learning about the adequacy of the model. This in turn may modify the criteria by which the model is judged and, through modification (or not) of the modelling language, may lead to the reformulation of a tentative model. This iteration continues until both a satisfactory model and accompanying assumptions are achieved. The broad arrow indicates that the content of the model is about the particular real-world situation, filtered by the set of assumptions.

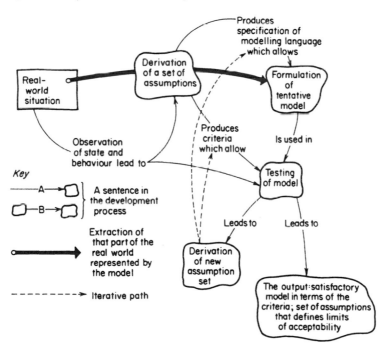

Figure A9. Model development process

318

It is frequently the case that, for a given situation or for a particular stage in the analysis, a single model is inadequate or too complex to be practicable. This feature is independent of the modelling language and can apply equally to models which are entirely conceptual as well as models which are analytic in form.

A way of overcoming this problem is to develop, not a single model, but a hierarchy of models. A concept that is crucial to this type of development is that of *resolution level*, or level of detail. The characteristic that differentiates one level in the hierarchy from another is the degree of detail with which the elements of the model are expressed. It is usual that the highest level in the hierarchy contains a broad description of the situation (low resolution) while the lower levels contain increasingly more detailed descriptions of less and less of the situation being modelled (high resolution).

The concept of resolution level is an essential consideration, though one which is difficult to apply, in the development of models of human activity systems (a particular class of conceptual models that we devoted considerable attention to in Chapter 2). In the meantime I will illustrate the concept by describing an application to a problem from the harder end of the problem spectrum.

The early designs of advanced gas-cooled reactor (AGR) nuclear power stations were required to contribute to the stability of the national electricity grid system. This meant that they had to be capable of changing power rapidly to compensate for loss of supply caused by other stations going off-line. The

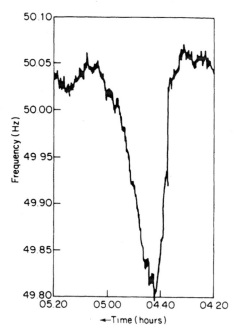

Figure A10. Variation of grid system frequency following the loss of a 500 MW set generating at 370 MW on 22 January 1969. [Reproduced by kind permission of M. J. Whitmarsh-Everiss, from Whitmarsh-Everiss (1969)]

indication that such a situation has occurred is a drop in the frequency of the electricity supply. This is controlled at a value of 50 Hz and ideally should not fall below 49.8 Hz. Such tight control provides the requirement that a station operating at 75% load should increase its output initially by 15% at the rate of 7.5% per second. This initial power response must be sustained and the output increased to 100% in 3–5 minutes. The kind of disturbance which gives rise to this specification is illustrated by Figure A10. The significance of this specification is that the station control systems must be highly sensitive to variations in grid frequency. Thus, any disturbances in grid frequency (due to faults or normal operation – see Figure A11) are rapidly transmitted to the reactor itself. This results in temperature cycling of the fuel pins within the core, leading to potential fracturing of the pin casing. In order to investigate the dynamic behaviour of a single pin as a result of grid frequency disturbances it is necessary to model at the level of the power station itself as well as at the detail of a single pin (which is 2 cm in diameter and 2 cm long in a core which is 10 m in diameter and 10 m high).

It is not feasible to consider this range of detail within a single model. A hierarchical model was used and developed as follows. First, a station model was constructed which contained analytic descriptions of the turbo-alternators and steam-raising units and a simple point model of the reactor core. The control systems for each unit were included and, for typical grid disturbances, the behaviour of the reactor controller was obtained. At the next level down the

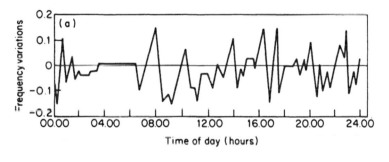

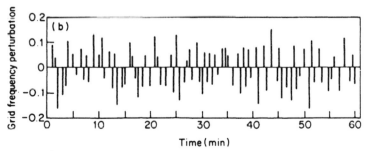

Figure A11. Grid frequency disturbances under normal operation: (a) variation in grid system frequency; (b) grid frequency noise. [Reproduced by kind permission of M. J. Whitmarsh-Everiss, from Whitmarsh-Everiss (1969)]

320

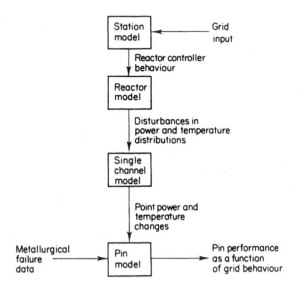

Figure A12. Hierarchy of models for fuel pin response investigation

reactor controller behaviour was taken as the input to a three-dimensional model of the reactor core in order to determine the disturbances in power and temperature distribution. At the next level, a single channel was selected which represented the worst situation in terms of power and temperature changes and these were used as the input to a single channel model in order to determine the position of the single fuel pin that was experiencing the maximum cycling. A single-pin model was then used to estimate the failure conditions given the cycling disturbances as input. Thus by using this particular hierarchy, the effects of grid disturbances on fuel pin failure behaviour were investigated. At each level major simplifying assumptions were made in order to reduce the complexity of the description for the components at the next level down. This complexity was then restored as the resolution level was increased but less and less of the power station was being considered. An illustration of the hierarchy is given in Figure A12.

Appendix II

A Systems Study of a Petrochemical Plant*

Gwilym M. Jenkins

Introduction

This paper describes a systems study of a petrochemical plant with the objective of reducing the cost per ton of manufacture of the product. The study showed that this could be most economically achieved by using a form of plant experimentation known as evolutionary operation, or EVOP. During the course of a six months' study, new process conditions were introduced which resulted in savings of £80 000 per annum at a total cost of £6000, including the cost of systems effort. To safeguard the security of the firm where the work was carried out, certain aspects of the description of the system have had to be disguised and all cost data has had to be changed. However, these changes should not detract from the main points brought out in the case study. The presentation follows closely the stages in the systems approach described elsewhere (Jenkins, 1967), namely:

1. *Systems analysis*
   1.1 Formulation of the problem
   1.2 Organization of the project
   1.3 Definition of the system
   1.4 Definition of the wider system
   1.5 Definition of the objectives of the wider system
   1.6 Definition of the objectives of the system
   1.7 Definition of the overall economic criterion
   1.8 Information and data collection

---

*This appendix was originally published as a paper of the same title in the *Journal of Systems Engineering*, 1, 90–101 (1969) and is reproduced with the permission of the publishers. Minor alterations have been made to maintain the style of the present volume.

2. *Systems design*
   2.1   Forecasting
   2.2   Model building and simulation
   2.3   Optimization
   2.4   Control
   2.5   Reliability

3. *Implementation*
   3.1   Documentation and sanction approval
   3.2   Construction

4. *Operation*
   4.1   Initial operation
   4.2   Retrospective appraisal
   4.3   Improved operation

Systems Analysis

Recognition and formulation of the problem

The problem started with a cry for help from the works manager in charge of manufacturing a group of products made in several plants under his supervision. The works manager had noticed that the daily production costs per ton (derived from accountancy data) of manufacturing one particular chemical seemed to vary a great deal, depending on process conditons. Changes in process conditions were usually made in an unsystematic way by the process operators, mainly to compensate for the effect of variations in feedstock quality on the purity of the main product.

As a consultant to the company, the author was called in to help with this problem. Discussions with the plant manager, responsible for the particular plant on which this chemical was made, brought out the following facts:

(1)   The plant manager agreed with the works manager that process conditions were not correctly chosen to achieve the lowest production costs.

(2)   The high production costs now being realized were partly due to the fact that, while product quality was appropriate for the end use of the chemical within the company, it was above the specification for the end uses of three external customers.

(3)   Discussions with the three external customers revealed that their original quality specifications were realistic and that, from their point of view, no special benefits resulted from their being supplied with a product whose quality was slightly better than they had specified.

(4)   Historically, what had happened was that the internal customer had been the dominant outlet for the product in the initial years, the outside market

having been developed only in the past few years. However, nothing had been done to modify process conditions – the objectives had changed but there had been no consequent change in the design of the system! It was established that a half-hearted attempt had been made to relax product quality for the outside customers about 18 months previously. However, the changes made then resulted in the production of some off-specification material and hence further attempts at relaxing quality requirements were abandoned.

(5) A very quick analysis of historical data suggested that if the best operating conditions achieved in the past had been realized at all times, savings of the order of £70 000 per annum might have been made.

To sum up, it had been established that:

(a) The works and plant managers both thought that the problem was an important one.

(b) It was sufficiently high on their list of priorities for them to give it sound backing.

(c) The qualitative opinion of the two managers, that there were substantial savings to be made, had been substantiated by a quick analysis of plant data – the problem was worth solving *now* and it was decided that manpower should be made available to enable the work to be carried out.

Therefore, it was decided that a systems study of the plant should be undertaken, its prime objective being to reduce production costs.

Organization of the project

The systems team

A small systems team was set up to carry out the investigation. This consisted of (a) the plant manager (part-time); (b) a chemist who had been involved in some relevant process research (part-time); (c) a chemical engineer who had considerable experience of process investigation and acted as team leader (full-time); (d) a statistician (part-time); (e) the author as consultant (part-time).

Scheduling the project

(1) It was agreed that the team should report back within 5 months to the works manager with some proposals for reducing production costs.

(2) Project planning was to be on the lines shown in Figure A13 and consisted of five stages. A rough project schedule, such as that given in Figure A13, is essential to instil *discipline* into the way that things are done. Otherwise, projects drift on aimlessly from one week to the next without much being achieved.

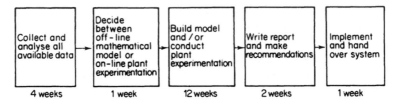

Figure A13. Rough project schedule for systems study of petrochemical plant

Stage 1 (4 weeks)

Four weeks were to be allowed for a preliminary investigation in which plant performance would be studied in detail by analysing historical data and collecting new data where necessary. In this way areas which were sensitive to cost could be isolated and then subjected to more detailed study.

Stage 2 (1 week)

By the end of stage 1 it was felt that enough facts should be available to make some important decisions about the future course of the project. In broad terms these were:

(1) What modifications to plant units, instrumentation and/or changes to process conditions would help to reduce production costs?
(2) How was the necessary optimization to be done? Possible methods of approach were: (a) using intuition and *ad hoc* methods; (b) building a mathematic model and using it *off line* to suggest plant modifications and/or changes to process conditions; (c) carrying out *on-line* plant experimentation.

Stage 3 (12 weeks)

It was felt that 12 weeks should provide enough time to produce worthwhile conclusions. During this period, the chemist and statistician could step up their involvement as required.

Stages 4 and 5 (3 weeks)

Ample time was allowed to write a report and to leave behind adequate documentation and instructions on how the improvements were to be realized by plant personnel.

After the rough project schedule was mapped out, a more detailed *critical path schedule* was constructed showing the various activities which had to be carried out in parallel if targets were to be met. Tasks could then be allocated to team members and their individual targets reconciled with the project schedule. Critical path schedules invariably have to be updated as the project proceeds, but again are essential if discipline is to be injected into the working of the team. Targets were reviewed at progress meetings held weekly and steps taken to ensure that the work of the team was adequately documented at each stage.

The system

A very simplified flow diagram of the plant is shown in Figure A14. The reaction scheme was

hydrocarbon + oxygen → organic oxide + carbon dioxide + fuel gas + heat.

The hydrocarbon feed, air, and recycle gas are mixed in a mixing vessel and then preheated before being fed to the reactor. The reaction takes place at a high temperature over a catalyst and the exist gases are cooled in a cooling system. This incorporates a waste heat boiler for producing LP (low pressure) steam which is fed into the LP steam main and is credited against the plant production costs. The cooled gases are then fed to an absorption column where the carbon dioxide is absorbed. The top product steam from the absorption column is first purged to avoid the build-up of nitrogen in the system and then recycled to the mixer. The bottoms product of the absorption column is fed to a separation system where the main product is separated from other gases which are used as fuel gas to heat boilers in another part of the site. Again the fuel gas is credited to the plant.

The wider system

The wider system consisted of a complex of plants on the same site, all making petrochemicals (Figure A15). The main hydrocarbon feedstock used in the petrochemical plant being studied was produced in another plant on the site. This hydrocarbon feedstock was also used by a number of other plants on the site. Approximately 60% of the oxide produced was fed on to a plastics plant on the same site and the remaining 40% sold to three outside customers.

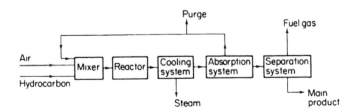

Figure A14. Simplified flow diagram of system studied

The objectives of the wider system

The overall production plans for plants on the site were produced on a computer using a linear programming model. The objective of this model was to increase the total profitability of all plants on the site and thus make a contribution to improving the company's overall profitability. The effect of the wider system on the chemical plant studied may thus be summarized as follows:

(1) The linear program specified how many tons of oxide were to be manufactured each month. It made due allowance for the fact that the plant was a competitor for its hydrocarbon feedstock with other plants on the same site. However, plastics manufacture took top priority whereas oxide manufacture for the outside manufacturers had much lower priority, relative to some other end uses of the hydrocarbon feedstock.

(2) Two different oxide quality requirements had to be met: (a) a higher quality for plastics manufacture; (b) a lower quality for the three outside customers.

The objectives of the system

The objectives of the system were dictated by the objectives of the wider system and could be summarized as follows: (a) to produce the specified amounts of oxide of appropriate quality for plastics manufacture and for the three outside customers, as determined by the site LP model; (b) to produce these specified amounts at minimum cost, thereby making a contribution to increasing the profitability of the whole site.

Definition of the overall economic criterion

Referring to Figure A14, the basic plant costs were (a) raw material costs and (b) utility costs, and there was also credit to the plant due to export of (c) fuel gas and (d) low pressure steam.

Suppose that, in a particular day,

N tons of oxide are produced.

M tons of hydrocarbon are needed at a cost of £x per ton.

P tons of LP steam are produced yielding a credit of £x per ton, this being the marginal cost of raising steam.

Q tons of fuel gas are produced yielding a credit of £z per ton, based on its calorific value.

Utility costs (steam water, electricity) are £u

Then the economic criterion used was the average cost per ton of manufacturing the oxide for this particular day, which is

$$0 = £\left(\frac{Mx + u - Py - Qz}{N} \right).$$

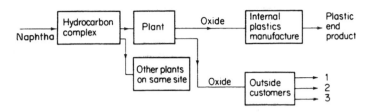

Figure A15. The wider system of which the plant studied was part

The objective was to minimize this economic criterion subject to the constraints that three different quality control criteria relating to product quality had to be satisfied. As mentioned earlier, two different sets of quality control criteria were used during the course of the study, one set relating to internal plastics manufacture and another set to the outside customers.

Although, on the face of it, the economic criterion looks very straightforward, great care is needed when an optimization is based on 'internal transfer prices' such as x, y, and z. Since the main objective here is to sell the plastic end product, there is a danger that if the transfer prices are unrealistic in the wider context of plastics manufacture, sub-optimization might result. However, a careful study of this aspect revealed that the economic criterion was realistic since (a) in times of hydrocarbon shortage, plastics manufacture received top priority, and (b) the cost £x per ton for the hydrocarbon was based on a careful breakdown of the marginal production costs in the complete cycle from naphtha to plastics.

Information and data collection

Referring to Figure A13, stage 1 of the project was devoted to collecting and analysing all available plant data. During this work, the following points were established:

(1) Irrespective of whether a mathematical model was to be built, or whether the problem was to be solved by plant experimentation, the plant *instrumentation* would have to be sufficiently accurate for measuring the variables which entered into the economic criterion. Hence the instruments used to measure the main process variables were checked and found to be adequate except for two. One instrument was found to be incorrectly calibrated and had to be recalibrated. Another instrument was found to be faulty and was replaced.

(2) There were no constraints on throughout requiring modifications to any of the plant units.

(3) Plant data over a period of 9 months was collected and analysed. An

empirical model was fitted to the data using *regression analysis* in which the cost per ton of manufacture was regressed against the main process variables. Although it is often dangerous to read too much into this form of analysis, based as it is on operating data in which no attempt has been made to make *planned changes* to the variables, it was concluded that only three or possibly four variables were important. Two were singled out on chemical engineering grounds, namely (a) the *reflux ratio* in the distillation column, which could be adjusted to affect the quality of the product, and (b) the recycle flow rate to purge flow rate ratio, which affected the materials loss in the system. Both of these variables would be expected to have significant effects on production costs.

(4) The plant had been bought under licence and hence there was little kinetic and thermodynamic data available to build a mathematical model which could be optimized *off line*. It was concluded that it would be too expensive, both in time and money, to develop such a model. Hence it was decided to use plant experimentation to search for better process conditions which might result in lower production costs. In the initial stages it was decided to make changes to the two variables (a) and (b) singled out in (3) above.

Systems Design

Forecasting

Production planning department reported that, for the foreseeable future, the plant would be operated with raw materials of the same quality as those used in the previous year. They also reported that the product-mix was expected to remain as at present, that is approximately 60% of the oxide would be used for plastics manufacture and 40% sold to outside customers.

Model building and simulation

Since on-line plant experimentation was to be used, any modelling would have to go hand-in-hand with the optimization of process conditions. By this is meant that the tentative models fitted to data obtained from plant experiments would then be used to adjust process conditions, leading to a new model, and so on until no further improvement was possible.

Optimization

In a further progress meeting, called to discuss the plant experiments, the following conclusions were reached:

(1) Large changes in the two variables mentioned above would have to be ruled out since (a) they might generate large dynamic disturbances which

would upset plant performance; (b) they might increase the risk of operating the plant near explosive conditions; (c) owing to the sensitivity of the separation system to relatively small changes in process conditions, they might increase the risk of producing off-specification material.

(2) Hence it was decided to make small changes in the two chosen variables about the existing operating values of 6.9 for the reflux ratio and 7.75 for the recycle to purge ratio. The changes were made according to the experimental plan or design shown in Figure A16. This technique was introduced by G. E. P. Box (Box, 1957; Box and Draper, 1969), who called this form of plant experimentation *evolutionary operation* (EVOP).

(3) Because of the need to avoid disrupting plant operation, the maximum permitted changes in the variables shown in Figure A16 were such that their effect on the output variables would be only of the same order as the inherent noise in the process. Hence, after conducting experiments at each of the five sets of conditions shown in Figure A16, it would be necessary to repeat the experiments a number of times and average the results until a definite pattern emerged in the costs of manufacture at each set of conditions. A decision could then be made as to the direction in which to move so as to achieve lower costs of manufacture.

(4) The changes to be made were sufficiently small for transient effects to have died out after 4–6 hours. A further 18 hours of steady operation would then be allowed to make measurements of the variables influencing cost. Hence a single experiment would take one day in all.

(5) Steps would be taken to arrange for flowmeters to be read at regular intervals and dips taken in storage tanks so that the amount of hydrocarbon used, the amount of product formed, the utility costs, and export of steam and fuel oil could be monitored.

(6) Quality was to be measured accurately during the course of the experiment. It was recognized that a point might be reached when the quality constraint for plastics manufacture would be violated. Hence

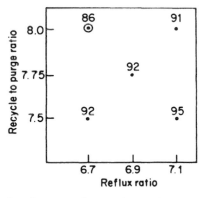

Figure A16. Conditions for plant experimentation and corresponding average costs per ton after five cycles or repetitions of phase 1 of the EVOP experimental plan

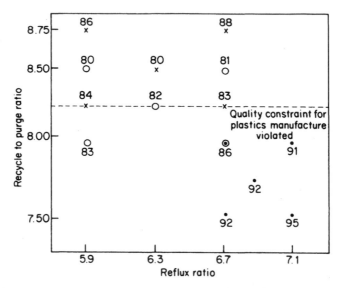

Figure A17. Average costs per ton at various process conditions during EVOP programme. *Key:* ●, phase 1 (cycle 5); ○, phase 2 (cycle 5); ×, phase 3 (cycle 4)

further experimentation would be possible only on those occasions when product for outside customers was being manufactured.

Simple statistical models were fitted to the data, obtained from the plant experiments, and the statistical analysis is summarized in Appendix A (see p. 334). After five cycles, or repetitions, of the experimental design, the average costs per ton shown in Figure A16 were realized. It was considered that enough information had now been obtained to suggest a move to better operating conditions. Hence a second phase of experimentation was started, centred at a reflux ratio of 6.3 and a recycle to purge ratio of 8.25, as indicated by the circles in Figure A17. During the course of this phase, the quality constraints for plastics manufacture were violated at certain process conditions. Hence, for these conditions, experiments were only possible thereafter when product was being manufactured for outside customers.

After five cycles of experimentation in the second stage, it was concluded that on optimum had been reached with respect to reflux ratio but not with respect to the recycle to purge ratio (see Appendix A). Hence a third phase of experimentation was centred at a reflux ratio of 6.3 and a recycle to purge ratio of 8.5, as shown by the crosses in Figure A17. The costs per ton after the end of phase 3 show very clearly that no further improvement is possible.

Conclusions

(1) The quality constraints were not violated during the course of the experimental programme for the product sold to outside customers. The

best operating conditions, corresponding to a reflux ratio of 6.3 and a recycle to purge ratio of 8.5, resulted in an average production cost of £80 per ton. This represented a reduction of £12 per ton from the figure of £92 per ton which was the value realized at the beginning of the experimental programme.

(2) The quality constraint was violated for the product used for plastics manufacture during phase 2 of the experimental programme. Some further experiments were run at a reflux ratio of 6.3 and a recycle to purge ratio of 8.0, yielding an average cost per ton of £83.5. This set of conditions was recommended for product to be used for plastics manufacture since it was sufficiently far away from the quality constraint boundary to make the risk of producing off-specification material acceptably small. (This point will be referred to later when we deal with the control stage.) In this case the cost of £83.5 per ton, corresponding to the new process conditions mentioned above, represented a reduction of £8.5 per ton from the previous figure of £92 per ton.

(3) Since plastics manufacture accounted for approximately 60% of the output of the plant, the expected average reduction in the cost per ton was

$$(0.4) £12 + (0.6) £8.5 = £9.9 \text{ per ton.}$$

Since the throughput of the plant was 10 000 tons per annum, this represented an expected saving of just under £100 000 per annum.

The experimental programme had taken 70 days, that is 14 cycles of 5 days, and was spread over a period of $4\frac{1}{2}$ months. The extended period was due to plant upsets and also to the fact that, from phase 2 onwards, experimentation was not possible on those days when the product was being made for plastics manufacture.

Control

The optimization, described above, resulted in the process being operated at conditions much closer to process constraints, namely quality constraints, than had been previously attempted. The experimental programme revealed that the existing control scheme for the reflux ratio, which varied the steam supply to the still reboiler, was not adequate to achieve the fine control required if the new operating conditions were to be maintained without violating the quality constraints. Accordingly, the control and instrumentation group were approached with a view to designing a better control scheme. As a result of their work, a new feedback controller was designed which resulted in much tighter control, without which it would have been necessary to operate the process at conditions much further removed from the quality constraint boundary for the two process variables. The resulting loss in profit would have far outweighed the cost of the controller (£500), so that the economic incentive for improved control could easily be justified in this case.

Reliability

No major reliability problems were foreseen except in relation to the functioning of the new controller. This was provided with a back-up alarm in case of malfunctioning so that manual control could then be reverted to.

Implementation

Documentation and sanction approval

A report was written which contained the following main recommendations:

(1) For product sold to outside manufacturers, process conditions should be set at a reflux ratio of 6.3, and a recycle to purge ratio of 8.5. This would result in an average reduction in the cost per ton of £12 from £92 to £80.

(2) For product used in plastics manufacture, process conditions should be set at a reflux ratio of 6.3 and a recycle to purge ratio of 8.0, yielding an average reduction in the cost per ton of £9 from £92 to £83.

(3) The expected savings from proposals (1) and (2) were approximately £100 000 per annum.

(4) Although two further process variables could have been introduced into the experimental programme, it was concluded that the likely reduction in production costs would be small and that systems effort could be used more effectively elsewhere.

(5) To operate near the quality constraints for proposal (2), it was necessary to install a new control scheme for the steam supply to the reboiler still at a cost of approximately £500. The increased profit which resulted from the improved control amply justified the cost of installing the controller.

(6) A better production planning scheme was needed so that longer runs could be achieved at each of the two sets of conditions now being recommended. Based on available storage and demand patterns, a scheduling scheme was proposed in the report.

(7) Proposals (1) and (2) had other implications for the running of the plant. These required that new operator manuals should be produced as separate documents.

(8) Finally a time table was proposed according to which the above recommendations could be completed in 5 weeks.

These recommendations were accepted in their entirely and approval was sought and obtained to implement them.

Construction

The only new hardware or software needed for this project was the feedback controller and its associated transducer, which were installed by the control and instrumentation department, at a cost of £500, and the repair and replacement of the instruments, mentioned earlier which cost £200.

Operation

Initial operation

The plant was run at the new conditions, together with the new scheduling system, for a period of 3 months. During this period the performance of the plant was monitored very carefully.

Retrospective appraisal

After the initial 3 month period of operating the new system, a three day exercise was mounted to compare economic performance during that period and the performance achieved during the year preceding the systems study. It was shown that

(1) After adjustments had been made due to increases in raw material costs and selling prices, the average cost per ton agreed with those predicted to within 1%.
(2) Based on performance during the three month period, the expected savings during the first 12 months were slightly under £80 000.
(3) The discrepancy between this figure and the predicted savings was due to the fact that (a) there had been a slight increase in the proportion of product made for plastics manufacture, resulting in smaller average savings per ton; (b) the increase in selling price to the outside customers had to be smaller than that which could be justified by subsequent increases in raw material costs. This latter was because customers maintained that they were now accepting lower product quality than previously and hence should pay a lower price.
(4) The economic performance of the plant was regarded as satisfactory and hence no further systems work was felt to be necessary.

Improved operation

Because the recommendations made during the original systems study had been substantiated by the retrospective appraisal, no further changes were made to plant operation. Hence this stage was not found to be necessary for this particular systems study.

Summary and Conclusions

The main costs of the systems study were:

Controller and instrumentation	£ 700
Salaries of systems team during period of study	£5250
Total	£5950

Hence the systems study had cost £6000 (excluding overheads) and had produced savings worth £80 000 per annum.

Appendix A: Statistical Calculations for EVOP Evaluation

(1)

Figure A18 shows the average costs per ton at the five points of the experimental plan after the end of the fifth cycle of phase 1 of the EVOP programme. The two process variables which were changed were: x_1, the reflux ratio in the distillation column, which could be adjusted to change the quality of the product, and x_2, the recycle to purge flow rate ratio, which affected the material loss in the system.

Effect	Estimate	95% confidence limits
1	4.0	± 3.1
2	− 5.0	± 3.1
12	1.0	± 3.1
CIM	− 0.8	± 5.8

Basic objectives of EVOP

The statistical calculations made in this appendix are described in greater detail in Box (1957) and Box and Draper (1969). The basic objective is to estimate the parameters θ_1 and θ_2 in the *linear model*

$$y = \theta_0 + \theta_1 x_1 + \theta_2 x_2, \tag{A6}$$

where θ_1 is the slope of the response surface $y(x_1, x_2)$ in the x_1 direction and θ_2 is the corresponding slope in the x_2 direction. Knowing the magnitudes and signs of θ_1, θ_2, the *line of steepest ascent*,

$$x_1/\theta_1 = x_2/\theta_2,$$

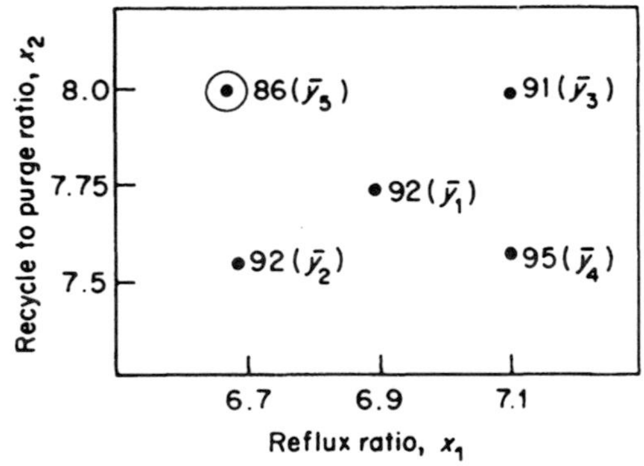

Figure A18. Analysis after fifth cycle of phase 1 of the EVOP programme

can then be calculated. This is the line along which greatest improvement in y can be expected. In practice, much cruder strategies are usually adopted, such as cornering the new design appropriate to a new phase at the best point of the design of the previous phase.

Model (A6) assumes that the surface is linear and this is a reasonable approximation initially when process conditions may be remote from the optimum. However, it is necessary to keep a check on whether a second order or quadratic model is required, which, together with the presence of small slopes would then indicate that the optimum was being approached. In addition, since the changes are of the same order as the noise level, the standard deviation of the noise must also be estimated so that due allowance can be made for its effect on the accuracy of the estimates. Thus, the model which is actually fitted to the data at the end of each cycle is

$$y = \theta_0 + \theta_1 x_1 + \theta_2 x_2 + \theta_{11} x_1{}^2 + \theta_{12} x_1 x_2 + \theta_{22} x_2{}^2 + e, \tag{A7}$$

where e is an error term representing the noise. Thus, the general strategy may be summarized: at the end of each cycle,

(1) The estimates θ_1 and θ_2 of the slope terms are calculated in a way to be described below.

(2) A check is kept on the size of the second order, or quadratic, terms $\theta_{11}, \theta_{12}, \theta_{22}$.

(3) An estimate of the standard deviation of the noise term e is obtained and then used to calculate statistical *confidence intervals* for the parameters. These are limits within which, with a given degree of confidence (say 95%), the true values of the parameters can be expected to lie.

In the initial stages of an EVOP programme, when one is likely to be far removed from the optimum, it would be expected that the slope terms θ_1 and θ_2 will be large relative to the second order terms θ_{11}, θ_{12}, and θ_{22}. However, as the optimum is approached, the second order terms will increase in value relatively until eventually a point will be reached when the slope terms are small compared with the second order terms and it can then be concluded that a point near the optimum has been reached.

Least squares estimates

Since there are only five distinct process conditions in the design of Figure A18 and six parameters in the model (A7), we can not hope to obtain unambiguous information about all six parameters. However, the design is so arranged that estimates of the important slope terms θ_1 and θ_2 are 'uncontaminated' with the estimates of the other parameters. The other estimates which can be obtained, apart from the θ_0 parameter which is not of any practical interest since it simply measures the average response at the centre point of the design, are those of θ_{12}

and $\theta = (\theta_{11} + \theta_{22})$. The term θ_{12} is usually referred to as the *interaction* term since it measures the extent to which improvements in y can be brought about by changing x_1 and x_2 simultaneously, and the term $\frac{4}{3}(\theta_{11} + \theta_{22})$ is usually referred to as the *change in the mean term* for reasons described below. Finally, it is customary in this work to quote twice the estimates $(2\hat{\theta})$ of the parameters rather than the estimates $\hat{\theta}$ themselves. The estimates $2\hat{\theta}$ are usually referred to as *effects* since, if we code, say, x_1, to lie between -1 and $+1$, then $2\theta_1$ measures the effect on y of a change in x_1 from -1 to $+1$.

The least squares estimates of the parameters may be obtained by minimizing the sums of squares of the errors e and the corresponding effects obtained by taking twice these values. Thus, for the data of Figure A18, the estimated effects are given by

$$
\begin{aligned}
1 = 2\hat{\theta}_1 &= \tfrac{1}{2}(\bar{y}_3 - \bar{y}_5 + \bar{y}_4 - \bar{y}_2) \\
&= \tfrac{1}{2}(91 - 86 + 95 - 92) = 4.0, \\
2 = 2\hat{\theta}_2 &= \tfrac{1}{2}(\bar{y}_3 - \bar{y}_4 + \bar{y}_5 - \bar{y}_2) \\
&= \tfrac{1}{2}(91 - 95 + 86 - 42) = -5.0, \\
12 = 2\hat{\theta}_{12} &= \tfrac{1}{2}(\bar{y}_3 - \bar{y}_4 - \bar{y}_5 + \bar{y}_2) \\
&= \tfrac{1}{2}(91 - 95 - 86 + 92) = 1.0,
\end{aligned}
$$

$$
\begin{aligned}
\text{CIM} &= \tfrac{4}{3}(\hat{\theta}_{11} + \hat{\theta}_{22}) \\
&= \tfrac{1}{5}(\bar{y}_2 + \bar{y}_3 + \bar{y}_4 + \bar{y}_5 - 4\bar{y}_1) \\
&= \tfrac{1}{5}(92 + 91 + 95 + 86 - 368) \\
&= -0.8
\end{aligned}
$$

The last estimate, CIM, or the change in the mean, may also be written $\bar{y} - \bar{y}_1$, where $\bar{y}$ is the mean of the five means $\bar{y}_1$, $\bar{y}_2$, $\bar{y}_3$, $\bar{y}_4$, $\bar{y}_5$. Hence the estimates of the second order, or curvature, terms $\theta_{11} + \theta_{22}$, are proportional to the difference between the mean of all experiments and the mean at the centre point of the design.

The calculation of the 95% confidence intervals is described in Box (1957) and Box and Draper (1969). It is necessary to emphasize that these confidence intervals are intended only to guide judgement in deciding when an effect is real or not. In no cases should it be necessary to wait until an effect is shown to be 'significant' in the usual statistical sense. Rather, a move to new conditions should be made once reasonable evidence has been obtained that the slope terms *1* and *2* are important.

Referring to Figure A18, note that the interaction and CIM terms are small compared with the slope terms. Hence a new phase was started with the new design cornered at the best point, £86 per ton, of the old design. It was also decided to double the spacing of the levels of variable *1* since it was felt that the original levels were too close.

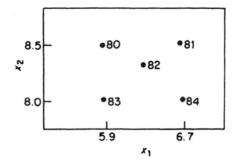

Figure A19. Analysis after fifth cycle of phase 2 of the EVOP programme

(2)

Figure A19 shows the results after the fifth cycle of phase 2. Note now that the effect of variable *1* has become much smaller, suggesting that an optimum is being approached with respect to this variable. Accordingly, the design for phase 3 was obtained by adding 0.25 to each level of x_2 in phase 1 and leaving the levels of x_1 unchanged.

Effect	Estimate	95% confidence limits
1	1.0	± 2.1
2	-3.0	± 2.1
12	0.0	± 2.1
CIM	0.0	± 4.0

(3)

Figure A20 shows the results after the fourth cycle of phase 3. Note now that the slope terms have changed sign from those given in Figure A19 but are

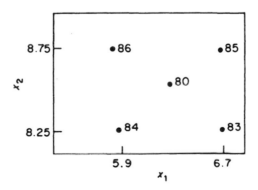

Figure A20. Analysis after fourth cycle of phase 3 of the EVOP programme

dominated by the change in mean term. Note also that no improvement has been achieved in this phase over the best point of phase 2. Accordingly, the conditions $x_2 = 8.5$, $x_1 = 6.3$ were accepted as being sufficiently close to the optimum to warrant stopping experimentation.

Effect	Estimate	95% confidence limits
1	− 1.0	± 2.9
2	2.0	± 2.9
12	0.0	± 2.9
CIM	3.6	± 5.0

Appendix III

Exercises

The exercises in this appendix are included to provide some practice in the application of the material in the book which is related to the use of human activity system concepts and associated methodologies. The exercises start at the level of basic systems thinking and progress to actual problem-solving assignments.

Picture Building

Systems thinking is about relationships as well as entities. Prose is particularly bad at conveying such relationships and it helps to clarify the thinking about a situation if it can be illustrated through a picture. The following passages are about problem situations. Convert them into pictures (this should not be done by merely writing statements in boxes and joining the boxes by arrows).

1 'Slimline', manufacturers of women's shoes, are hoping to improve their performance with a new range called 'Carefree'. (From a peak return on capital of 22% three years ago they have fallen to 15% then 11%.) The Managing Director discovers that the Production Department have introduced a new glue for sticking soles to uppers. This eliminates a sole-roughening process and has enabled them to achieve an 8% reduction in production cost. The MD, however, when investigating a Marketing Department complaint that for the last two weeks they have been 20% down on the supplies of shoes they need from Production, finds that the glue is in short supply. It seems that the Purchasing Section of the Production Department has cancelled an arrangement by which they received a 35% discount on bulk supplies of the original glue (this discount could not now be reinstated) and are buying the new glue ('STIX'), at the same price as the old glue, from a different supplier who has failed to meet delivery promises.

2 The workers at the Edge Hill (Liverpool) factory of Airfix Industries are currently staging a 'work-in'. This particular factory produces Meccano, the traditional construction toy, and Dinky toys, which range from model cars

through all varieties of vehicles to agricultural implements. The situation has arisen because Airfix have stated their intention to close the factory, making some 940 workers redundant. It is claimed that the factory has been losing money for years, mainly through the use of inefficient and antiquated production processes. Very little investment in new machinery has taken place over the last 50 years, resulting in production methods which are time consuming and labour intensive. For example, the individual pieces of Meccano are hand loaded on trays in the enamelling plant and each piece is then turned by hand. The workers claim that they have a viable product and, given the opportunity, they intend running the factory as a workers' co-operative. This would require financial support from the Government and a meeting has been arranged between local union officials and representatives of the Department of Industry to discuss the situation.

Meccano, which has been a household name in toys for most of this century, was invented by J. F. Hornby, a Liverpool businessman in 1893. As the business developed he added model trains and Dinky toys, all three products being highly successful. After the last war they suffered severe competition from other manufacturers, such as Lego, Triang, and Matchbox toys, resulting in the decline of the Meccano share of the market. Fifteen years ago, Hornby Trains Ltd was bought out by Triang, leaving the two product lines currently produced at Edge Hill.

3   The international airline Gulk Air, owned by the four Gulf States of Bahrain, Qatar, Oman, and the United Arab Emirates, has achieved a remarkable growth record. Over the past 5 years its turnover has climbed from 2 million dinars per annum to 100 million dinars per annum. During this period the board of the company, which consists of representatives of the four States, was under the chairmanship of H. E. Shirawi, a Bahrainian. He accepts that this growth rate cannot continue and that the company needs to assess how it acquires and uses its assets in a way which will consolidate their position and provide a stable base from which to undertake future development. Planning has been non-existent and the rapid growth is a result of *ad hoc* decision-making, at senior management level. There has been massive investment in aircraft, computers, housing, hotels, and recreation facilities. Expertise and computer packages have been bought in, mainly from other airlines, in order to provide an operating capability; a practice which cannot continue. Four-State ownership introduces constraints. Each State wishes to be seen as an equal partner; hence there are international airports in each State even though it is difficult to justify them on the grounds of the traffic to the Gulf. It is hardly a tourist area, and therefore the traffic consists mainly of businessmen, through-tourists, freight, and mail.

It has been argued that continued use of the airline will depend on it being seen as attractive in terms of the other amenities that are provided; hence the investment in hotels and recreation facilities. The housing was acquired mainly to attract personnel from other airlines, though most ex-patriots view their association with Gulf Air as temporary. We have been asked by the Chairman to provide help during the difficult transition phase that they are now in.

Hierarchical Description

The system chosen as relevant to an area of concern will represent one level of resolution from a number of possible levels in associated hierarchies. These examples are included to initiate thinking about what those associated hierarchies might be.

1   Take a petrol pump (in a garage forecourt) to be a system. To what hierarchies of systems does it belong?

2   Place the activity 'writing a letter to a personal friend' in two different systems hierarchies. What characterizes these hierarchies?

3   Take the activity 'control quality' in a manufacturing company and place it in relevant hierarchies.

Root Definitions and Conceptual Models

These two concepts represent the most important tools of the systems thinker in relation to problems at the soft end of the problem spectrum. These exercises, first of all, examine the structure of 'given' root definitions and then provide practice in modelling them.

1   Analyse the following root definition in terms of CATWOE elements:

A Company owned system concerned to develop and produce particular polymers with selected saleable properties so as to continuously exploit market opportunities world-wide with a performance acceptable to the Company and to the Group of which the Company is a part.

2   A well known event that takes place in most organizations immediately prior to Christmas is the office party. Take this to be a human activity system and produce two root definitions that could define its nature.

3   Take the 'Miss World' contest to be a human activity system and produce two root definitions that could express its nature. Discuss the CATWOE elements of each definition, and develop broad level conceptual models (10 activities).

4   Make an activity model which could form the basis of an information system used to control the provision of sterilizable (i.e. non-disposable) surgical instruments in a teaching hospital with a number of operating theatres.

5   The following extract is taken from the *Code of Practice for the Use of Management Consultants by Government Departments*, produced by the Civil

Service Department. It represents the set of activities to be undertaken under the heading 'How to select a consultant'. Produce a root definition of the human activity system of which this is a model and comment on its completeness as a model.

(a) Prepare a statement of the problem of the organizational unit concerned, its functions, and the objectives of the assignment, and submit that statement to the Civil Service Department for advice as to who should carry it out.

(b) Assuming that it is decided that the assignment is appropriate to an outside consultancy, invite an agreed short list of three firms to interviews to discuss the proposed assignment, bringing at least one member of their staff who would be closely concerned with the assignment. The invitations should be as informative as possible, especially with regard to such matters as the functions of the departmental areas concerned and the main individual officers, and should assume that the consultant is not familiar with departmental shorthand and jargon. They should encourage the seeking of clarification, if necessary, and name an officer to handle enquiries.

(c) At the interview discuss the firm's approach to the assignment, the way it might be manned, the timing of it, and the machinery for liaison with the department. It is usually appropriate to invite the firms to supplement the discussions by undertaking a short unpaid preliminary survey of the problem.

(d) Ask the firms to provide, in confidence, relevant information concerning previous assignments, particularly those carried out by the individuals who would be allocated to the proposed work.

(e) Ask for proposals to be submitted to show the method of approach, stages in the assignment, reporting arrangements, staff to be employed (including details of their experience), possible start dates, duration, cost, and terms of business.

(f) Evaluate the proposals and interview the firms again to make a final choice and agree terms of reference and financial arrangements. At the interviews the firms would be expected to introduce at least the 'resident' consultant (i.e. the consultant who will bear the brunt of the field work in the department) and his supervisor.

Initial Systems Analysis

The purpose of this exercise is to put together the elements so far discussed and hence it is concerned with an initial exploration of a problem situation. The exercises are presented first and then an example of the kind of analysis required is given.

The following problems are intended to illustrate areas in which there are statements of concern. Your task is *not* to try to solve the problems as stated, but to derive systems models as a means of exploring these concerns. For each example:

(a) Express the problem situation pictorially.

(b) Choose a system relevant to the problem and derive a root definition for it.

(c) Develop a conceptual model at a broad resolution level.

1   The University Catering Manager is concerned with maintaining quality and preventing price rises. How would you examine this?

2   A new Management School building on the University site, now under construction, is 6 + weeks behind schedule. We want to get it on schedule and yet meet agreed budgeted cost. What should be looked at and how?

An example

The following is an example showing how to approach the two questions above.

Statement of concern

At the AGM the local choral society always has difficulty in obtaining nominations for its officers and commitee. As it is a performing society a number of non-choral tasks must be managed. How could this problem by examined?

Problem situation expressed

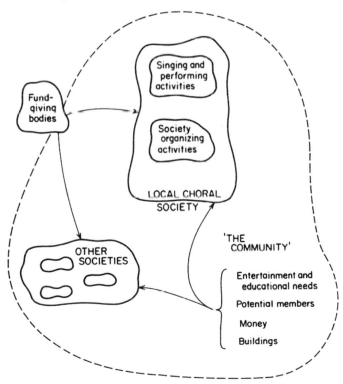

A root definition

> A partially supported, society owned and operated system which seeks to promote a greater appreciation of choral music within the local community through continuing participation in, and presentation of, a range of choral works

Analysis of this root definition may be expressed as follows:

C   The local community
A   The society
T   Promotion of greater appreciation
W   Choral music is worth appreciating
O   The society
E   Supporting bodies, the community (including other societies)

Conceptual model

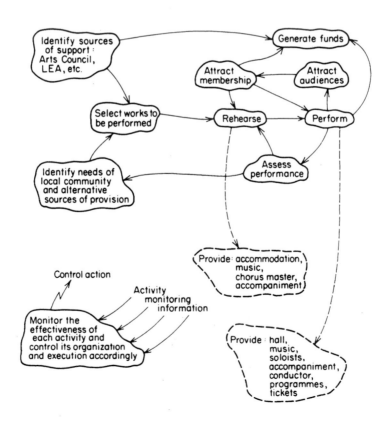

Methodology Related to the Nature of the Problem Situation

The following examples represent statements of problems related to particular situations. How would you tackle them? You cannot solve the problems, but you are required to derive the activities in a problem-solving system which is specific to the nature of the situations.

1 A multi-product packaged food company has two manufacturing sites, one in Manchester and one in Glasgow. It has traditionally operated with a large number of distribution depots, spread over the country, to deliver its products to supermarkets and major grocery stores in England, Scotland, and Wales. In a move towards higher efficiency, the company wishes to rationalize its depots. How should it decide how many depots it should have and where these should be located?

2 Anglo-American owns a copper refinery at Rhokana in Zambia. The refining process takes in copper concentrates and, through the unit processes of smelting, converting, and refining, produces the copper product. The smelter area, in which the unit processes take place, has traditionally been managed by European personnel. However, due to the policy of Zambianization, these managers are being replaced by Zambian nationals. It has been suggested by a small London-based contracting company (as a means of establishing themselves in the African market) that, as a result of this changeover, communication systems to provide personnel control are required. The processes to be managed are five smelters, six converters, four refiners, and five overhead cranes (which transfer ladles of material between units), plus the overall control of the smelter area. How would you analyse this problem?

3 An oil company wishes to improve oil production through better maintenance of its off-shore oil rigs. The planning and doing of maintenance, the provision of materials and labour, and the longer term feedback to rig design and modification, rely on the provision of information. How could the company decide what information systems were essential, so that it could improve production?

4 A local philanthropist wishes to donate a large sum of money for the establishment of a community centre for the benefit of an under-privileged sector of the community in a specific area of Liverpool. He wishes to remain anonymous and has appointed a director to manage its setting up. How should he decide: (a) what to provide and (b) how it should be managed? (A picture of the situation relevant to this problem was given in Chapter 1 of the book.)

Problem-solving

The following three exercises are concerned with the use of systems concepts and methodologies for carrying out an investigation of concerns expressed by particular actors in a problem situation. The exercises are artificial and are necessarily constrained in terms of the amount and type of information available. In a real situation you would be able to collect more information. That is not the case here but don't be frustrated by that. Do the best systems analysis that you can with the data provided. Similarly, you will not be able to obtain any feedback on the ideas that you are pursuing. Use your own judgement in deciding what to recommend.

Chocktree & Sons

This is an exercise in formulating an approach to analysis at the start of a project. Thus the material tells you something about the problem content system; you are required to think about an appropriate problem-solving system.

The handouts are in three categories: (a) letters from the organization regarding the setting up of the project; (b) information on the company; (c) data provided by the organization.

Your remit is as follows: An ISCOL consultant will be visiting the company to discuss our approach and to submit a proposal for the study. The company have provided some information on which the proposal will be based. Use systems concepts in deriving the content of your problem-solving system and report on your findings to the consultant. Since he will formulate his proposals from your initial analysis, it will be necessary to convince him of the soundness of the approach that you have adopted in reaching your recommendations. NB: Take the year to be 1984.

Correspondence

Chocktree of Stoke,
Agate House,
Clayton
8th November

Professor P. B. Checkland,
ISCOL Ltd,
University of Lancaster,
Bailrigg,
Lancaster

Dear Professor Checkland,

Thank you for visiting us last week and for putting us in the picture so clearly on what you are trying to do at Lancaster. I am very sympathetic towards universities and in particular to those who are trying to bridge the gap between the university and the outside world. I am sure that we can find a problem suitable to one of your students and I will discuss this with my colleagues.

As you know, my nephew has recently completed a similar course in business analysis and it might be interesting to attach your student to him in order to test out your methods on someone who has recently completed a course in modern management methods.

Yours sincerely,

C. Chocktree
Managing Director

Chocktree of Stoke,
Agate House,
Clayton
23rd March

Professor P. B. Checkland
ISCOL Ltd,
University of Lancaster,
Bailrigg,
Lancaster

Dear Professor Checkland

With reference to our telephone conversation of yesterday I would like to confirm that we will take one of your postgraduate students, starting on 1st May. Sir Charles and I have discussed possible projects and I think that one in the production area would be most suitable considering the time scale involved.

We have recently commissioned a new finishing line and I believe that the whole production process is now capable of greater profitability, and this could form the basis for the project.

I take your point regarding the need to look more widely at the company but I think that the student will have enough on his plate to examine the production system.

Sir Charles has suggested that the project manager at our end should be Samuel Chocktree and I would have no objection to this even though the project is unlikely to be in his area.

I suggest that once the student has assimilated sufficient background knowledge we get together with Samuel to decide exactly what needs to be done.

Yours sincerely,

M. Chocktree
Technical Director

Chocktree of Stoke,
Agate House,
Clayton
10th April

Professor P. B. Checkland,
ISCOL Ltd,
University of Lancaster,
Bailrigg,
Lancaster

Dear Professor Checkland,

I have been asked by Sir Charles to act as organization manager to your student when he arrives in May.

I have persuaded Sir Charles to widen the remit for the project so that we can look at more than production. We can now examine the marketing aspects of the company in order to determine where our business lies in the next ten to fifteen years.

The market for quality goods is fairly static and our performance over the last five years has also been fairly static. If we look at the potential market for mass-produced items there would appear to be every opportunity for rapid growth if we were to enter now.

I am sure that you would agree that the only sensible thing to do is to produce a study which will point out to the board the need for a change in thinking so that the company can take advantage of this expanding market.

I look forward to meeting you in May and to the prospect of having a fruitful project experience.

Yours sincerely,

Samuel Chocktree
Promotions Manager

ISCOL Limited,
University of Lancaster,
Bailrigg,
Lancaster LA1 4YR
17th April

Sir Charles Chocktree,
Managing Director,
Chocktree & Sons Ltd,
Agate House,
Clayton,
Staffs

Dear Sir Charles,

ISCOL Project

I have now heard from Matthew and Samuel Chocktree, and I am writing to confirm the arrangements for the ISCOL project commencing in May. I have sent a separate copy of this letter to Matthew and Samuel.

I now see the project as being directed towards improving the profitability of the production process in relation to the markets which it can best serve. We hope, therefore, to be able to make, in the first instance, recommendations for improvements in the profitability gained from your existing operations. We should also be able to indicate which other markets you could serve with your existing resources, and further, other opportunities which may require additional capital in the longer term.

This emphasis is somewhat broader than that of our earlier discussions. During the first few weeks of the project it will become obvious if it is over-ambitious, but at that stage we will be in a better position to agree priorities and establish a work plan.

In order to clarify the situation further, I am arranging to send one of our full-time consultants to make a presentation to you. He will show how we would apply our problem-solving approach to the data you have already sent to us. He will contact you directly.

I hope this is satisfactory to you, and I look forward to meeting you again during the summer.

The project manager and student will report to Mr Samuel Chocktree at 9.00 a.m. on 1st May.

Yours sincerely,

P. B. Checkland

cc. Mr Matthew Chocktree
    Mr Samuel Chocktree

Background to the company

Chocktree Ltd was established in 1780 by Sir Josiah Chocktree, a member of a family noted at the time for wood-carving. By 1850 they were sufficiently well known as the makers of high class pottery for Queen Victoria to order a 200 piece dinner service for her personal use. Following the highly successful reception of the completed service, Josiah Chocktree was allowed to use the 'By Appointment to Royalty' motif. In the early days the products varied from statuettes and other *objects d'art* in black basalt and alabaster to delicate tea sets in hand-painted bone china.

By the turn of the century the company had 100 employees and was firmly established in the high quality end of the ceramic market. During the depression and the two world wars, the company was badly hit, both in terms of loss of market and also in terms of loss of designers.

The traditions of the company have been maintained by the family management. Their current turnover is in the region of £0.50 million per annum.

They are in a very specialist market, a large proportion of which is in North America. They rely to a large extent on their reputation as the means of generating orders, though a small amount of advertising is taken in such magazines as *Sphere, London Life, Harpers Bazaar,* etc.

Their main expertise lies in the quality of design and durability of the glaze. This particular glaze has been in use for many years and exhibits a high resistance to detergents and food acids.

The personnel complement of the company has been fairly static over the past ten years. There has been virtually no turnover in design staff but it has been difficult recently to recruit unskilled workers due mainly to the better wages paid by other nearby industries. This problem was the main incentive for the introduction of a modest amount of automation of the finishing processes.

Company management structure

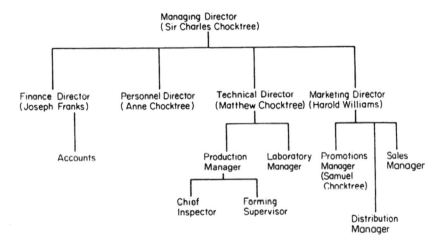

Biographical details of top managers

Managing Director (Sir Charles Chocktree)   Grew up with the firm, is now Senior Executive and major shareholder. Age 70 – should have retired but still has a highly active mind and is keen to keep control of the business. He is a very shrewd business man and has a sound knowledge of the market at the high quality end of the business.

Finance Director (Joseph Franks)   Trained as an accountant before joining the firm in the Accounts Department. Has been Finance Director for twenty-three years.

Personnel Director (Anne Chocktree)   Is the daughter of Sir Charles and was trained originally as a graphic art designer. First joined the company as decoration supervisor in order to learn the business. Forming and decoration are still regarded as the key craft areas and hence, as Personnel Director, she occupies a key role with regard to the quality of personnel employed in the company.

350

Technical Director (Matthew Chocktree)   Younger brother of Sir Charles. Has no formal training but joined the firm as an apprentice former and has worked his way to the board by showing considerable competence in all aspects of quality ceramic production. Was Production Manager for fifteen years and has been on the board for the last ten.

Marketing Director (Harold Williams)   Is a relatively new appointment and has been a member of the board for three years. He has been in the ceramics industry for ten years and was previously sales manager for a large pottery company in the medium quality area. It is believed that he was appointed to fill the gap between the retirement of Sir Charles and the future appointment to the board of Samuel Chocktree, the son of the Technical Director.

Promotions Manager (Samuel Chocktree)   Age 35. Has generally been a disappointment to the family. Intention was that he should succeed his father as Technical Director and was sent to university to obtain a degree in mechanical engineering. Having failed this he transferred to the Department of Geography and obtained a lower second. After a short period abroad in the Foreign Office of the Civil Service he returned to do a one-year course in business studies and received an MA. He joined the company in the sales office five years ago and was appointed to the position of Promotions Manager by the new Marketing Director.

Production sequence

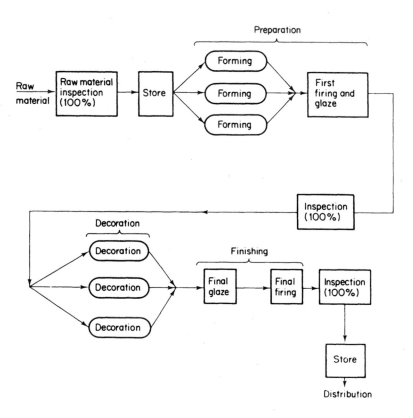

Production processes

The production process consists of the three basic operations of preparing, decorating, and finishing. The first two of these are hand operations whilst the finishing process has been automated.

A high quality raw material is used as the input to the preparation stage. Here the articles are hand thrown and the craftsmen who perform this operation are known as formers. When each article has reached its final shape they are assembled on trays and dipped into a primary glaze. Each tray is fed into a kiln for the first firing. At this stage surface defects occur such as surface cracking, distortion due to inhomogeneity in the clay, and imperfect glaze. The acceptable pieces then go on to the decoration stage.

The company prides itself on the quality of designs produced and this is all due to the skill of the decorators. As many as 10 000 brush strokes can go into a single design and hence the quality of craftsmanship employed at this stage is just as critical as at the previous stage.

The finished articles are again assembled on trays which are fed on to a conveyor. The conveyor transfers the trays via a glaze vat to the kilns for final firing. The speed of the conveyor determines the throughput but a balance must be maintained between the residence time in the glaze and the firing time in the kiln.

The finished articles are inspected and stored prior to distribution. Surface faults are the main cause of rejects and can be due to underglazing or underfiring (resulting in a non-uniform glaze) or overfiring (resulting in a surface covered in hairline cracks).

A completely new finishing line was installed and commissioned in 1982 and after some initial faults in the conveyor system and glaze flow control, it has operated without serious breakdown.

Prior to the installation of the automated finishing line, all the handling and transporting of trays and pieces between the various parts of the prrocess had been carried out manually.

Production information

Finishing line

Conveyor speed	0.60 ft/hr
Glaze flowrate	8 gall/min
Firing temperature	1200 °C.
Inspection	100%
Average throughput	6 times/hr
Rejection rate	11.4%

Commissioning tests (under manual operation)

Conveyor speed (ft/hr)	Kiln temperature (°C)	Pieces fired	Rejects
0.2	1000	75	8
0.3	1000	49	5
0.2	1100	67	7
0.3	1100	59	6
0.2	1200	58	6
0.3	1200	40	4
0.2	1250	51	5
0.3	1250	60	6

(Cont.)

Commissioning tests *(Cont.)*

Conveyor speed (ft/hr)	Kiln temperature (°C)	Pieces fired	Rejects
0.4	1100	79	8
0.5	1100	38	4
0.4	1000	68	7
0.5	1000	56	6
0.6	1000	42	5
0.7	1000	33	4
0.7	1100	50	6
0.6	1100	52	6
0.4	1200	30	3
0.5	1200	87	9
0.6	1200	27	3
0.7	1200	43	5
0.4	1250	99	10
0.5	1250	76	8
0.6	1250	44	5
0.7	1250	32	4
0.7	1300	30	4
0.6	1300	48	6
0.5	1300	70	8
0.4	1300	63	7
0.3	1300	82	9
0.2	1300	44	5

Glaze control

The control system installed maintains the glaze flowrate proportional to conveyor speed so that

$$G_f = KV_c + 5.6,$$

where G_f = glaze flowrate (in gallons per minute), V_c = conveyor speed (in feet per hour), and K = proportional control gain.

The glaze flowrate controller uses integral action and goes unstable for flowrates below 7 gall/min.

Typical operating records

The conveyor speed control (automatic) is shown by the following diagram (controller set-point is 0.6 ft/hr):

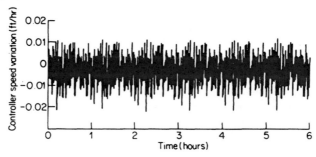

The glaze flow control (automatic) shows the following responses subsequent to initial disturbance: (a) at a flowrate of 7.0 gall/min

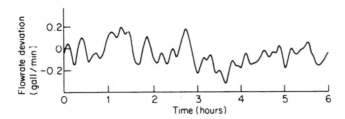

and (b) at a flowrate of 8.0 gall/min

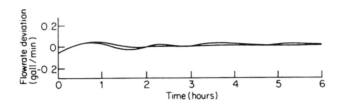

Statistics related to hand-decorated products

Average throughput	30 000 items/year
Order book	6 months
Number of decorators	50
Number of formers	10
Range of products	
Tea services	two standard designs, 34 pieces
Coffee services	one standard design, 15 pieces
Dinner services	two standard designs, 41 pieces
Individual designs produced to order	

Sales statistics 1983

		Revenue
Dinner services	17 special	£18 500
Dinner services	250 standard	£75 000
Tea services	200 standard	£40 000
Coffee services	300 standard	£27 500
Miscellaneous items	6450	£300 000

Customer statistics 1983

25 large (67% orders) }
200 small (33% orders) } 78% export

Company performance over past 5 years

Year	1979	1980	1981	1982	1983
Turnover (£m)	0.46	0.50	0.52	0.55	0.5
Pre-tax profit (£m)	0.043	0.055	0.055	0.03	0.047
Raw material cost (£m)	0.044	0.049	0.052	0.056	0.053
Total production and distribution costs (£m)	0.373	0.396	0.413	0.464	0.400
Production losses due to scrap (£m)	0.020	0.0218	0.0226	0.024	0.026

Polyglot

Inspection of a plastic product

You are a member of a team set up by your Works Manager in a company producing a plastic product. He tells you that the whole business of inspection at the factory is unsatisfactory because, although a lot of product is rejected, the number of complaints received is still much too high. Thus the economics of the process suffer while time, effort, and money are expended on dealing with the complaints. He asks you to look into the matter and let him have an analysis of the situation.

Use a systems approach to analyse the problem situation which has been outlined. Assume that the year is 1984.

Background information

The study is concerned with one type of polymer chips in a variety of grades.

Production situation

There is fierce competition in the industry and this has a number of effects:

(a) For products made for general sale there are continual external pressures to produce to higher and higher standards.
(b) For products made for certain customers there is a continual need to make more grades of product to very tight specifications.

It is increasingly difficult to dispose of material which only just fails to meet specifications. At the same time the need to maximize production puts pressure on the quality control organization to pass such material. It need hardly be said that the requirement for high standards puts a greater strain on both the plant and production personnel.

Management aspects

Specifications

Manufacturing specifications are set up by sub-committees of a Specification Committee. Both are made up of representatives of Production, Engineering, Research, and Technical Services Departments. Because of their close contact with the market, the Technical Services Department takes a predominant part.

Inspection

Inspection is the responsibility of a chief inspector who is responsible to the Works Manager. Inspection is carried out by Works Laboratory shift inspectors. In general, they have no discretionary powers. Reliance is on laboratory personnel for test results which are 'rubber stamped'.

The inspection organization is independent of the Technical Services Department and the Production Department, in order to eliminate bias. They are under pressure from both sides and bear the odium of any decision which is unwelcome to one or other.

It is probably fair to say that the department concerned spends much more of its time on achieving high standards where they are necessary, *than on lowering standards where they are not necessary.*

Standards and complaints

While manufacturing standards exist it is not possible to keep them all completely up to date. Furthermore, production for special requirements cannot be covered by standard specifications.

Standard specifications are analytical and numerical (objective), or visual (subjective). It may happen that although one property is out of specification it may not matter to some customers. Strictly speaking, if the product is out of specification, the inspector must take the responsibility of passing it.

Complaints fall into two categories:

(a) They may concern gross variations which have not been detected by 'quality control'.

(b) They may concern a material which is found to be unsuitable for a specific application although it is on specification. They may happen even when the grade of product is correctly specified for the application.

The number of complaints received is too high. These too cost money to investigate and put right. They also occupy the time of personnel who could be more productively employed.

On the whole it is felt that standards are often incorrectly set and relations between the departments concerned may become strained if there is disagreement over a particular decision.

356

OUTLINE PROCESS FLOW CHART

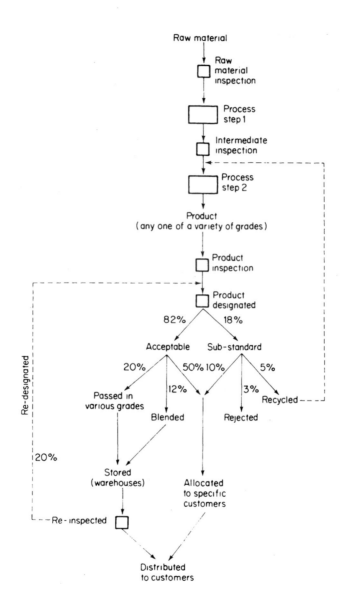

ORGANIZATIONAL CHARTS

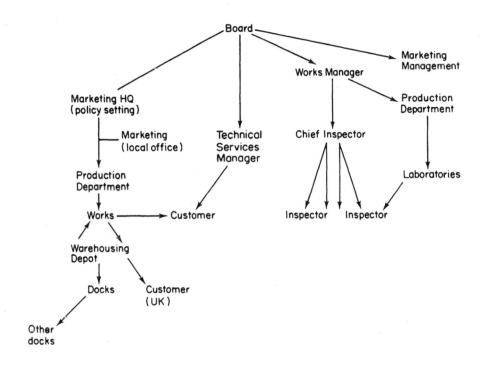

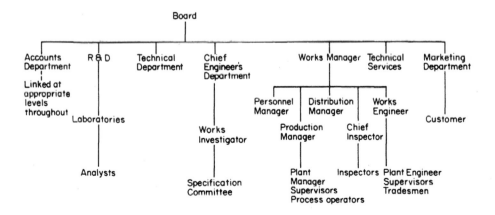

ORDER INFORMATION FLOWS

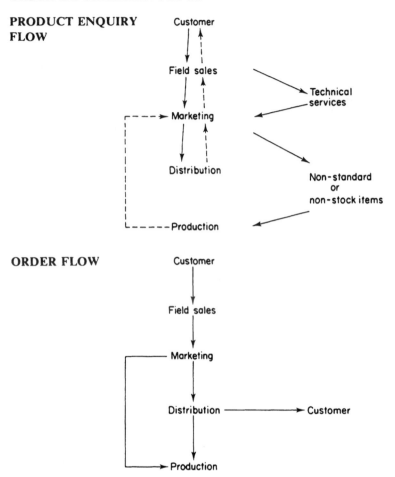

PRODUCT ENQUIRY FLOW

Customer → Field sales → Marketing → Distribution → Production

Technical services

Non-standard or non-stock items

ORDER FLOW

Customer → Field sales → Marketing → Distribution → Customer

Production

Financial data – 1984

Orders and Sales

Grade	up to 100 ton orders (pence/lb)	Over 100 ton orders (pence/lb)	Sales (tons/year)
1	11	10.5	10 000
2	10.5	10.0	15 000
3	10	9.5	50 000
4	9,5	9.0	15 000
5	9	8.5	10 000
			100 000

Costs expressed as a percentage of sales price (SP)

	%		%
Recycle cost	11	Cost of raw material	45
Second value	= next grade down	Cost of raw material inspection	0.5
Cost of insepction	2.5	Cost of process 1	25
Cost of processing	38.5	Cost of inspection	1
		Cost of process 2	10
		Cost of inspection	1
		Cost of storage	1
		Cost of re-inspection	0.5
		Cost of distribution	2
		Profit	14
			100

The costs above all include overheads

Degree of the level of inspection

1% raw material
1% sample for 1st inspection
3% on final inspection
1% on re-inspection

Details of customer complaints – 1984

(1)	Faults overlooked by inspection	64%
(2)	Faults arising in transit	1%
(3)	Failure to achieve effect while meeting specification	20%
(4)	Level of service, e.g. packing and general	5%
(5)	Delivery dates	10%

Total number of orders

2000 from small customers (20% of production)
800 from large customers (80% of production)

Marketing information

70% of customers are small but in total account for 25% of sales by weight.
30% of customers are large and in total account for 75% of sales by weight.
25% by weight small customers UK and Overseas.
75% by weight large customers UK and Overseas

self-used 60   for resale 40%

Subject to economic/political climate, especially overseas.

360

Sales

Large
    Home one-fifth of large sales
    Overseas four-fifths of large sales
Small
    Home four-fifths of small sales
    Overseas one-fifth of small sales

	Standard specification	Loose specification (mix of grades)	Specials
Regular production	80%	20%	–
Occasional production	20%	10%	70%

Company performance

The company performance tabulated below contains the following assumptions: (a) selling price constant; (b) grade pattern constant; (c) processing cost constant.

	Year				
	1980	1981	1982	1983	1984
Production (million tons)	0.12	0.12	0.125	0.12	0.10
Total revenue (£m)	26.8	26.8	28.0	26.8	22.4
Pre-tax profit (£m)	5.28	4.88	4.60	3.98	2.64
Raw material cost (£m)	10.7	11.1	12.0	12.0	10.2
Total process and istribution costs (£m)	10.2	10.2	10.7	10.2	9.0
Total inspection costs (£m)*	0.62	0.62	0.70	0.62	0.56
Recycle and rejection costs (£m)*	2.14	2.14	2.24	2.14	1.79
Number of complaints*	140	145	135	170	200
Cost of complaints (£m)	0.050	0.049	0.050	0.055	0.050

The three items asterisked in this table are included in the overheads.

Albion group

The information given below describes a situation in a company in which you, as a consultant, have been asked by the managing director of the group to carry out an investigation.

The main purpose of this exercise is to give practice in analysing a problem situation, using the concepts and methodologies described previously in this book (or of your own design). The required output is argued, recommendations for the managing director in line with the last paragraph of his letter L.1.

Information is available in the form of (a) a set of internal letters about the problem in the firm; (b) some additional notes prepared recently for other consultants; (c) other information relating to certain areas of activity in the group.

The description of the process technology of the group is intended to provide a plausible sequence of events leading from raw material to finished product which is adequate for the purpose of the exercise. It does not purport to be a detailed description of actual processes in any industry. Similarly, the market forecasts and sales figures do not apply to any specific product.

Assume the year to be 1984.

Internal letters

L.1.

To the Systems Consultant

I am enclosing a letter in which our Stockist Manager is applying for £642 000 for warehouse expansion. Clearly, there is some case for increasing our finished stocks. but since they have managed reasonably well for the last few years with the present stock policy, I wonder whether the application is unnecessarily large. Bearing in mind the low profitability of the Division, I am unwilling to inject capital into our Stockist without more understanding of the situation.

I am enclosing copies of letters which I think you should see. Let me know if you require any further information.

I would like to have advice on what action to take over this request for expansion and stock increase. Please investigate this and let me know what course of action you recommend and the approach you have used in arriving at it.

H. C. Symonds

Managing Director
The Albion Group

L.2.

The Director of Finance,
Albion House,
St Jame's Square,
Kingston-upon-Hull

Surface Stockists,
Treated Hardboard Surface
Specialists,
Ealing, London W5

Dear Mr Turner,
   You will remember that we discussed the possibilities of extension to our present premises when I last visited you. I would now like to make a formal claim for £642 000 for extensions to our warehouse facilities.
   There are two reasons for this extension. Firstly, we need to increase our overall stock level by 30% to assure our customers of a better delivery. Secondly, we anticipate an increased demand for finished board, especially in the heavy gauges.
   I enclose the T.11 expenditure application form.

Yours sincerely,

J. R. B. Small

T.11	

Expenditure sought:
   £642 000
	Applicant:
	   J. R. B. Small, Manager, Surface Stockists

Reason for expenditure:
   To extend Warehouse 3 to increase volume storage by 30%
	Details:
	   Purchase 5 acres 500 000
	   Building adjacent to Warehouse
	   No. 3 42 000
	   Handling equipment 100 000
	            ———
	            642 000

L.3.

J. R. B. Small,
Stockist Manager,
Surface Stockists,
(Albion Group Ltd),
Ealing

Tower Construction Co.,
97 Conduit Street,
London EC5

Dear Sir,
   We have surveyed the land adjacent to your No. 3 Warehouse and have studied your requirements.

We estimate that the building extension 120ft × 75ft × 35 ft will cost   £42 000
Handling equipment and storage racks will cost   £100 000
         ———
         £142 000

   We await your further instructions.

Yours faithfully,

pp. R. T. Thomas
Estimates Manager

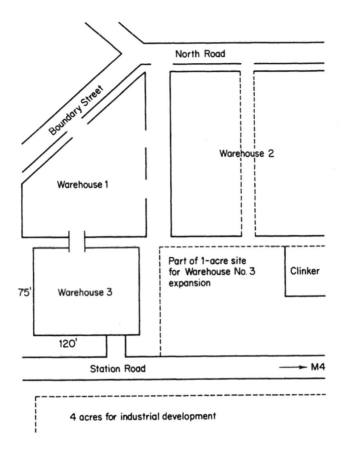

North Road

Boundary Street

Warehouse 1

Warehouse 2

75'

Warehouse 3

120'

Part of 1-acre site
for Warehouse No. 3
expansion

Clinker

Station Road

→ M4

4 acres for industrial development

L.4.

Thomas Mander Limited,
Plant Handling Division,
Sheffield 4.

J. R. B. Small,
Surface Stockists,
(Albion Group Ltd),
Ealing

Dear Mr Small,

Our Engineer has inspected your premises and reports that we should be able to increase your effective warehouse volume by 40% without increasing floor area used, or roof height.

We have devised a handling system based on a pallet unit (the pallet size can be specified by you). This has been installed with great success in G.K.L., Mudlow steel, Farles Bearings and many other companies.

We would increase your present storage matrix to a height of 28 ft (from the original 20 ft). This would involve substantial strengthening of certain key frames and floor covers. A fork lift truck based on a Warwick Climax design is used in our system, and to handle the throughput you require, six trucks would be necessary.

The total cost of increasing your stockholding and accessing capabilities would be:

	£
Increasing racking volume (including strengthening of structure)	175 000
Palletization of containers	67 000
Six 'Highlift' Mander trucks	84 000
	326 000

Yours sincerely,

N. T. Cross
(Thos. Mander)

L.5.

ALBION GROUP LTD

MEMO

From: Chief Accountant
To: Finance Director

I have looked into the question of the alternative investments – new buildings, or improvements to existing ones – at Surface Stockists.

I am afraid I will not be able to let you know my recommendations before mid-January.

It seems as if an investment analysis of some sort should be made – but the pressure of work near the end of our Financial Year makes it impossible for me to give you an immediate reply.

The purchase of land in that part of London is always a good investment, and the purchase price of £500 000 for 5 acres is not excessive for the area. It is unfortunate that Real Estate Holdings are unwilling to sell the 1-acre site on the north side of the Station Road by itself. However, I believe that as we have the ability to purchase at such a good price we should accept, and develop the excess 4 acres through our own financial expertise. The rate of return on investment should easily exceed that on our current operations.

I enclose some information about land prices and rents in this area. Also depreciation rates for our plant and buildings. I feel that from these we should be able to evaluate the worth of the alternative investments to us in the future.

M. R. Thompson

L.6.

ALBION GROUP LTD

MEMO

From: Chief Accountant
To: Finance Director

I enclose a letter from the Ealing Borough Surveyor, showing what an attractive proposition the 5-acre site is. The opportunity of receiving a high return on investment is extremely attractive.

Here is some other information promised you:

Building Depreciation
On Offices   (NBV £1 600 000) 10% straight line
On Factory   (NBV £1 250 000) 14% straight line
Plant Depreciation(NBV £3 870 000) 20% straight line.

Our present activities are producing profits of 6% after tax on turnover, and this percentage has decreased from 9%, 7½%, 7% over the past three years.

M. R. Thompson

L.7.

Ealing Borough Surveyor's
Department,
Hermitage Road,
London W5

The Chief Accountant,   Our Ref: EB/83/G03
The Albion Group Limited,
Albion House,
Hull

Dear Sir,
With reference to your request for a survey of the 5-acre plot on both sides of Station Road, we have completed the inspection and find:

(i)   that the land is well supplied with drainage, sewerage and other utility systems;
(ii) that the density of packing of clinker on the north-west corner is sufficient for supporting heavy structures;
(iii) that no road development scheme is planned to cross the land.

Officially, you must realize that we cannot comment on the value of the land, but similar plots have sold at up to £120 000 per acre in Ealing, and Station Road, with its easy access, is an attractive site. Land prices have increased by 17%, 20% and 22% over the past three years and rents for buildings in the centre of Ealing have increased by 18% per year over the past five years.
We will send a detailed survey to you soon, and invoice you then.

Yours faithfully,

T. Cockraine
(Chief Surveyor)

NOTES FOR PREVIOUS MANAGEMENT CONSULTANTS

N.G.10

To Systems Consultant

Information about the Albion Group
I am afraid that it will be impossible for you to meet the senior members of the Group due to their previous commitments and to the tight time scale on this project. Therefore, in order to give you some background into our activities, I am enclosing notes made about the Group, which were prepared recently as a preliminary exercise for a Management by

Objectives study (items N.G.1, N.S.11, 34, and 25, N.B.31, 32, 43, 44, and 35, and N.A.10). I have not had time to look at these myself, and would be interested to hear if you find them useful as a basis for assessing the Group, with particular reference to improving our overall profitability. It will be most helpful to me to receive fairly specific suggestions, rather than general comments like the desirability of improving communications within the Group.

H.C. Symonds

Managing Director,
Albion Group

N.G.1

THE ALBION GROUP

To: MBO Consultant
From: Managing Director, Albion Group

I have only just been appointed to this position by J. F. Tyzacks, Merchant Bankers, who have recently acquired a majority shareholding in the Albion Group.

My terms of reference have not been detailed yet; hence a start by this Management by Objectives exercise.

Clearly, I have to increase the profitability of the group as a whole. There is a seemingly healthy spirit of competition inside the group, which I feel must be essential for good overall profitability.

Albion Mills is the oldest part of the group and I feel that I should look more closely into their management methods, as there is, without doubt, a lot of management of the 'old school'. Betterfinish and Surfaces I think present less of a problem, and I am quite happy, initially, to let them manage themselves until I have a better feel for the group situation. We have a lot of experience in the preparation of prefabricated building sections and surfaced boards. It is this experience which Tyzacks wish to use to best effect in the improving market situation which is foreseen during the next few years. My success in directing the group must be measured in terms of overall group profits, and our ability to grow to meet the market demands.

H. C. Symonds

N.S.11

SURFACE STOCKISTS

From: Stockist Manager

We hold finished board from Betterfinish, and some untreated board from the Albion Mill. We have facilities for trimming and specialist packing, but our main task is one of supplying customers with a reliable delivery.

Thirty per cent of our throughput goes to three major customers – 'English Lily', 'Elizabethan', and 'Ramply' – for kitchen furniture construction. Prefabricated Bathroom Units are taking 15% and a further 15% is exported.

Why we want to expand

We are keen to explore possibilities of expansion, even if this means stocking non-Albion Group board. An independent market survey, arranged by us, showed a large increase in demand for the heavier gauge products, and enquiries to us from customers suport this finding. However, at present, we are unwilling to quote a delivery date on heavy gauge products for new customers, because delivery from Betterfinish is particularly poor. From our point of view, it is very profitable to sell, but we must be known as a reliable deliverer. We would be uncompetitive if we offered delivery in excess of three weeks, yet with heavy gauge it has often taken over three weeks on average to fulfil an order, and so rather than risking being unable to supply from stock, we are placing specific orders on Betterfinish and requesting immediate delivery to help ensure that specific customer needs are met. We are especially keen to do this for our major customers.

How we aim to expand

An AIC consultant looked into this last March and found that the mean lead time (i.e. the time from placing the order with Betterfinish to receiving the board from them) for our orders from B is 16 working days. Thus, we are aiming to increase our stock to the 20 working days level, which will entail increasing our stockholding volume by approximately 30%. When we have this larger safety stock, we will be able to offer for delivery from stock and can then start to place larger, more economic orders with Betterfinish. The Warehouse Foreman, who is in charge of a lot of the re-ordering, storage and despatch, will, I hope, take charge of the re-organization of stock when the expansion is complete, as I regard him as a key figure in our ability to maintain reliable delivery.

N.S.34

SURFACE STOCKISTS

From: Warehouse Foreman

Although we are not completely full at present, I understand what the manager means when he says we must expand our stocks, and agree that the only way to ensure a reliable delivery is to have a good safety level. Besides, I'm keen to take on more modern machinery, more warehousing and more responsibility. As far as my job is concerned, it's my duty to trigger off a re-order for a fresh delivery from Betterfinish. We've got a system here whereby I order three weeks' worth of board and re-order another three weeks' worth when it arrives, as the delivery from Betterfinish takes about three weeks. I think it is a fairly good system because it is easy to operate. Even my leading chargehand can operate it, and he often has to if I'm away.

N.S. 25

From: The Manager, Surface Stockists

Information available for the sales of board over the period 1978–83 is given in the following two graphs:

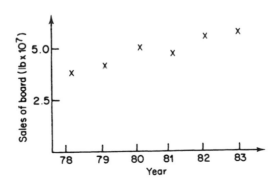

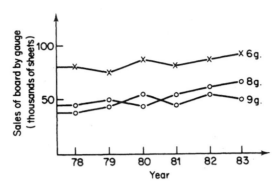

In 1983 we sold 95 000 sheets of 6 g board, 65 000 sheets of 8 g board, and 50 000 sheets of 9 g board.

N.B. 31

BETTERFINISH FABRICATORS

From: The Works Manager

The production process carried out at Betterfinish is illustrated by the diagram below. About 80% of the mill board used is supplied by Albion Mills; the remaining 20% is bought from other suppliers.

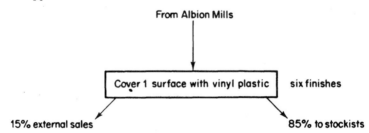

(1) Nearly all our lighter gauge board comes from Albion. Most of our bought-out board is the heavy gauge board.

Thickness	1″	0.75″	0.50″
Gauge	9 g	8 g	6 g
From Albion (sheets)	15 000	60 000	85 000
From outside (sheets)	35 000	–	5000

(2) We use five covering machines, multipurpose 'Challands Applicators', with a machine throughput rate of 350 ft/hr. We work two 8-hour shifts per day, 5 days a week, 50 weeks per annum.

(3) There are six different finishes, one of which is rolled continuously on one machine. Other machines are scheduled with a sequence of different finishes.

(4) The boards are coated with Bestick 'Permabond' and then fed through the 8′ wide rolls together with the coating sheet (vinyl or plastic).

(5) Machine throughput rate is governed by the machine design (limited by dryers at back end of machine). Machine changeover times are 35 minutes for gauge change and 15 minutes for a finish change. Roll alignment is critical; overpressure causes the

surface to spring and underpressure creates no adhesion at all. We feel quite happy that the machines are of the best on the market for quality and throughput.

N.B. 32

BETTERFINISH FABRICATORS

From: Sales Manager

85% of our production goes straight to our stockists. We have a couple of large contracts for direct supply of heavy gauge, negotiated recently, but our profitability on these is not large. In general, we are keen to stay off the heavy gauge.

Although we know we could easily sell more heavy gauge, we are having difficulty in producing sufficient to meet our existing orders. As the contribution on each heavy board is relatively low (£13.24 on 9 g, £13.76 on 8 g, £14.28 on 6 g), there is no incentive, given our present output problems, to manufacture more heavy gauge. (Contribution is the difference between the selling price and the material costs (board + surface).) We buy much of the heavy gauge board from Laver Bros., since they charge £1.12 less per board than Albion, but supplies are limited and we have a long lead time on delivery. We are trying to negotiate with Albion for a reduction in price since their delivery is better. They may reduce their prices marginally, but fear that their profit margin might suffer.

Our stockist is tending to place a lot of small orders, and we are trying to come to some arrangement over increasing order size. We are told that until the stockist's holding capacity is increased, orders will remain small.

N.B. 43

BETTERFINISH FABRICATORS

From: Works Manager

The Production Director at Albion House has emphasized the need for autonomy so much that I feel under obligation to no-one to manufacture surfaced board if it is not profitable for us to do so. I'm afraid that recently (because it would reduce our machine utilization, which is always a good excuse) we have been delaying our production of heavy gauge. My production scheduler has my full authority to do this, especially on these small orders of heavy gauge, so that we can recover our overheads and reach the profit targets I am setting.

The case is quite clear. We obtain a contribution per board of £13.24 on 9 g, £13.76 on 8 g, and £14.28 on 6 g. If we have to buy 9 g from Albion (and we have to take 15 000 of these from them in order to meet demand), we have an even lower contribution of £12.11 per board. We must negotiate with Albion for better terms, and if we cannot manage this, we will try to negotiate for bulk rates with Laver Brothers. For a given gauge, all six finishes bring in the same contributions.

N.B. 44

BETTERFINISH FABRICATORS

From: Production Scheduler
Attached please find a (partially complete) specimen production schedule for machines 1 and 5.

BETTERFINISH FABRICATORS														
Production Schedule (Week no. 17)														
Machine 1			2			3			4			5		
Gauge	Finish	Ft	Gauge	Finish	Ft	Gauge	Finish	Ft	Gauge	Finish	Ft	Gauge	Finish	Ft
9	5	600										9	2	280
8	6	720										9	2	320
6	4	240										6	2	160
9	1	80										9	2	1120
6	1	920										6	2	960
6	3	160										8	2	340
6	6	840										8	2	280
8	3	140										6	2	40
6	1	540										6	2	180
8	1	140										6	2	80
6	3	60										8	2	40
												6	2	960
												9	2	640

N.A. 10

ALBION WORKS

From: Production Manager

(1) Albion Works have for the past 20 years produced a high quality Mill Board with excellent waterproofing and fire resistant qualities. It uses long-fibre softwoods imported from Scandinavia and shipped into Hull, together with special resins. Production processes are shown diagrammatically below:

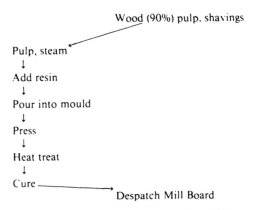

(2) Our pulping and steam injection equipment is capable of handling 10^8 lb per year on 2×8-hour shift working, 5-day week.

(3) Resin added varies with thickness of sheet to be produced.

(4) The moulds are standard area, $8' \times 20'$, and of three thicknesses, Each mould is slid on to a frame which holds 10.

(5) Three are five ovens and three frames per oven. One frame for loading, one for firing (55 min in the oven), and one for cooling and unloading for curing. Ovens are worked on 2 × 8-hour shifts. Recently we have been considering increasing our oven capacity and buying a new oven specially for the lighter gauges.
(6) Curing is the final drying and resin bonding period of three days.
(7) Our board is near enough the same density as the wood we start from, i.e. 30 lb/ft^3.

N.B. 35

BETTERFINISH FACBRICATORS

From: Production Scheduler

I work from the list of orders which I receive from Surface, and also from the few direct orders from outside the group.
I know there have been complaints about our delivery performance on heavy gauge. It's hardly my fault – I'm under pressure from the Factory Manager to limit the amount of 9 g we process. I try to rush through all the lighter gauges as quickly as possible to help Surface.

ADDITIONAL INFORMATION

The Albion Group Structure

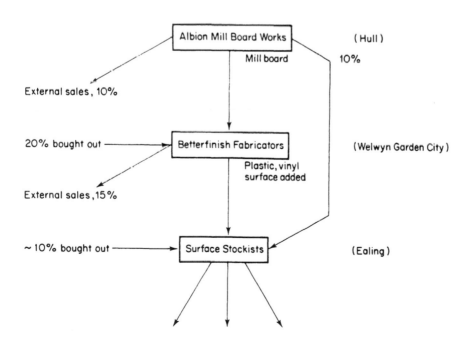

Management structure of Albion Group

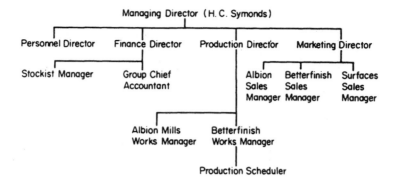

Albion Group policy

(1) There has always been encouragement for strong autonomy of Albion Mills, Better finish, and Surface Stockists. Internal competition is good for us, keeps us efficient and on our toes.
(2) The Works Managers at Albion and Betterfinish, and the Stockist Manager at Surface Stockists, are responsible for the profitability of their companies.
(3) The Board meets monthly to discuss technical, personnel, and expenditure plans. Prior to this there is a Management Meeting, at which the senior management report to their functional Director.
(4) There is a quarterly report on revenue and expenditure, that is, on Sales and Firm Orders, and on material and production costs.

Independent market forcast for finished board (Market Research Inc.)

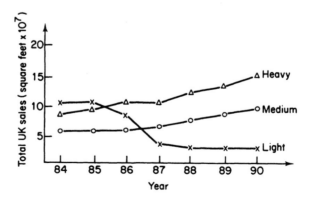

(1) The main increase in heavy gauge is due to the increase in fitted kitchens and bathrooms, and prefabricated wall sections. Light gauge is decreasing because of the increasing use of plastic surfaces.
(2) 80% of the market is in the London area, and Surface Stockist have good contacts with a large number of outlets in this area. From a distribution point of view they are ideally situated.

Sales pricing information

From: Sales Manager, Albion Mills

Gauge	9 g	8 g	6 g
Thickness	1″	0.75″	0.50″
Wood cost (£ per 10^3 lb board)	100	100	100
Resin cost (£ per 10^3 lb board)	21.2	19.6	18.0
Overheads per 10^3 lb of board (recoverable on same basis as direct material cost is incurred)	11.2	11.2	11.2
Cost to us (£ per 10^3 lb board)	132.4	130.8	129.2
Sells at (£ per 10^3 lb board)	139.6	140.8	145.6
Percentage profit (%)	5.5	7.7	12.7
No. of boards produced (× 000)	55	60	85

Clearly, our bread and butter line is the light gauge board, with a profit margin of nearly 13%. The margin on the heavier gauge is much less, and we are losing out at this end of the market. We suspect that several competitors are selling heavy board at a loss in order to maintain their sales of lighter board. Our own light board is selling very competitively, on both price and quality.

Surface Stockist contribution from finished board

From: Manager, Surface Stockist

The contribution from finished board (per cubic foot) are

9 gauge	8 gauge	6 gauge
£0.992	£0.90	£1.168

Contribution here is the difference between selling price and buying price.

Survey carried out by Market Research Inc.

MARKET RESEARCH INC.

Survey of Environment Surrounding the Albion Group Carried out for J. F. Tyzacks, Merchant Bankers

These are the key points taken from our main report, which will be published in January, 1985:

(1) Our Market Survey, already in the possession of the Albion Group, shows a healthy increase in demand for heavy gauge board.

(2) The market is shared by seven companies; the Albion Group is in the top three of these for turnover and assets.

(3) Each company has a stable nucleus of customers (who take approximately 70% of output). The users of board are fairly conservative, and will only change their supplier if there is good reason to do so (e.g. consistently poor delivery performance).

(4) It is hard to assess price competitiveness as contracts are on a highly personalized basis. However, each company appears to maintain the same price to each of its customers.

(5) The customers, who are in the prefabricated kitchen and bathroom furniture and wall section trades, appear to request a range of delivery times. The majority of orders require delivery in under three weeks, and it is uncompetitive not to meet this delivery service requirement.

(6) Surfaces have contact with some of the largest users: 'English Lily', 'Elizabethan', and 'Ramply'. Prefabricated Bathroom Units are also fairly large users. In the main, customers are fairly small.

(7) The technology in the board-producing industry is fairly static, and no developments are anticipated in future (apart from instrumentation improvements in the control of the pulping process). However, it is known that in the application of surfaces to boards, 'Starboard', a subsidiary of Laver Bros., is developing a new process. In this process the board is given a surface by spreading liquid plastic over the board in a controlled manner.

LIBRARY EXERCISE

The following documentation refers to a fictitious company in which a study of the in-house library services is being undertaken. Your role in this study is as a member of a consulting team which has been invited in to study the information requirements of the library services. the consulting company to which you belong is called ISCOL Ltd and the senior consultant who has been involved in the negotiations is called Andrew Taylor.

A number of roles are referred to in the documentation and their attitudes to an internal information study are illustrated through a number of memoranda. An organization chart is included which illustrates the formal responsibility relationships that exist within this part of the company.

It is your task in this exercise to use this documentation as the source of a number of view points (Ws) and having identified these to develop a number of root definitions and conceptual models relevant to the primary task of the library services. This is a similar process to that illustrated earlier within the text in relation to the Olympic Games. This will be the first stage in the analysis which leads to a definition of the information requirements. Use the information requirements analysis methodology which has been described in the text and identify the information processing procedures which you feel need to be designed to provide adequate information support. Because the study is being undertaken by ISCOL Ltd you will find that this methodology is referred to as the ISCOL methodology. In real life you would have access to the various managers who hold these roles and you would be in a position to ask them further questions. Unfortunately this is not the case in an artificial exercise such as this but I hope you find it a useful vehicle for testing your understanding of the approach described in Chapter 6.

STRICTLY CONFIDENTIAL

From: Gemma Pitt
         Director, Administrative Services

To: Information Planning Consultants (ISCOL Ltd)

ORGANIZATION AND INFORMATION REQUIREMENTS
STUDY OF LIBRARY SERVICES

I have recently been discussing your study with my management group, and I feel I should know more about it than I currently do. There are possibilities that it may have impact on some organization issues that I am currently pondering. I should stress that I have no wish to directly influence your study – it should be most definitely focused on John Barth's organization. Nonetheless, I have some concern that issues extending beyond the Library Services Department, and John's perceptions of its activities, should be taken into appropriate consideration.

In particular, there are three points I would like to stress to you:

(1)   John has only recently been appointed as Head of Library Services. We are waiting to see whether he can establish, in a coherent manner, a new and more comprehensive role for Library Services. We particularly need a clearer understanding of the costs and benefits of the services his Department is offering. I would welcome your views and assessments.

(2)   The company is in the middle of a cost cutting exercise. It is also apparent that the Departments reporting to me are changing their respective roles – and much of this change is due to developments in technology. I suspect that existing departmental boundaries no longer make sense and am contemplating a rationalization and perhaps separating out the computer operations as a separate profit centre; maybe combining some of the services of Library and Computer Applications areas – that sort of thing anyway. While I don't want you to worry about this directly (even if it happens, it won't be in the short term), any issues you throw up concerning co-ordinated developments and future integration should be handled carefully.

(3)   I know that Jerry Bray in Computer Applications is very eager to be involved with your study. I think at this stage that this would not be such a good idea. Obviously I would like to see a reduction in the amount of money being spent on bureau services. On the other hand, a competitive element is beneficial for ensuring the commercial awareness of the computing department. For the short term, then, I think I had best talk with Jerry myself and I'll pass on his concerns to you myself.

I trust you will be in touch with me soon.

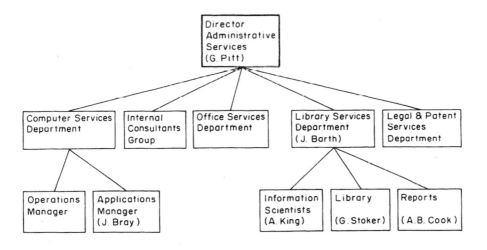

CONFIDENTIAL

From: John Barth
            Head of Department
            Library Services Department

To: Andrew Taylor
            ISCOL Ltd

INFORMATION REQUIREMENTS OF LIBRARY SERVICES DEPARTMENT

Dear Andrew,

I am pleased to be able to tell you that I have obtained approval for the proposed Information Planning study of my Department.

Enclosed is the report produced by Andrew King, my senior Information Scientist, discussing Computer Systems Development in the Library Services Department. I have considerable sympathy with the principles underlying his recommendations. In particular, I believe that if my Department is to contribute to the Division's activities in any significant way we must become more *active* in our provision of services.

As they stand, however, the current recommendations are unacceptable and would only serve to polarize the various groups in my Department, and exacerbate relations between them. If we are going to successfully introduce new technology, I need the support of all groups involved. The report, in fact, shows insufficient consideration of the genuine needs of the Library and Report Groups – I am also sending you copies of the responses to the report from the heads of those groups. I hope, and trust, that applying your approach will help to establish priorities and plans for information provision which are both understandable and acceptable to all parts of my Department.

In conclusion I would like to emphasize two problems which it is essential that your study should consider.

Firstly, ever since I became responsible for this Department three months ago, and having had no previous experience of the workings of Library Services Departments, I have been trying to establish in my own mind how effective and efficient this particular Department is. This is proving to be impossible. Not only are there no procedures for assessing the relative use and user satisfaction of the various services which the Department offers, but, moreover, it is difficult to establish where our financial resources are being expended. In short, I trust that any Information System developments will include satisfactory 'management information' as a major priority.

Secondly, I have to admit to some apprehension about the necessary decisions concerning computer technology. Neither I, nor anyone in my Department, are in a position to select any particular technical option. In addition, we must be seen to be moving in directions consistent with developments in the Computer Services Department. Having said this, what I *am* in a position to insist upon is that my need for information are not distorted or ignored in favour of expedient technical solutions. Obviously, it will be necessary for you to present your findings to the Computer Services Department – they are to be joint sponsors, with me, of your study.

I look forward to hearing from you soon.

                                          Yours sincerely,

                                          John Barth

COMPUTER SYSTEMS DEVELOPMENT IN LIBRARY SERVICES DEPARTMENT

CONTENTS

1 INTRODUCTION AND SCOPE OF STUDY

This report has been prepared in response to a request from the Manager, Library Department, for an evaluation of the way in which the Divisional Library Services should develop in the future, and the consequent requirement for new computer systems.

The issues now facing the library are threefold. Firstly, library users within the division are becoming increasingly aware of the advantages to be gained from making full use of the library's facilities. This increase in utilization is reducing the level of service that can be provided to each individual user and adverse comments on the quality of service are now being received. This situation is likely to become worse in the future as out-of-date library systems, which are currently almost entirely manual, are stretched further and further.

The second problem is closely related to the first. Not only are more people using the library, they also now expect more from it. The growth in the availability of published material continues to accelerate, and users can no longer hope to keep themselves up to date. Instead of simply obtaining items requested by users, the library is now seen as having the function of keeping users informed on its own initiative. There has, of course, been considerable innovation in information-gathering technologies, including computer-based text handling facilities and vastly improved communications. This represents the third major issue to be confronted – what action should the library in this Division take in response to these new technological opportunities?

Recommendations made in this report for the development of new services are not only based on consideration of weaknesses in current provision, but also take into account the advances in library science being made by other Divisional libraries within the group. It is felt that the current review should not be restricted to means of achieving satisfactory levels of performance in the immediate future, but should also plan for investment which would restore this Division to its traditional place of being a leader in library provision within the company.

Consequently, this report is presented in three main sections:

1. A review of the organization and current activities of the library and their associated problem areas.
2. A discussion of possible future developments in services.
3. Recommendations for computer systems to support such future developments.

2 SUMMARY

Service provided to users is currently unsatisfactory, and this situation is likely to deteriorate as increased demands are placed on the library as a central information-provider. This will be an active role, with the Library Services Department initiating a wide range of current awareness, abstracting and similar activities on behalf of users.

Out of date and cumbersome systems for both information provision and administration currently render these new developments in services impracticable. Major investment in new technology will enable Library Services Department to vastly improve services to users, as well as internal administration, whilst at the same time reducing staffing levels in line with company policy.

3 CURRENT ORGANIZATION AND ACTIVITIES

For the purposes of this description, it is convenient to consider the activities of the library under three headings, corresponding to the current library organization depicted below.

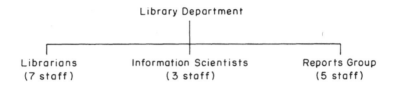

3.1 Librarians

The Librarians' section is responsible for what might be described as the traditional functions generally associated with the operation of a library. This includes the purchase and cataloguing of books and periodicals, the day-to-day administration of loan procedures and general assistance to users wishing to do their own research in the library. It is helpful to examine these functions by taking books and periodicals separately.

3.1.1 Administration for books

Books are currently purchased for the library on the basis of user requests (subject to the approval of the Head of the Librarians' section) or following receipt of publishers' circulars or published reviews which indicate that the work is likely to be of general relevance to the Division. Details of books received are entered in a card-index catalogue, indexed alphabetically by title and author, with a separate classified index. The catalogue now contains entries for more than 10 000 volumes.

The book loans procedure is based on the 'card and pocket' system familiar in many public libraries. A pocket is glued into each volume, into which is inserted a card containing title, author and classmark information. Each user has a set of tickets identical to the book pocket showing name and department, into which the book card is placed when it is loaned, showing who the borrower is. A return date is stamped inside the book, but no sanctions are imposed against readers who do not return it by the stated date, and no procedure exists for following up overdue books unless the title has been requested by another reader. This practice has given rise to a number of difficulties.

Firstly, books are frequently 'lost' as either the book card becomes separated from the user ticket, or both get mislaid. Alternatively, long overdue volumes are finally chased as the result of another reader's enquiry, only to find that the original borrower has been transferred to another division, or has left the company, presumably still with the book. If the title is sufficient interest, this generally means that a new copy has to be purchased, or obtained on inter-library loan, both of which are expensive and time-consuming.

Furthermore, the ability of users to 'browse' in a particular subject area is severely curtailed. It is not possible to find out what material is held in a given area, firstly because the relevant books are missing from the shelves and secondly, there can be no guarantee that a title appearing in the card catalogue is not irretrievably lost, and a long wait for re-acquisition or external loan may be unacceptable.

It should be noted, however, that if a title has, in fact, been obtained on inter-library loan, efforts are made to ensure its timely return, because the library itself would otherwise be subject to penalty. Staff indicate, however, that their current workload is too great to make routine recall of all overdue items a practical possibility.

One further point is worthy of mention. The absence of any comprehensive financial information concerning the operation of the library frequently means that budgets run out early in the year. When this happens, demands for new acquisitions cannot be met until new funds are available in the following year. This introduces a further element of unreliability into the service provided.

3.1.2 Administration for periodicals

Procedures for handling periodicals are necessarily more complex than for books because of two additional requirements:

1. Circulating current periodicals to interested readers.
2. Agreeing and implementing the renewal or deletion of titles from year to year.

In principle, periodicals are obtained only in response to user requests. This means that librarians regularly send to each responsible departmental manager a list of periodicals currently received in that area of interest, requesting that the manager indicate which are to be maintained and which deleted. The librarians also notify users when new journals appear, to determine whether or not a subscription should be taken out. In practice, many departmental managers do not respond at all, with the result that the head of the Librarians' section has to take these decisions herself.

When periodicals arrive, they have to be distributed, and this involves the preparation of circulation slips daily, weekly, monthly, quarterly or annually, depending on the frequency of the publication. Maintaining and producing accurate circulation lists manually would clearly be an arduous and time-consuming task. There is, however, a company-wide periodicals administration system designed to cope not only with circulation lists but also with the maintenance of a periodicals index and the ordering and renewal arrangements described above. In fact, this system now causes at least as much work as it saves.

The periodicals administration system is an old, mainframe computer system intended for use in all the company's Divisional Libraries. It is very old, poorly maintained and inadequately updated. This results in records being kept for publications no longer received, and for publications which have changed their name. In the latter case, there is likely to be no record at all for the publication under its new name. The system is also designed to cancel automatically subscriptions to any publication whose continuance is not verified six weeks before the due date, which causes immense problems if, as described above, users do not notify requirements.

3.1.3 Assistance to users

Many readers use the library as a source of reference material and undertake their own research in the library itself. The librarians are frequently called upon for assistance, in terms of their specialist knowledge of both the cataloguing and layout of library contents. Some support is available from the library indexing system, which is simply a file of author and title details for the library's holdings, which can be asked to produce a listing according to key words in the title. This key word search facility is a batch system provided by Computer Services Department seven or eight years ago, and is obviously very limited in scope.

Given these continuing requests for assistance from visitors to the library, which occupies much of the librarians' time, it is clear that both the periodicals and book administration procedures are now totally inadequate to cope with demands from users.

3.2 Information Scientists

Whereas the Librarians' Section is basically responding to user requests, the task of the Information Scientists is principally to act on their own initiative, making users aware of new material which is likely to fall within their area of interest. There are currently only three staff in this section, reflecting the low priority given to this activity in the past, but it is the area most likely to expand in the future in response to new demands from users.

The limited staff is further hampered by limited resources. The existing current awareness service is based on a very restricted profile of user interests, and to a small number of users. Despite the widespread use of abstracting databases elsewhere, this Division continues to rely on printed abstracting journals, and the only technological support is from a small word-processing system.

Access is available to abstracting data bases through a bureau service, but no budget has been allocated to fund this resource. Occasional use is made of it, but for major searches this extremely efficient option is only available if paid for by the user. Otherwise, searches are performed manually.

3.3 Reports Group

The Reports Group is a separate section of the library dealing exclusively with documents produced within the division. It has been in existence, with largely the same procedures, for more than fifty years. These procedures are as follows. Three copies of every report issued for distribution are sent to the Reports Group. Their first task is to agree a security classification with the originator. One copy of the report is kept in the archives, and the other two are kept for reference in the library. These copies are generally available, except for security category A items, which are released at the discretion of the Head of the Reports Group. Certain highly important documents (e.g. patent specifications) are also kept on microfiche for additional security. Some of the oldest items have also been microfilmed to reduce storage costs.

Completely separate author, title and subject indexes are kept for material handled by the Reports Group, and no information is made generally available on the holdings, except for a recently introduced alphabetical title index suggested by the Information Scientists' Section. This is of very limited value, however, since the full listing has more than 30 000 entries, and even the listing restricted to the last five years has more than 4000. The Reports Group argue that there is no real need for any further details to be disseminated, since the producers and users of these reports tend to be limited to a few departments whose need for access is well understood. In any case, the Reports Group see their prime concern as being the security of company information, and would consequently regard it as inappropriate to encourage wider distribution of such material.

4 DISCUSSION AND ANALYSIS

4.1 Problem areas

From this study, the key problem areas now facing the Library Department seem clear-cut, and can be summarized fairly simply as follows. There is no doubt that users will come to expect that in the short-to-medium-term the library should provide them with a greatly improved service. This will require the library not only to maintain existing services, such as journal distribution and book loans, but will also demand a wide range of new functions in which the library's role will be that of an active information-provider rather than a passive information-storer as at present. This will have to be achieved with fewer staff, since it is unreasonable to expect that the library will be able to avoid the company-wide manpower cuts now being imposed. The likelihood is that the library's establishment will be reduced from 15 to 10 as a result of this programme. This outlook poses two major questions:

1. What is the source of the new information which will need to be provided and how can it most effectively be disseminated?
2. How can the library streamline its operations in order to meet the greater performance expectations with fewer people?

The answer to both these questions lies in the reshaping of the library's operations as a result of the effective implementation of new technology.

4.2 Future developments

From all the preceding discussion it is clear that it is the information science aspect of the library which is certain to experience the greatest growth in the future, and the library's traditional librarian functions which offer the most scope for rationalization.

The quality of information provided to library users can only be as good as the quality of information obtained, and this is a major weakness in the division at present. This can be seen in the areas of both access to information and structuring of information. At present, effective access to information is severely limited, even for the library's own stock. Many of its books are left out on indefinite loan, and there is a whole class of material – that managed by Reports Group – which is scarcely available at all because of unrealistic claims as to the need for security. If the library is to be seen in the future as the central provider of information as a business resource, we cannot afford for knowledge generated within the division to be so closely guarded as it is currently. Apart from highly secret material, all the library's holdings – books, periodicals and reports – should be catalogued, indexed and made available as a whole, and not segregated on the basis of outmoded distinctions.

If access to the library's own stock is poor, it is an insignificant problem when related to the access available to external material. In this Division, it has not yet been realized that 'published material' now includes electronically stored and transmitted items of all kinds. Access to computer files made available by public information providers is now an indispensable element in the resources of any library. Our current provision in this respect is effectively nil.

This leads on to considerations of information structuring. A wide range of separate unwieldly, out-of-date card index catalogues offers no realistic basis for a comprehensive information service. Even if it were possible to list all books and journals catalogued under a particular classmark, it would be impossible to read through them all to find the items relevant to one specialist field of interest. In addition, items published in other works classified elsewhere which may be highly relevant would be missed entirely. Modern abstracting databases and sophisticated text-handling computer facilities overcome this problem by allowing truly comprehensive searches for items in highly specific, tightly defined subject areas. The saving in library department and user time and effort using this approach is enormous.

Exactly the same features of new systems allow highly sophisticated current-awareness services to be offered. Each user can have a properly constructed detailed interest profile, so that searches can be made for items of interest about which he should be notified. The relevant abstract can be provided also, thus saving the time and cost of providing a reprint of an article not really relevant, which happens frequently when only the titles of pieces are given, as at present.

4.3 Internal systems

The case for investing in powerful new technology to provide innovative information services is clear, but this will also provide benefits to the internal organization of the department. Maintenance of a fully-integrated library catalogue would be vastly simplified with on-line updating. Indexes could be made available not only by authors' name with cross-references to joint authors, and title, but also in detail by subject, and with no additional effort. Administration of loans would be computerized with automatic reminders being despatched for overdue books and the production of warning lists to monitor non-returns. Periodical administration would be vastly simplified, with the company system scrapped in favour of a system tailored to the needs of the Division.

These new procedures should enable Librarians to shed three staff, and the integration of the Reports Catalogue into the main library catalogue should reduce the Reports Group staff by two. This would meet the Division's demand for a reduction in numbers to ten overall. The Information Scientists group, being already the smallest, could not safely be cut, given the increased responsibilities which would fall on them.

A major spin-off from these developments would be the ability to acquire a financial management package to monitor the library's performance against budget. It is well known that at present no financial monitoring is available, with the result that the only form of expenditure control exercised is that services are provided until the budget is exhausted, which may occur as quickly as six or seven months into the year. The uncertainties created by this situation, which results from pressure of work, would be greatly eased by the introduction of new technology.

5 SYSTEM RECOMMENDATIONS

These recommendations are based on a major review of recently published articles in the professional journals covering the area of library and information management. They represent the consensus of opinion on the most technologically-advanced and effective technology now available.

5.1 Text handling

In order to provide the services to users discussed above, the most sophisticated text-handling software is required. The preferred package for this Division is IBM's STAIRS. Marketing Department are actively considering purchasing this package as a means of structuring and accessing product and competitor information and market surveys. If the Library Department were also to request this facility, its purchase would be ensured. The library could also become a centre of expertise in using this software, which would provide benefits both to Marketing and to other Divisions, in addition to improving the library's services in general.

The STAIRS package would run on the Division mainframe computer, as would obviously be required for optimum processing and communication with other systems providing relevant databases. A number of terminals with access to the system would be required both in the library and possibly also in selected user areas. Mainframe processing of text, in addition to offering the most powerful operational solution, is also likely to prove cheaper than working through a bureau as at present, given our experience of the considerable costs of on-line access which we have had amply demonstrated in the past.

This recommended system would also solve the cataloguing and indexing problems within the library. The required details could simply be input as text, and the required catalogue formats produced using the text-editing facilities provided by the package. This was the additional attraction that no additional programming support need be requested from Computer Services.

5.2 Library administration

The sophisticated mainframe package described above is not, of course, suitable for internal library administration. Maintaining loan records, periodicals distribution lists, etc., will require local computing facilities within the library, and for this purpose it is recommended that an IBM 8100 machine be acquired.

This system will meet all foreseeable requirements in the following areas:

- book, periodical and reports loan records
- production of overdue notices and control reports
- book and periodical purchase administration – printing purchase orders, progressing payments, etc.
- administration of periodicals distribution
- financial monitoring.

Some applications development may be required in some areas if suitable packages cannot be found, but once again it is desirable that Computing Department involvement be minimized.

Note that this system will allow proper financial records to be kept, and will provide Library Department with a powerful word processing facility. Thus the investment will not only provide for improvements in existing tasks, but will offer completely new facilities within the Department for little or no extra expenditure.

MEMO

From: G. Stoker
        Library Group
        Library Services Department

To: J. Barth
        Head of Department
        Library Services Department

(Copies to: A. M. King, Information Science Group
               A. B. Cook, Reports Group
               J. B. Bray, Computer Services Department)

COMPUTER SYSTEMS DEVELOPMENT IN LIBRARY SERVICES DEPARTMENT

You have asked us for our comments on the report by Andrew King. Unfortunately it is difficult for us in the Library Group to find much with which we can agree in the report.

The assessment of the current activities of the library fails to mention that the Library Group supports the most widely used services of the Department, and does this with considerable success despite its lack of resources and outdated computer systems.

This point aside, the recommendations of the report would still be inappropriate, given the current priorities of the Department.

(1) If we are to cut staff and improve our service, the *basic* procedures must be considered as first priority for computerization. This implies a need for:

   * a computerized Book Ordering system
   * a computerized Book Catalogue system
   * a new, computerized Periodicals Handling system.

   Procedures already exist which could support each of these systems; all we require is a computerization of the manual, and outdated computer, information handling of these procedures.

(2) No consideration is given to the alternative option of a mini-computer system similar to those used by other libraries in our company. The advantages of such an option would be that:

   * it is cheap;
   * software to support the three systems mentioned in (1) above is being developed in other divisions, so the systems could be implemented quickly;
   * it would be compatible with the general approach of other library services in the company.

In conclusion, I would like to stress that, in the opinion of our group, the requirements for information systems development in the Department are quite clearly specifiable and we should pursue these before considering such adventurous long-term developments as are detailed in the report.

MEMO

From: A. B. Cook
        Reports Group
        Library Services Department

To: J. Barth
        Head of Department
        Library Services Department

COMPUTER SYSTEMS DEVELOPMENT IN LIBRARY SERVICES DEPARTMENT

I have read this report, as you requested, and have to say that I find its recommendations completely unacceptable.

It is essential that we maintain our control over access to the information in the Reports Group. Our good relations with our major users depends on this security.

Moreover, the services we provide are specifically adapted to the personal requirements of our users, with whom we have established good understanding. I believe, therefore, that attempts to computerize these services would either be impossible due to their complexity or would lead to a deterioration in the service we offer.

A. B. Cook

MEMO

From:

    J. B. Bray
    Applications Development Manager
    Computer Services Department

To: A. D. Taylor
        ISCOL Ltd

Dear Andy,

I'm glad to see that your study into Library Services Department's information requirements is going ahead. I'm looking forward to working with you in the application of the ISCOL methodology.

Some points you should know:

(a) Marketing Department are, in fact, unlikely to invest in the STAIRS package; so if LSD go ahead with their idea of using the package, they'll have to find the money for it.

(b) There are, not surprisingly, discussions within LSD about the possibility of a 'mini' solution. I'm sure I do not need to tell you of the advantages of a mainframe development (reliability, software support and expertise, etc.). And it is, in fact, the policy of our CSD to stick to mainframe options wherever feasible.

(c) There are all sorts of political problems with the use of bureau services. We would have to look very carefully at any arrangements along these lines.

I reckon that I'm effectively saying that if you can sort out what they actually *want* in LSD, my department can handle the problem of how best to give it to them.

Good luck, hope to see you soon.

                                             Jerry Bray

References

Ackoff, R. L. (1962). *Scientific Method: Optimizing Applied Research Decisions*, John Wiley, New York.

Anderson, O. D. (1976). *Time Series Analysis and Forecasting*, Butterworth, London.

Anderton, R. H. (1970). Industrial dynamics: can it provide a unifying systems language? Paper presented to an Institute of Measurement and Control symposium on 'Control Engineering Approach to Management Systems', January 1970.

Anthony, R. N., and Dearden, J. (1976). *Management Control Systems, Texts and Cases*, Irwin, Toronto.

Beer, S. (1966). *Decision and Control*, John Wiley, New York.

Beer, S. (1979). *The Heart of Enterprise*, John Wiley, New York.

Bishop, D. M. (1979). Control of management systems: a control engineer's view. Paper presented to an Institute of Measurement and Control symposium, March 1979.

Blake, R., and Mouton, J. (1978). *The New Managerial Grid*, Gulf Publishing, Houston.

Bleazard, G. B. (1976). *Program Design Methods*, NCC, Manchester.

Bowen, P., and Wilson, B. (1971). A systems approach to the design of management control systems for production plant. *Journal of Systems Engineering*, **2** (2).

Box, G. E. P. (1957). Evolutionary operation: a method for increasing industrial productivity. *Journal of Applied Statistics*, **6** (1).

Box, G. E. P., and Draper, N. R. (1969). *Evolutionary Operation*, John Wiley, New York.

Box, G. E. P., and Jenkins, G. M. (1970). *Time Series Analysis, Forecasting and Control*, Holden-Day, San Francisco.

British Computer Society (1977). Report of the British Computer Society Data Dictionary Systems Working Party. *Data Base*, **9** (2).

Burnstine, D. C. (1979). *The Theory behind BIAIT – Business Information Analysis and Integration Technique*, BIAIT International, Petersburg, N.Y.

Campbell, D. P. (1958). *Process Dynamics*, John Wiley, New York.

Carlson, W. M. (1979). Business information analysis and integration technique (BIAIT) – the new horizon. *Data Base*, **10** (4).

Checkland, P. B. (1971). A systems map of the Universe. *Journal of Systems Engineering*, **2** (2).

Checkland, P. B. (1979). Techniques in soft systems practice. 2. Building conceptual models. *Journal of Applied Systems Analysis*, **6**.

Checkland, P. B. (1981). *Systems Thinking, Systems Practice*, John Wiley, Chichester.

Checkland, P. B., and Wilson, B. (1980). Primary task and issue-based root definitions in systems studies. *Journal of Applied Systems Analysis*, **7**.

Chestnut, H. (1965). *Systems Engineering Tools*, John Wiley, New York.

Chestnut, H. (1967). *Systems Engineering Methods*, John Wiley, New York.

Coyle, R. G. (1978). *Management System Dynamics*, John Wiley, New York.

Dermer, J. (1977). *Management Planning and Control Systems*, Irwin, Toronto.

Downs, E., Clare, P., and Coe, I. (1988). *Structural Systems Analysis and Design Method: Application and Context*, Prentice Hall, Hemel Hempstead Herts.

Forbes, P. E., and Checkland, P. B. (1987). Monitoring and control in systems models.

Internal Discussion Paper, Department of Systems and Information Management, University of Lancaster.

Forrester, J. W. (1961). *Industrial Dynamics*, MIT Press, Cambridge, Mass.

Forrester, J. W. (1968). Industrial dynamics, after the first decade. *Management Science*, **14**.

Foresrester, J. W. (1969). *Urban Dynamics*, MIT Press, Cambridge, Mass.

Foster, M. (1972). An introduction to the theory and practice of action research in work organizations. *Human Relations*, **25**.

Gane, C., and Sarson, T. (1979). *Structured Systems Analysis*, Printice-Hall, Englewood Cliffs, NJ.

Gass, S. I. (1964). *Linear Programming*, McGraw-Hill, New York.

Gifford, V. A. J. (1970). Corporate design for control: the dynamics of funds flows. Paper presented to an Institute of Measurement and Control symposium on 'Control Engineering Approach to Management Systems', January 1970.

Goode, H. H., and Machol, R. E. (1957). *Systems Engineering*, McGraw-Hill, New York.

Hall, A. D. (1962). *A Methodology for Systems Engineering*, Van Nostrand, Princeton, NJ.

Herzberg, P., Mausner, B., and Snyderman, B. (1959). *The Motivation to Work*, John Wiley, New York.

Horovitz, J. H. (1979). Strategic control: a new task for top management. *Long Range Planning*, **12**.

Huse, F. E. (1980). *Organization Development and Change*, West Publishing Co., St. Paul, MN.

IBM (1975). Business Systems Planning, Information Systems Planning Guide IBM publications Aug 1975 GE20-0527-1.

Jackson, M. A. (1975). *Principles of Program Design*, Academic Press, New York.

Jenkins, G. M. (1967). Systems and their optimization. Inaugural Lecture, Lancaster University, Lancaster.

Jenkins, G. M. (1969a). The systems approach. *Journal of Systems Engineering*, **1** (1).

Jenkins, G. M. (1969b). A systems study of a petrochemical plant. *Journal of Systems Engineering*, **1** (1).

Kanter, J. (1970). *Management Guide to Computer System Selection and Use*, Prentice Hall, Englewood Cliffs, NJ; see Chapter 3.

Kerner, D. U. (1982). Business information control study methodology. *The Economics of Information Processing*, Vol. 1, *Management Perspectives*, John Wiley, New York.

Lefkovits, H. C. (1977). *Data Dictionary Systems*, Q.E.D. Information Sciences, Wellesly, MA.

Levin, R. I. (1978). *Statistics for Management*, Prentice-Hall, Englewood Cliffs, NJ.

Lundeberg, M., Goldkuhl, G., and Nilsson, A. (1979). *Information Systems Development A First Introduction to a Systematic Approach, Information Systems*, Vol. 4, pp 1–12, Oxford, Pergamon Press.

McGregor, D. (1960) *The Human Side of Enterprise*, New York, McGraw Hill.

Mallen, G. L. (1970). Control theory and decision making in organisations. Paper presented at an Institute of Measurement and Control symposium on 'Control Engineering Approach to Management Systems', January 1970.

Maslow, A. H. (1954). *Motivation and Personality*, Harper, New York.

Meadows, D. H. *et al.* (1972). *The Limits to Growth*, Universe Books, New York.

Methlie, L. B. (1979). Systems requirements analysis – methods and models. Paper presented at an International Federation of Information Processing conference, Bonn, June 1979.

Misshauk, M. J. (1979). *Management Theory and Practice*, Little, Brown, Boston.

Nolan, R. L. (1979). Managing the crisis in DP. *Harvard Business Review*, March–April 1979.

Orsey, R. R. (1982). Methodologies for determining information flow. *The Economics of Information Processing*, Vol. 1, *Management Perspectives*, John Wiley, New York.

Pitt, F. L. (1968). Management – a control system. Paper presented to the 3rd UKAC

Conference on Advances in Control of Systems, April 1968.

Popper, K. R. (1963). *Conjectures and Refutations*, Routledge, London.

Pritsker, A. B., and Kiviat, P. S. (1969). *Simulation with GASP II*, Printice-Hall, Englewood Cliffs, NJ.

Quade, E., and Boucher, W. I. (eds) (1968). *Systems Analysis and Policy Planning: Applications in Defence*, Elsevier, Amsterdam.

Rockart, J. F. (1979). Chief executive define their own data needs. *Harvard Business Review*, March–April 1979.

Schoderbek, P. P., Kefalas, A. G., and Schoderbek, C. G. (1975). *Management Systems*, Business Publications, New York.

Shearer, J. L., Murphy, A. T., and Richardson, H. H. (1967). *Introduction to Systems Dynamics*, Addison-Wesley, Reading, Mass.

Sisson, R. L. (1969). Simulation – uses. *Progress in Operations Research*, 3.

Smyth, D. S., and Checkland, P. B. (1976). Using a systems approach: the structure of root definitions. *Journal of Applied Systems Analysis*, 5 (1).

Statland, N. (1982). The relationship between data flow and organization management. *The Economics of Information Processing*, Vol. 1, *Management Perspectives*, John Wiley, New York.

Swann, W. H. (1967). Optimum design of an acrolein plant by process model building and optimization. MA thesis, University of Lancaster, Lancaster.

Taylor A. (1982). Private Communication.

Tocher, K. D. (1964). *The Art of Simulation*, English Universities Press, London.

Tocher, K. D. (1969). Simulation – languages. *Progress in Operations Research*, 3.

Warmington, A. (1980). Action research: its methods and its implications. *Journal of Applied Systems Analysis*, 7.

Welbourne, D. (1965). *Analogue Computing Methods*, Pergamon Press, London.

Whitmarsh-Everiss, M. J. (1969). Design and control of a generating station. Paper presented at the IEE Summer School on Systems Engineering, September 1969.

Williams, T. J. (1961). *Systems Engineering for the Process Industries*, McGraw-Hill, New York.

Wilson, B. (1971). Design and improvement of process systems. *Journal of Systems Engineering*, 2 (2).

Wilson, B. (1979). The design and improvement of management control systems. *Journal of Applied Systems Analysis*, 6.

Wilson, B. (1980a). The Maltese cross – a tool for information systems analysis and design. *Journal of Applied Systems Analysis*, 7.

Wilson, B. (1980b). Systems engineering at Lancaster and the lessons learned. Paper presented at the International Conference on Systems Engineering, Coventry, September 1980.

Wilson, B. (1987). A systems methodology for information requirements analysis. *Systems Development for Human Progress* (Eds. Klein and Kumar) pp. 176–195, Elsevier, North-Holland, Amsterdam.

Zachman, J. A. (1982). Business systems planning and business information control study: a comparison. *IBM Systems Journal*, 21 (1).

Index

Index compiled by Annette Musker